Particle Physics

ADVANCED
PHYSICS
READERS

Christopher Bishop

JOHN
MURRAY

© Christopher Bishop 2002

First published in 2002
by John Murray (Publishers) Ltd
50 Albemarle Street
London W1S 4BD

Layouts by Wearset Ltd
Illustrations by Wearset Ltd
Cover design by John Townson/Creation

Typeset in 11.5/13pt Goudy by Wearset Ltd, Boldon, Tyne and Wear
Printed and bound in Spain by Bookprint, S.L., Barcelona

A catalogue entry for this title is available from the British Library

ISBN 0 7195 8589 9

Contents

Chapter 6 Fundamental interactions, symmetries and conservation laws

Chapter 7 The Standard Model of particle physics

Chapter 8 Particle physics and the early universe

Introduction

The 20th century has been a period during which great strides have been made in attempting to understand how nature builds a relationship between matter and energy. The emergence of particle physics as a scientific discipline that can explain the 'mechanism' of forces has met with many notable successes, including a theory of the origin of the universe.

This book has been written primarily for A-level students who are studying Particle Physics as part of their Physics syllabus. Much of the content assumes an understanding of 'core physics' concepts such as basic Newtonian mechanics, heat and properties of matter, wave motion, radioactivity and atomic structure. The mathematical skills required are those of GCSE Higher level with the addition of some further concepts, particularly that of *symmetry*, which is a key guiding principle in theoretical particle physics.

Particle physics has relied heavily on quantum theory for its development and description. A proper discussion of quantum mechanics would deserve a book in its own right and I have not attempted this here. I have introduced quantum ideas only where they are relevant to the subject and to help the reader understand how they are used in particle physics. As well as providing necessary information for the A-level student, my aim is to give the reader an idea of what the subject is like post A-level. To this end, some of the subject matter goes beyond the requirements of A-level, including a more in-depth introduction to Special Relativity.

At the end of each chapter is a summary of the main points, followed by a number of self-assessment questions which allow students to test their understanding. Past examination questions from popular examination papers are included. Some of the questions are more searching but students will learn valuable lessons by attempting all of them. Answers with outline solutions are provided at the back of the book, together with some useful tables of mathematical and physical data.

It would be all too easy to write a book that simply listed particles and their discoveries together with physical formula, all quoted in a rote fashion. In *Particle Physics* I have attempted to explain the subject both in a historical as well as a conceptual context, so that the reader can get a feel for how the various ideas of particle physics have developed. I have also tried to enlighten the reader as to the origin and meaning of the colourful and bizarre terminology in which the subject abounds!

Particle physics is a fast-moving subject with new discoveries constantly being made, and, for this reason, an associated website at www.ph.surrey.ac.uk/starbase/pp has been created. This is maintained and run by the Physics Department at the University of Surrey, and is under continual review. As well as finding here an abridged version of the text, students will be able to 'hot link' to sites of current interest in particle physics that will keep them abreast of the latest developments. It is hoped that student and teacher alike will find this book plus the website to be a valuable resource for their understanding of the exciting subject of particle physics.

Christopher Bishop

Acknowledgements

Many people have been involved in the production of *Particle Physics*. It gives me pleasure to express my thanks to the following: Katie Mackenzie Stuart (Science Publisher) who encouraged me to start writing educational physics texts; Jane Roth (Project Manager) whose excellent editorial and production skills ensured that the book saw the light of day and cajoled me to deadlines in the nicest possible way; Geoff Amor who edited the manuscript and made many valuable suggestions concerning the presentation of text and illustrations; Jon Butterworth of University College London, who read through the text and checked it for technical accuracy; Liz Moore who researched the photographs and discovered some I never knew existed.

I am also grateful to the following for their advice during the preparation of the manuscript: Dr Keith Fuller, formerly of Guildford High School for Girls; Steven Beith and Kathryn Jarman of the Arnewood School New Milton, who made many valuable comments and suggestions to the text from a teacher's perspective; Professor Phil Walker of the Physics Department, University of Surrey, for useful comments on explaining the Heisenberg Uncertainty Principle and solutions to some of the self-assessment questions; Professor David Owen, Ben Gurion University, Israel, for many interesting discussions about symmetry and conservation laws; also Professor Ron Johnson, Dr Stuart Vincent and Dr Jim Al-Khalili of the Surrey Physics Department for clarifying some points in nuclear physics.

I would also like to thank the following colleagues at the University of Surrey for their support: Professor Bill Gelletly, Head of the School of Physics and Chemistry, Professor Jeff Tostevin, Head of the Physics Department, and Dr Alan Emsley, Administrator of the School of Physics and Chemistry.

I am grateful to Mieke van den Bergen of Fermilab for permission to reproduce details of the Tevatron from their own publicity material.

Thanks also to my wife Sally for putting up with the decorating not being done, and to Ciaran, for being patient when Dad commandeered the computer. Finally, this book is dedicated to my three-year-old daughter Charis – a very strongly interacting particle who has plenty of 'charm' and knows how to use it.

Christopher Bishop

In memory of my mother, Patricia Catherine Bishop, 1923–2002.

Examination questions

Examination questions have been reproduced with kind permission from the following boards. The answers provided have been written by the author, and the examination boards bear no responsibility for their accuracy.
Edexcel (formerly ULEAC)
OCR (incorporating OCSEB, UODLE)

Source acknowledgements

Thanks are due to the following for permission to reproduce copyright extracts:
Figure 3.22 Institute of Physics Publishing, from *Advancing Physics* course materials, reproduced with permission; **p.184** John Gribbin, *Q is for Quantum*, Weidenfeld and Nicolson (1998); **p.210** © Feynman Lectures by the Division of Physics, Mathematics and Astronomy at the California Institute of Technology, Pasadena, CA 91125, USA

The following are sources from which diagrams have been adapted:
Figures 1.13, 8.5 W. J. Kaufmann, *Universe*, W. H. Freeman (1968); **Figure 1.15** R. A. Millikan, *Phys. Rev.* 7, 355 (1916); **Figures 2.1, 2.9** Halliday, Resnick, Walker, *Fundamentals of Physics* (1997); **Figures 3.19, 3.21** Institute of Physics, *Advancing Physics* course materials; **Figure 4.2** Caro, McDonell, Spicer, *Modern Physics*, Edward Arnold (1971); **Figure 4.18** C. R. Kitchin, *Astrophysical Techniques*, Institute of Physics Publishing (1998); **Figure 4.21** F. Close, M. Marten, C. Sutton, *The Particle Explosion*, Oxford University Press (1987); **Figures 4.24, 4.27** Institute of Physics/The Open University, *Teachers' Guide*, Institute of Physics Publishing (1992); **Figures 6.6, 6.10** The Open University, M203 *Introduction to Pure Mathematics*, Unit 1 Symmetry (1978); **Figures 6.12, 6.16** The Open University, S354 *Understanding Space and Time*, Block 3 Units 8, 9; **Figures 6.15, 8.9** J. Silk, *A Short History of the Universe*, W. H. Freeman (1997); **Figure 8.1** The Open University, S281 *Astronomy and Planetary Science*, Block 4

Sources of copyright photographs are listed on page 312.

Early ideas about matter and energy

1

What is particle physics?

Have you ever walked along a seashore and seen the waves rolling in, the sun shining brightly, the wind blowing sand along the beach, or perhaps watched a spectacular thunderstorm in progress with dramatic displays of forked lighting (Figure 1.1)? Have you ever wondered what the basic forces are that govern all the different phenomena that you can see? Why do magnets attract and repel each other? What causes electricity? What is the source of energy that makes the stars shine? And what exactly is this force called gravity that holds us to the Earth? To answer these questions, we need to understand how nature builds a relationship between matter and energy. **Particle physics** is the study of matter and energy at its most fundamental level. **Particle physicists** seek to understand the basic structure of matter and the fundamental forces or **interactions** that govern the workings of the universe.

Figure 1.1 Examples of the forces of nature at work

Particle physicists use machines called **accelerators** to accelerate beams of sub-atomic particles to very high kinetic energies and then make them hit a target material or collide them together to produce more particles. Using special detectors and powerful computers, the experimenters hope to learn more about the various physical relationships that connect different kinds of particles and how the particles can combine together to make more complex forms of matter. You can imagine particle physics as being rather like smashing a watch with a hammer and deducing how it works by examining the wheels, cogs and springs that emerge in the debris. This is why particle accelerators are popularly called 'atom smashers'.

The particles that emerge from accelerator experiments have enabled particle physicists to construct a theory of matter and forces called the **Standard Model**. The Standard Model has been very successful in explaining many of the physical phenomena of nature as well as how matter is organised at the sub-atomic level. Particle physicists have discovered many different kinds of particles. The protons and neutrons in the atomic nucleus are now understood to be composed of elementary units called **quarks** arranged in different combinations. Particles that cannot be broken down into smaller units are called **elementary**, which is why particle physics is also known as **elementary particle physics**, and you have almost certainly heard of the elementary particle known as the electron.

These are exciting times in the history of physics. In this book, you will see how a process of successive discoveries over the past 100 years has led the way in formulating our modern understanding of matter. Particle physics involves some very advanced mathematics, but you do not have to be a mathematical wizard to understand the basic ideas involved. As we will see, a key concept that particle physicists use is that of **symmetry**, or a deep order in the structure of matter that enables them to make sense of the diverse range of particles and their interactions with each other (Figure 1.2). Of great significance is when the symmetry of nature is broken, that is, when the regular order is violated in some way, and particle physicists have found that **broken symmetries** provide important clues to understanding the underlying processes of nature. Within the last 50 years we have come to see that understanding the very small leads to an understanding of the very large, including the question of how the universe began and what may happen to it in the future. Yet how far will particle physicists get in understanding the nature of matter? Are they close to a complete description of the physical universe in terms of its basic building blocks?

Figure 1.2 Symmetry is inherent in nature, such as in these diatoms

Well, some believe that we will be able to obtain a complete theory of matter and forces and that then we will have reached the end of the road – and for physics, there will only remain the development of new technologies. Others think that there will always be something infinitely smaller in the structure of matter and, just like a Russian doll, there may still be elementary particles waiting to be discovered. Finally, there are those who think that particle physics will be forced to recognise its limitations due to insuperable technical difficulties in building ever larger accelerators and the complex mathematical problems that will need to be solved theoretically, and this will limit the frontiers of knowledge.

Whichever of these views is right, there are still many questions that remain unanswered about the nature of our physical existence, and particle physics is firmly at the forefront of attempts to answer them. Whether or not we will ever be able to construct a 'Theory of Everything (TOE)' that we can pass on to our descendants is still uncertain but, whatever the outcome, you can be sure that particle physics will continue to remain one of the most exciting and challenging fields of human endeavour for many years to come! Before we describe the progress of particle physics to date, let's first take a look at some early ideas about the nature of matter and energy, including some key discoveries that helped lay the foundations of modern particle physics.

Early ideas about the nature of matter

The ancient Greeks were among the first to speculate about the nature of matter, and many of their philosophers formed schools where fundamental questions of nature and morality would be debated. We begin with Democritus of Abdera, who lived from about 470BC to about 380BC (Figure 1.3).

Figure 1.3 Democritus (c.470–380BC) explaining that matter consists of indivisible *atomos*

For an ancient philosopher, Democritus was remarkably up to date. For example, well before the telescope was invented he believed that the Milky Way was a vast collection of stars, and even our modern system of government, *democracy*, is named after him. Democritus held the view that all matter consisted of particles that were so tiny they could not be broken down into smaller pieces. He coined the word *atomos* for these particles, which literally means 'indivisible'. In Democritus' view, these **atoms** were physically different from each other, so that 'atoms of water' were smooth because water flowed and had no discernible shape, 'atoms of fire' were thorny, which was why fire gave you painful burns, and 'atoms of earth' were rough and jagged so that they stuck together to form hard materials. The observed changes in matter arose due to the separation of different atoms and their re-forming in new patterns. But while the idea of atoms having surface features like this may seem odd to us today, thanks to modern accelerators and detectors we can see that Democritus' essential idea of indivisible particles was very nearly right.

Another influential philosopher at that time was Empedocles (c.430–390BC). He combined the earlier ideas of Thales (c.640–540BC), who thought that the basic element of matter was Water, his contemporary Anaximenes (611–546BC), who thought it was Air, and Heraclitus (c.540–475BC), who thought it was Fire. To this Empedocles added one of his own: Earth; thus was born the notion of all matter being made out of differing amounts of Air, Fire, Earth and Water (Figure 1.4).

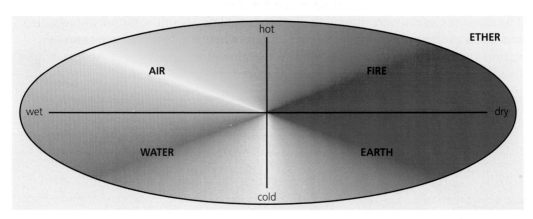

Figure 1.4 The 'elements' of Aristotelian physics. Matter was regarded as being composed of four elements: Air, Fire, Earth and Water. Aristotle added a fifth element, the *Ether* (or *Aether*), from which the heavens were made

The most renowned of the Greek philosophers was Aristotle (Figure 1.5), who lived a little later (384–322BC). He founded the famous school of philosophy in Athens called the Lyceum. His lectures on nature were collected into many volumes in which he entirely rejected Democritus' idea of atoms. Aristotle perpetuated Empedocles' view of matter and his thinking dominated philosophical thought until the Scientific Revolution of the 16th and 17th centuries, which overturned much of the old Aristotelian physics and brought us back to Democritus' idea of atoms.

Figure 1.5 Aristotle (384–322BC) in discussion with members of the School of Athens

Dalton's atomic theory and the Periodic Table of the elements

Sir Isaac Newton's (1642–1729) scientific treatise *Principia Mathematica* of 1687 heralded the age of reason whereby physical phenomena, including the dynamics of moving bodies, could be explained by careful observations and the correct application of mathematics. Newton's own experiments with light, for instance, led him to propose a particle theory in which light was composed of 'corpuscles' – an interesting idea that anticipates the modern concept of the photon. But it was Newton's legacy of the **scientific method** – observation, theory and measurement – that was the foundation stone of all contemporary physics.

Figure 1.6 John Dalton (1766–1844)

Our modern understanding of matter may usefully be traced back to the atomic theory of John Dalton (1766–1844), an English chemist (Figure 1.6). Dalton was interested in the properties of gases, particularly the experiments of Robert Boyle (1627–1691), from which came the familiar Boyle's Law. Dalton thought that all gases consisted of tiny particles, and he contributed to the theory of gases by founding a law that we know today as **Dalton's Law of Partial Pressures**. This states that the pressure of a mixture of gases is the sum of the partial pressures of its constituents. Dalton, however, went further than this and asserted that *all* matter and not just gases must consist of

5

small particles. Dalton at once saw the similarity of this to the atomic theory of Democritus 21 centuries earlier and, using Democritus' own term, he called his tiny particles of matter *atoms*. Whereas Democritus' idea was a purely philosophical one, Dalton's was based on observation and measurement. Dalton's atomic theory held that all substances are composed of atoms of the elements in different proportions. This idea was supported by Joseph Proust (1754–1826), a French chemist who showed that copper carbonate always contained precise proportions by mass of copper, carbon and oxygen, regardless of how it was prepared, that is, always 5 parts of copper, 4 parts of oxygen and 1 part of carbon. From this, Proust generalised that all samples of a chemical compound contain the same elements in the same proportions. This is known to chemists as **Proust's Law** or the **Law of Definite Proportions**.

Dalton's theory also stated that one substance could change into another by breaking a particular arrangement of atoms and forming a new one. All the atoms of one element, said Dalton, were exactly identical, and the atoms of each element were different from the atoms of every other element. Furthermore, the atoms differed from each other only in mass. Now, using the scientific method, this last feature was something that could be measured experimentally, and, from the properties by mass of the chemical elements known at that time, Dalton attempted to calculate the relative masses of the different atoms. He even drew up a table of atomic masses, which in this respect preceded the modern Periodic Table of the elements.

However, it was Dmitri Ivanovich Mendeleyev (or Mendeleev) (1834–1907), a Russian chemist, who brought order to Dalton's atomic theory. When Mendeleyev placed the elements in order of atomic mass, he noticed that the property of *valence*, or the ability that each type of atom has for combining with other atoms, had a periodic nature. For example, the first atom in Mendeleyev's list was lithium, which had a valence of 1. Beryllium has a valence of 2. Next is boron with a valence of 3 and then carbon with a valence of 4. Mendeleyev found that he could arrange all the 63 elements then known in a table in order of their atomic mass and get periodic rises and falls of valence, 1, 2, 3, 4, 3, 2, 1, etc. He then arranged them in rows under each other so that elements with the same valence all fell into a vertical column. Mendeleyev saw that when the elements were placed in this way, the columns of elements tended to show similar chemical properties. It is because of this periodicity of valence that Mendeleyev's table is called a *periodic* table. Mendeleyev published his table in 1869, and this was an important first step in bringing classification and order to matter. In 1955, a newly discovered element number 101 in the Periodic Table was named Mendelevium in his honour. The modern Periodic Table is given on page 290.

Electric and magnetic forces

Natural magnetism began to be noticed when lumps of iron ferrite or **lodestones** were first discovered. Thales wrote about them, and the ancient Greeks gave the stones the name *magnes* (magnets) as they could be abundantly found on the island of Magnesia in the Aegean Sea. Magnetic compass needles are known to have been in use by mariners by AD1100, although we cannot be sure when they were first used for navigation.

Thales in his writings had also noted that a fossil resin, which we call amber, had the ability to attract feathers, cotton threads and bits of fluff when it was rubbed against fur. This was the first inkling of the existence of **static electricity**. William Gilbert (1544–1603), an English physician and physicist, began the first serious investigation of magnetic and electrostatic phenomena. He wrote a book *De Magnete* ('Concerning Magnets') in which he showed that a compass needle points in a northerly direction and also dips towards the Earth. From these observations, Gilbert suggested that the Earth was a giant spherical magnet and that the compass needle pointed towards the poles. He also studied the attractive properties of amber and found that other substances such as rock and a variety of gems behaved in a similar fashion. All the 'rubbing' substances he grouped under the name 'electrics' from the Greek word for amber, *elektron*.

By the mid-1600s, the German physicist Otto van Guericke (1602–1686) had mechanised Gilbert's 'electrification by rubbing' phenomenon by producing the first electric machine that could generate static electricity by friction. In 1765 at the University of Leiden in Holland, it was noticed that when one of these machines was connected to a **Leyden jar** (this was a metal-lined glass jar with a conducting rod inserted in it via a cork), large amounts of charge could be stored. The jar could be discharged by touching it with the hand, or when metal was brought near a spark could be seen to leap across the air gap accompanied by a crackling noise. This fact was noted by the American statesman and scientist Benjamin Franklin (1708–1790), who wondered if lightning and thunder involved the storage of charge in a manner rather like a giant Leyden jar. In order to test this hypothesis, Franklin flew a kite with a pointed wire on it in a thunderstorm with a long silk thread attached to a metal key. As the lightning flashed the key became electrically charged and Franklin was able to charge up a Leyden jar with it, demonstrating that lightning is indeed an electrical phenomenon.

Safety note: Don't try this! Flying kites during thunderstorms is extremely dangerous, involving a high risk of electrocution and death. Franklin was most fortunate not to kill himself.

Franklin carried out other experiments with static electricity. He observed that two amber rods (as did two glass ones) repelled each other when they were rubbed and electrified. However, an electrified amber rod *attracted* an electrified glass one. To explain this, Franklin supposed that electricity was a kind of 'fluid' that could exist either in an excess or as a deficit. Two electrified substances containing an excess of fluid repelled each other, as did two that had a deficit. On the other hand, an electrified substance with an excess and one with a deficit attracted each other – the 'fluid' flowed from one into the other, as it were. Franklin called an excess of fluid 'positive electricity' and a deficit 'negative electricity'. Although Franklin's concept of electricity as a fluid was incorrect, he was the first to anticipate that electricity possessed the property of **polarity**, that is, positive and negative charge.

Figure 1.7 Michael Faraday (1791–1867)

It was Michael Faraday (1791–1867) (Figure 1.7) who in a series of classic experiments showed that there was a link between electrical and magnetic phenomena. Faraday was intrigued by the experiments of Hans Oersted (1777–1851), a Danish physicist, who showed that an electric current could deflect a magnetic needle. Faraday constructed the first electric motor and thereby showed that magnetic and electric forces could produce mechanical movement. However, Faraday wanted to see if the symmetry of Oersted's discovery held – could magnetism give rise to an electric current? In a famous experiment Faraday showed that this was indeed the case and this led to the invention of the first generator.

Faraday was also aware of the work of André Marie Ampère (1775–1836), a French physicist, who demonstrated that magnetic force circled a wire carrying an electric current – you will have seen this when iron filings are sprinkled on to a sheet of paper perpendicular to a current-carrying wire. Faraday had no mathematical training and began to imagine the magnetism stretching out around the electric current as 'lines of force' (Figure 1.8). Lines could be drawn through points where the force had the same magnitude, rather like a contour map. It seemed to him that the iron filings followed these lines of force and, by using the iron filings in this fashion, he could work out the pattern of force lines for other sources of magnetism such as bar and horseshoe magnets. In

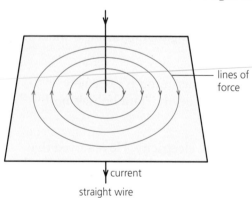

lines of force

current
straight wire

Figure 1.8 Faraday's 'lines of force', constituting a magnetic field pattern

this way Faraday, as well as establishing a link between electricity and magnetism, introduced the idea of a magnetic **field**. As we shall see later in this book, the idea of a field is an essential key concept in modern particle physics.

Faraday also investigated the process of **electrolysis**, which is the chemical decomposition of certain substances when an electric current is passed through the substance in a dissolved or molten state. His **Laws of Electrolysis**, which related the chemical decomposition to the magnitude of the electric current, provided strong evidence that electricity might be particulate in nature, although Faraday himself was not an enthusiastic believer in atoms.

Faraday went on to make further discoveries in **electromagnetism**, ushering in our electricity-dependent civilisation of today.

Maxwell's theory of electromagnetism

Figure 1.9 James Clerk Maxwell (1831–1879)

The Scottish physicist James Clerk Maxwell (1831–1879) worked out a theory that put Faraday's lines of force on a firm mathematical basis. Maxwell (Figure 1.9) showed that a few simple equations could neatly encapsulate all the known phenomena of electricity and magnetism. In Maxwell's theory, electricity and magnetism are inextricably bound together. An electric charge that is oscillating will radiate an **electromagnetic field** which moves outwards from the source at a constant velocity. This velocity, according to the equations, has a value equal to the reciprocal square root of the product of two physical constants, that is, an electric force constant $\epsilon_0 = 8.85 \times 10^{-12} \, \mathrm{F \, m^{-1}}$ (the permittivity of a vacuum, which appears in **Coulomb's Law** for the force between charged particles, $F = (1/4\pi\epsilon_0)Q_1Q_2/r^2$ and a magnetic force constant $\mu_0 = 4\pi \times 10^{-7} \, \mathrm{N \, A^{-2}}$ (the permeability of a vacuum, which appears in the SI definition of the ampere):

$$\text{velocity of the electromagnetic field} = \frac{1}{\sqrt{\mu_0\epsilon_0}}$$

Maxwell noticed that this value was remarkably close to the velocity of light as previously measured by two French physicists, Armand Fizeau (1819–1896) and Jean Foucault (1819–1868). Could it not be, he said, that the essence of light was due to an oscillating electric charge? Furthermore, according to his theory, the electric charge could oscillate at *any* frequency so there should exist an **electromagnetic spectrum** of electromagnetic radiation, of which visible light is but a small part. Although ultraviolet and infrared radiation were already known, Maxwell predicted that electromagnetic radiation should exist far beyond both infrared and ultraviolet frequencies. This was experimentally confirmed by the German physicist Heinrich Hertz (1857–1894), who demonstrated the existence of radio waves with his spark-gap apparatus.

At this time the accepted view was that light consisted of a transverse wave motion, which propagated through a luminiferous (light-carrying) **ether** that occupied all space, since it was taken as self-evident that any wave needed a medium to travel through, although no one had any idea what the ether might be made of. The ether was considered to be 'at rest' with respect to space, and Maxwell believed that Faraday's magnetic lines of force were distortions through the ether, suggesting the idea of 'action by contact' rather than 'action at a distance'. However, while Maxwell's theory was an accurate mathematical description of electromagnetism, it was soon shown that the idea of the ether as a propagating medium could not be right.

The Michelson–Morley experiment

Albert A. Michelson (1852–1931) and Edward W. Morley (1838–1925), two American scientists, showed that the concept of a universal ether permeating through all space must be wrong. Maxwell had determined the value of the speed of light c, and 19th century physicists believed that this was its speed with respect to the ether. The concept of the ether had been invoked because light and associated phenomena such as interference effects were understood as a wave motion – and all known waves propagated through a medium. But, could the ether itself be detected?

It was imagined that the ether flowed past the Earth rather like a river flowing past a small island. If this were so, then light from a source travelling in the same direction as the Earth's relative motion through the ether ought to travel more rapidly than light travelling perpendicularly to it. Michelson and Morley used an optical arrangement called an **interferometer** to look for a shift in the interference fringes that would be produced by the different beams having different effective path lengths due to their differing velocities (Figure 1.10).

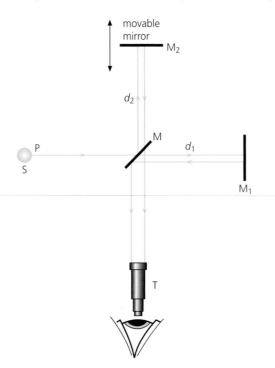

The apparatus is essentially an interferometer, which splits the path of light originating at a point P of an extended source S. Mirror M splits the light into two beams, which reflect from mirrors M_1 and M_2 back to M and then to a telescope T, where the observer sees a pattern of interference fringes. The path length difference for the two light beams when they recombine at T is $2d_2 - 2d_1$. The Earth's motion through the ether (if it existed) would cause a change in the effective path difference and a shift of the interference fringes when the entire instrument was turned though an angle of 90°. No such shift was observed.

Figure 1.10 Michelson and Morley's apparatus

Within the bounds of experimental error, Michelson and Morley could detect no such shift in the interference fringes, whatever the orientation of the interferometer. The speed of light seemed to be constant in whichever direction it was travelling relative to an observer, and at a stroke their result demolished the notion of the ether. Some other explanation, therefore, was needed to explain the propagation of light and its constant velocity to all observers.

The Special Theory of Relativity

It was the Irish physicist George F. FitzGerald (1851–1901) who in 1895 put forward a novel explanation for the null result of the Michelson–Morley experiment. His idea seems to go against all intuition, but nonetheless has been shown to be true experimentally.

FitzGerald proposed that all matter *contracted* in the direction in which it was moving and the amount of contraction increased with velocity. According to FitzGerald, Michelson and Morley's interferometer had contracted in length in the direction of the Earth's motion by an amount equal to the difference in distance that the light beam had to travel. So it now seemed that nature did not recognise the concept of 'absolute motion' or 'absolute space', which had been a central tenet of the physics of Newton.

A similar phenomenon occurs with mass. The Dutch physicist Hendrik A. Lorentz (1853–1928) was interested in the dynamics of charged particles. He was unaware of FitzGerald's paper but suggested a similar idea; in addition, he proposed that the mass of a charged particle would increase as its length decreased. He reasoned that, assuming the mass of a charged particle was due to the potential energy of its own charge, if the particle was compressed into a smaller space due to its length contraction then its potential energy (which is a function of the reciprocal of the distance from the charge) would increase, and hence its mass would increase. In fact the mass of a particle is related not to its charge but, as Einstein went on to show, to its total energy. Nonetheless Lorentz's mass increase is found experimentally to be true and was put on a firmer theoretical basis by Einstein.

Albert Einstein (Figure 1.11) pointed out that without the ether it was not possible that one could regard *anything* as being at absolute rest or in absolute motion. All the motion we observe, said Einstein, must be viewed from the observer's point of view or **frame of reference**. This is arbitrarily chosen (usually for convenience) and all measurements are made relative to it. Taking the result obtained by Michelson and Morley, Einstein made two key assumptions or *postulates* which he incorporated into his **Special Theory of Relativity** of 1905.

Figure 1.11 Albert Einstein (1879–1955)

Postulate I   *The laws of physics are the same in all inertial frames of reference.*

Postulate II   *The speed of light is the same in all inertial frames of reference.*

By an **inertial frame of reference** we mean a reference frame that is not accelerating, that is, it is at rest or moving with uniform velocity.

Einstein showed that if the constancy of the speed of light was assumed and all motion was considered relative, Michelson and Morley's result could be explained, Maxwell's equations of electromagnetism remained unchanged, and the length contraction and mass increase equations of FitzGerald and Lorentz could be derived. Note that Maxwell's expression for the velocity of an electromagnetic field $1/\sqrt{\epsilon_0 \mu_0}$ contains only physical constants, which according to Einstein's first postulate must be the same in all reference frames and whose values do not depend on motion.

The Special Theory also led to another bizarre effect – the rate at which time passed varied with the velocity of motion. Einstein's equations tell us that moving clocks run slower than those at rest. A consequence of this 'time dilation' is that all notions of **simultaneity** have to be abandoned. We can no longer say that two events separated in space are simultaneous if the rate at which time passes depends on the relative motion of an observer's clock in a particular reference frame. Space and time are no longer separate entities but linked together in a **space–time continuum**.

These effects of length contraction, mass increase and time dilation with velocity are real, but we do not normally observe them in everyday life, since they only become pronounced at velocities approaching an appreciable fraction of the speed of light. The particles in particle accelerators, however, routinely travel at near light speeds and, as we will see later, relativistic effects need to be taken into account when making measurements on them. We will take a closer look at the effects of Special Relativity in Chapter 3.

Energy and mass

Another important result from the Special Theory of Relativity is its assertion that energy and mass are *equivalent*. It is expressed as the **mass–energy relation**:

$$E = mc^2$$

where E is the energy, m is the mass and c is the speed of light. Since c is such a large number, a small amount of mass converts into a large amount of energy. The mass–energy relation tells us that mass and energy are different aspects of a single phenomenon. While we still have the law of conservation of mass and the law of conservation of energy for most everyday situations, there is now a greater generalisation, the **conservation of mass–energy**. The mass–energy relation is very important in particle physics as it explains how matter can be created from energy and vice versa; it is the key to how particles can be 'manufactured' in particle accelerators.

The Special Theory was published in 1905 and marked the beginning of a fundamentally new way of looking at the universe in which the 'absolute' nature of space and time as assumed in Newtonian physics was gone forever.

The quantum nature of matter

Figure 1.12 Max Planck
(1858–1947)

In 1900, the German physicist Max Planck (Figure 1.12) put forward a revolutionary idea that electromagnetic energy is absorbed and emitted in discrete 'packets'. Planck was concerned with **thermal radiation**, particularly the properties of 'blackbodies'. In thermodynamics, a 'blackbody' is an ideal object that absorbs *all* radiation that falls upon it; such an object does not reflect any radiation, and hence it would appear black. We would also expect a blackbody to be an ideal *emitter* of radiation. By 'ideal' here we mean that the amount of radiation emitted by a blackbody and the relative intensities present depend only on its *temperature* and not on any other properties such as its chemical composition.

The total radiation emitted by a body when it is heated was investigated by the Austrian physicist Josef Stefan (1835–1893) and is formulated as **Stefan's Law:**

$$P = \sigma A T^4$$

where P is the power radiated by a body of surface area A at temperature T and σ is **Stefan's constant**, whose value is $5.7 \times 10^{-8}\,\mathrm{W\,m^{-2}\,K^{-4}}$. As the temperature of a blackbody rises, the peak intensity of the radiation shifts towards shorter wavelengths. To see this, imagine an iron bar being heated to increasingly higher temperatures (Figure 1.13). At first it will start radiating in the infrared, then it will glow dull red, followed by cherry red, then orange, then yellow-white and finally (if we could keep it from completely vaporising) it would glow blue-white.

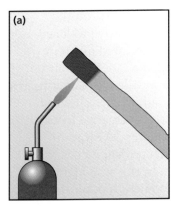
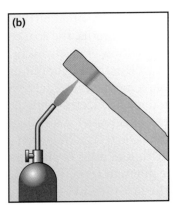
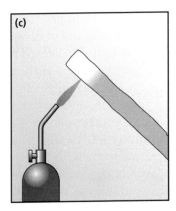

Figure 1.13 Heating a bar of iron: as the temperature of the bar increases, the amount of energy it radiates increases and the dominant wavelength of light emitted decreases, causing the glowing iron to change colour. This phenomenon can be explained using the Planck radiation theory

The distribution of radiant energy with wavelength emitted from a blackbody has a characteristic curve (see Box 8.2, page 248). In order to explain this shape and the relative intensities of the energy when a blackbody is heated, Planck introduced the notion that the energy is radiated in the form of particles with zero mass. He called these particles **quanta** from the Latin meaning 'how much'. The size of each of these quanta, or the 'quantum' of energy E, was directly proportional to the frequency f of the radiation, as given by the **Planck Relation**:

$$E = hf$$

where h is the **Planck constant**, which is now recognised as one of the fundamental constants of the universe. Its preferred value is $6.62606876(52) \times 10^{-34}$ J s (the brackets indicating uncertainty in those decimal places).

The Planck Relation is a very important equation in particle physics as it gives rise to the concept of the photon. It is sometimes expressed in terms of the wavelength λ of the emitted radiation:

$$\text{since} \quad c = f\lambda \quad \text{then} \quad E = h\frac{c}{\lambda}$$

and this explains the form of the electromagnetic spectrum with high-energy short-wavelength gamma rays at one end and low-energy long-wavelength radio waves at the other.

While Planck did not suggest that light, or other forms of electromagnetic energy, existed as quanta (which we now call photons), his radiation theory marked the beginning of the **quantum theory** of matter and radiation, which is the basis for all our modern understanding of particle physics.

Einstein's explanation of the photoelectric effect

In 1905, Einstein applied Planck's quantum idea of energy to the **photoelectric effect** (Box 1.1). It was first noted by Hertz that, when ultraviolet light shone on the negative terminal of his spark-gap apparatus, the spark was more easily produced than with visible light. This stimulation of electrons by light, declared Einstein, occurred because a particular wavelength of light corresponded to a fixed quantum of energy ejecting an electron from a metal atom. Light with a wavelength larger than a certain critical value would produce no electron emission at all, whereas light of more energetic quanta (shorter wavelength) would cause more energetic electrons to be emitted. (The discovery of the electron is discussed in Chapter 2.) Einstein referred to these light quanta with zero mass as **photons**. Such critical values of wavelength would of course be different for different metals. In this way Planck's quantum theory was used for the first time to explain a phenomenon other than blackbody radiation. Einstein received the Nobel Prize for Physics for this work rather belatedly in 1921.

Box 1.1 The photoelectric effect

That light was quantum in nature was suggested by Albert Einstein in 1905 as an explanation for the **photoelectric effect**. Figure 1.14 shows the basic apparatus needed to demonstrate photoelectricity. The cathode C, consisting of a photosensitive metal of large surface area, together with a collector of electrons A, is enclosed in a vacuum. Light of wavelength λ is incident on the metal surface and electrons are emitted. Some of these electrons strike the collector and an electric current flows between C and A.

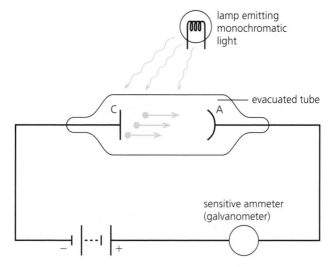

Figure 1.14 Simplified apparatus for demonstrating photoelectric emission

The maximum kinetic energy of the emitted electrons KE_{max} is found to be independent of the intensity of the incident light, and this is contrary to what we would expect from a classical description of light as a continuous wave motion. We would expect that increasing the intensity would simply increase the kinetic energy gained by the individual electrons. However, this is *not* what is observed. The maximum KE of the emitted electrons is the *same* for a given λ, no matter how intense the light is. Einstein suggested that this could be explained if light consists of particles called **photons** which travel as packets of energy of value $E_{photon} = hf$, where h is the Planck constant and f is the frequency of the incident light (or, using $c = f\lambda$, this can also be written as $E_{photon} = hc/\lambda$, where c is the speed of light).

The intensity of a beam of light is proportional to the number of photons per square metre per second, but the *energy* of a photon is proportional to its frequency and does not depend on the light intensity. A photon interacts with an electron in the metal and the photon disappears, giving all its energy to the electron. If ϕ is the minimum energy needed to remove an electron from the surface of the metal, then the maximum KE of the emitted electron is given by

$$KE_{max} = hf - \phi$$

This expression is known as **Einstein's photoelectric equation**. The quantity ϕ is called the **work function** and has different values for different metals.

Figure 1.15 shows a graph of KE_{max} versus frequency for a particular photoemissive metal (sodium). The graph is a straight line whose slope is equal to h. Photons with frequencies less than a certain **threshold frequency** f_t, with a corresponding **threshold wavelength** $\lambda_t = c/f_t$, do not have enough energy to eject an electron from the metal. By setting KE_{max} to zero, the threshold frequency can be found from the photoelectric equation as $f_t = \phi/h$ and the threshold wavelength as $\lambda_t = hc/\phi$. Typical values for work function energies are a few electronvolts (see page 29 for the definition of the electronvolt, eV).

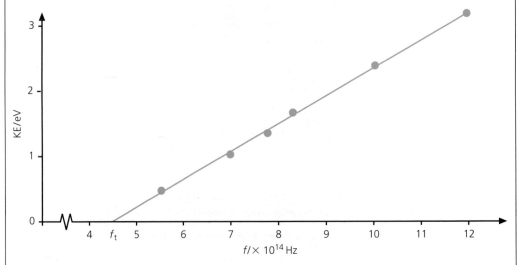

Figure 1.15 Kinetic energies of photoelectrons ejected from sodium by light of different frequencies. f_t marks the threshold frequency for sodium

---

WORKED EXAMPLE 1.1

Sodium is a metal that has a work function of 2.0 eV. Calculate the threshold frequency for which electrons are emitted.

At the threshold frequency $KE_{max} = 0$. Using the photoelectric equation $f_t = \phi/h$ (where $h = 6.6 \times 10^{-34}\,\mathrm{J\,s}$) and substituting $2.0\,\mathrm{eV} = 2.0 \times 1.6 \times 10^{-19}\,\mathrm{J}$,

$$f_t = \frac{2.0 \times 1.6 \times 10^{-19}\,\mathrm{J}}{6.6 \times 10^{-34}\,\mathrm{J\,s}} = 4.8 \times 10^{14}\,\mathrm{Hz}$$

Cathode rays and X-rays

By the middle of the 19th century, physicists were trying to send electrical discharges through evacuated glass vessels, or **Geissler tubes**. These were glass tubes that had been reliably evacuated of air to low pressures and which physicists could readily use for electrical experiments. They are so named after their inventor, the German glass-blower Heinrich Geissler (1814–1879).

Sir William Crookes (1832–1919), an English physicist, worked with a Geissler tube in which the **cathode** (or negative electrode) emitted a luminous glow when the tube was subjected to a strong electric potential (Figure 1.16).

Figure 1.16 Crookes (1832–1919) investigating cathode rays

In 1874 Crookes invented an improved Geissler tube in which this cathode radiation could be more effectively studied. In this **Crookes tube** he found that the **cathode rays** cast shadows when small objects inside the tube were placed in their path. In addition, cathode rays could turn the vanes of a small wheel when they were incident on one side, suggesting that the rays had mass. Crookes also showed that they could be deflected by a magnetic field and from this concluded that he was dealing with charged particles moving along in straight lines.

Wilhelm Konrad Röntgen (1845–1923), a German physicist (Figure 1.17), was working with a Crookes tube and was interested in the luminosity that cathode rays produced in certain chemicals. Since the luminosity was faint, Röntgen enclosed the Crookes tube in black cardboard and darkened the room so that he could observe it more clearly. To his surprise, he found that a sheet of paper coated with a chemical called barium platinocyanide, placed some distance from his apparatus, was glowing. It was glowing even though the cathode rays were blocked by the cardboard and could not possibly reach the paper. Röntgen found that this still happened when the Crookes tube was enclosed by several thicknesses of paper and even thin metal sheets. He concluded that some kind of penetrating radiation must be emerging from the tube. Since he didn't know what the radiation was, Röntgen named them **X-rays**. For this discovery, Röntgen was the first person to be honoured with the Nobel Prize for Physics in 1901.

Figure 1.17 Wilhelm Röntgen (1845–1923)

Radioactivity

Antoine Becquerel (1852–1908) was a French physicist who was particularly interested in the phenomena of fluorescence and phosphorescence – the absorption by some forms of matter of light at one wavelength and emission at another. Becquerel was aware of the discovery of X-rays by Röntgen and wondered if any of his fluorescent materials could be emitting them. He was working with a fluorescent compound called potassium uranyl sulphate which contained atoms of uranium. After inducing fluorescence using sunlight, he placed the compound on top of some wrapped photographic plates. Sure enough, the plates when developed were found to be exposed, showing the compound's physical outline. Becquerel therefore concluded that X-rays must be present in the fluorescence as radiation was passing through the wrapping. There then followed a period of cloudy weather when he could not continue his fluorescence experiments. There were some crystals of uranyl sulphate kept in a drawer in the laboratory which had not been exposed to sunlight for some days. They were lying on top of some photographic plates which Becquerel decided to develop to see if there had been any weak residual exposure. He was astonished to find that a plate was very strongly exposed, showing clearly the outline of the crystals. Becquerel at once realised that this could not be due to any fluorescent or X-ray effect and that the crystals were giving off radiation of their own account (Figure 1.18).

The Polish–French chemist Marie Curie (1867–1934) knew of Becquerel's unknown radiation (then called 'Becquerel rays') and gave them the name **radioactivity**. Marie Curie studied these emissions and found that the radioactive rays ionised the air and caused it to conduct electricity. It was also found that there were three distinct rays: alpha, beta and gamma. Together with her husband Pierre Curie (1859–1906) she experimented with different uranium compounds and showed that the intensity of the radioactivity was proportional to the amount of uranium they contained. This indicated that the radiation was coming from uranium atoms and, since some of it carried electric charge, provided the first evidence that the atom had an internal structure and might contain charged particles. Becquerel and the Curies received a share of the 1903 Nobel Prize for Physics for their work.

Figure 1.18 Becquerel (1852–1908) and his discovery of radiation

Einstein and gravity

The one force of nature that we are aware of all the time is **gravity**. Newton in his *Principia* had postulated his **Law of Universal Gravitation**, *'that the force between two masses* M_1 *and* M_2 *is inversely proportional to the square of the distance r between them'*, $F = GM_1M_2/r^2$. The constant of proportionality, G, is regarded as a fundamental constant of nature and its preferred value is $6.673(10) \times 10^{-11} \, \mathrm{N \, m^2 \, kg^{-2}}$. Newton, however, did not describe any mechanism of gravity. In fact we still don't know exactly what gravity is, although particle physicists are much closer now to finding out how it works. In 1915, Einstein published a paper referred to as the **General Theory of Relativity**, which dealt with the more general case of masses undergoing uniform *acceleration*, whereas the Special Theory considered only the motion of masses with uniform *velocity*. The Special Theory is in fact a specific case of the General Theory. Einstein presented a set of equations called the **Field Equations of General Relativity**, which describe how the motion of a body with mass is affected in the presence of a gravitational field. The mathematical form of these equations is beyond the scope of this book, but the important point is that the field equations do away with the concept of force and instead rely on a *geometrical* description of space and time where bodies move along paths where space has been 'curved' by the presence of mass. The larger the mass, the greater the curvature and deflection of the body's motion (Figure 1.19).

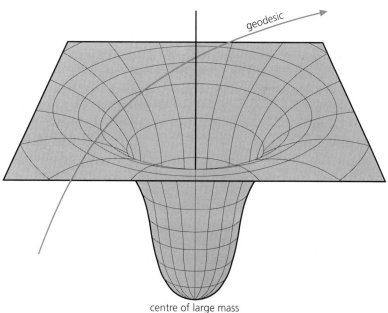

Figure 1.19 Gravity according to Einstein's General Theory of Relativity: matter curves space–time. The concept of force as found in Newtonian mechanics is not used. This diagram represents a two-dimensional analogy of the geometry of space–time near a massive object. Such an object moves along a path called a geodesic (see Chapter 8)

geodesic

centre of large mass

The General Theory of Relativity has been very successful in explaining a number of physical phenomena and has been verified experimentally many times. One consequence is that light should be deflected by a strong gravitational field (Figure 1.20). This was put to the test during the total solar eclipse of 19th March, 1919. A number of bright stars were in the vicinity when the Sun was in totality and, if the

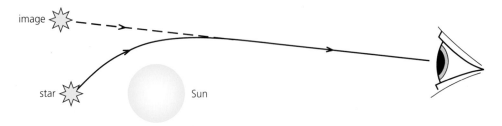

Figure 1.20 'Bending' of starlight by the Sun, as proposed by the General Theory of Relativity

theory was correct, their apparent positions should have altered slightly from those they occupied six months previously when they were observed in the night sky. This was due to the light photons from the stars being deflected as they passed close to the Sun. An expedition was sent to observe the eclipse and found that the altered positions were exactly in accordance (within experimental error) with the General Theory. Newton's Law of Gravitation is not however invalidated, as it turns out to be just a special case of the General Theory when the degree of curvature of space–time is not significant.

An important result of the General Theory is the **Principle of Equivalence**, which states that '*you cannot distinguish between a uniform gravitational field in a non-accelerating inertial reference frame and a uniformly accelerating (non-inertial) one*'. Here Einstein is saying that a free-fall acceleration can 'cancel out' the effect of a gravitational field. The Principle of Equivalence is used to train NASA astronauts in weightless conditions using a specially adapted aircraft nicknamed the 'Vomit Comet' (Figure 1.21). The pilot flies the aircraft over the top of an arc and for a few minutes the occupants are in a state of free-fall as the aircraft accelerates towards Earth under the influence of gravity. During this time, inside the aircraft, all objects mysteriously lose weight (including food in the stomach), the effect of gravity is neutralised and you can float about just as if you were in space!

In the last years of his life, Einstein tried to unify gravity and electromagnetism, but this task eluded him. The unification of gravity with the other forces of nature is still one of the outstanding problems of particle physics today.

Figure 1.21 Inside the 'Vomit Comet', free-fall is indistinguishable from the lack of a gravitational field

Classical and non-classical physics

We have now come to the dawn of the era of modern particle physics. In this chapter we have traced the development of key discoveries and introduced some of the individuals who have laid the foundations essential to our modern understanding of matter and energy. Starting with the philosophical speculations of the ancient Greeks, we have progressed to the scientific renaissance of Newton and then finally to the quantised and relative world of Planck and Einstein. The deterministic world view of Newton, in which time is an absolute entity and physical variables take an infinite range of values, is shown to break down under Special and General Relativity and the Planck radiation theory. However, Newtonian mechanics is still very useful, and in many cases provides a perfectly adequate description of nature. It is simply a special case in situations where quantum and relativistic effects are unimportant. For this reason, physicists sometimes talk about **classical** descriptions where quantum and relativistic effects can generally be disregarded when making measurements on physical systems, and **non-classical** ones where such effects are significant and must be included. This definition is far from exact, and some historians of physics regard all physics before 1900 as classical. Nowadays, non-classical physics is usually referred to as 'modern physics' and, for our purposes, any physical description of matter and energy in which measurements are allowed only discrete quantised values, such as those appearing in particle physics, is very definitely non-classical and modern!

As in every science, the way of discovery has been by speculation, careful observation, accurate measurement and sometimes just fortuitous circumstance. This process is still continuing, and in the rest of this book we will be looking at the modern development of particle physics and its extraordinary successes in explaining the fundamental nature of matter and energy in our universe.

Summary

◆ The ancient Greeks were among the first to speculate about the nature of matter. In particular, Democritus first put forward the idea of **atoms** as fundamental particles of matter. Democritus' ideas were superseded by the philosopher Aristotle's concept of matter being made out the 'four elements' Air, Fire, Earth and Water, which held sway until the scientific renaissance of the 16th and 17th centuries.

◆ The properties of magnetism were investigated by William Gilbert. Benjamin Franklin investigated static electricity and found that it possessed the property of **polarity**. Using the scientific method of Sir Isaac Newton based on observation and measurement, the atomic theory of John Dalton restored the concept of the atom, and Dmitri Mendeleyev introduced order and classification among the chemical elements by drawing up the first **Periodic Table**. Newton also stated in his **Law of Universal Gravitation** that every object attracts every other object.

◆ Michael Faraday, in a series of experiments, established a link between electricity and magnetism. He also introduced the idea of a **field**. James Clerk Maxwell provided a mathematical theory for Faraday's discoveries and predicted the existence of the **electromagnetic spectrum** consisting of **electromagnetic waves** propagating through a medium called the **ether**, which was at rest relative to all other objects in the universe.

◆ The **Michelson–Morley experiment** disproved the existence of a universal ether and established the constancy of the speed of light for all observers. George FitzGerald proposed **length contraction** to account for Michelson and Morley's observations. Hendrik Lorentz put forward **mass increase** as a consequence of length contraction.

◆ Einstein formulated the **Special Theory of Relativity**, which described the motion of bodies moving close to the speed of light. Newton's 'absolute' concepts of space and time were abandoned and measurements needed to be relative to a **frame of reference**. A consequence of the theory is that the passage of time depends on velocity and events take place in a **space–time continuum**. He also showed that energy and mass are equivalent.

◆ Max Planck proposed that electromagnetic radiation was particulate in nature and put forward the idea of the 'quantum' of radiation. Albert Einstein used Planck's radiation theory to explain the **photoelectric effect** and introduced the quantum of light called the **photon**.

◆ Wilhelm Röntgen discovered **X-rays**. Antoine Becquerel discovered **radioactivity**. The Curies investigated radioactivity and discovered that it comes in three forms, alpha, beta and gamma. They deduced that the atom may have an internal structure.

◆ Einstein formulated the **General Theory of Relativity**, in which gravity is explained in terms of the curvature of space by massive objects, and introduced the **Principle of Equivalence**. He developed the **Field Equations of General Relativity**, which describe how the motion of a body is affected in the presence of a gravitational field.

◆ Since the beginning of the 20th century, physicists have made a distinction between the continuous, deterministic description of matter and energy based on Newtonian mechanics, or **classical physics**, and the quantum **non-classical physics** of Planck and Einstein, which these days we call 'modern physics'.

Questions

1 Outline the main features of the Aristotelian model of matter. What observations led Dalton back to an atomic theory similar to that of Democritus?

2 The magnetic field strength B in teslas at a distance r around a current-carrying conductor may be described using the equation

$$B = \frac{\mu_0 I}{2\pi r}$$

where $\mu_0 = 4\pi \times 10^{-7}\,\text{NA}^{-2}$ is a magnetic force constant called the *permeability of free space*. If a current of 2 A flows in a conductor, what is the magnetic field strength at
a 1 m, **b** 2 m, **c** 3 m?
Plot a graph of B against r.

3 **a** How many electrons does 1 coulomb of charge correspond to?
[Take $e = 1.602 \times 10^{-19}\,\text{C}$.]
b Using Coulomb's Law

$$F = \frac{1}{4\pi\epsilon_0}\frac{q_1 q_2}{r^2}$$

calculate the force between a charge of $q_1 = +25.0 \times 10^{-9}\,\text{C}$ and a charge of $q_2 = -7.20 \times 10^{-9}\,\text{C}$ that are separated by a distance of $r = 6\,\text{cm}$.
Is the force attractive or repulsive?
c What charge q placed 12 cm from a charge of $-30.0 \times 10^{-8}\,\text{C}$ will produce an attractive force of 0.18 N?

4 The gravitational force between two masses m_1 and m_2 separated by a distance r is described by Newton's Law of Gravitation

$$F = G\frac{m_1 m_2}{r^2}$$

a Calculate the gravitational force between a proton and an electron in a hydrogen atom separated by a distance $0.5 \times 10^{-19}\,\text{m}$. [Use mass data for proton and electron given in the Tables of Particles on page 284.]
b What is the electrostatic Coulomb force between a proton and an electron in a hydrogen atom?
c What do you conclude about the effects of gravity on sub-atomic particles?

5 One result of Maxwell's equations is that the speed of light c may be written as the equation

$$c = \frac{1}{\sqrt{\mu_0 \epsilon_0}}$$

where $\epsilon_0 = 8.854 \times 10^{-12} \, C^2 N^{-1} m^{-2}$ is an electric force constant called the *permittivity of a vacuum*. The constant $\mu_0 = 4\pi \times 10^{-7} \, NA^{-2}$. Confirm that the result is equal to $3.0 \times 10^8 \, ms^{-1}$. Show also that the equation is homogeneous with respect to units.

6 What do you understand by the *quantum theory*? What is meant by 'classical physics'?
   a How many photons per second are emitted from a red light bulb with an output power of 60 W? [You may take the wavelength of red light to be 630 nm.]
   b In the photoelectric effect, the maximum kinetic energy of electrons emitted from a metal is $1.6 \times 10^{-19} \, J$ when the frequency of the emitted radiation is $7.5 \times 10^{14} \, Hz$. What is the minimum frequency of radiation for which electrons will be ejected from the metal? [Take the Planck constant as $6.6 \times 10^{-34} \, Js$.]
   c The release of energy E in a nuclear reactor is through the conversion of mass m into energy via Einstein's mass–energy relation $E = mc^2$. If the mass of fuel is converted into energy at a rate of $1 \, \mu g$ ($10^{-9} \, kg$) per hour, what is the rate of energy production in megawatts (MW)?

2 The nuclear atom

'It was quite the most incredible event that had ever happened to me in my life. It was as incredible as if you fired a 15-inch shell at a piece of tissue paper and it came back and hit you.'

Ernest Rutherford (1871–1937), New Zealand-born British experimental physicist

The discovery of the electron

The first elementary particle to be discovered was the electron. In Chapter 1 we saw how Crookes' experiments with evacuated tubes in the 1870s determined the properties of cathode rays. He found that cathode rays could be deflected by a magnetic field, but it wasn't until 1897 that the English physicist J. J. Thomson (1856–1940), using highly evacuated tubes, succeeded in deflecting them with electric fields and established beyond doubt that cathode rays consisted of tiny particles with negative electric charge. Thomson's apparatus was essentially a **cathode ray tube**, not unlike those found today in television sets. Electrons were emitted via **thermionic emission** at the **cathode** and accelerated towards the **anode** by a potential difference. They then passed through a pair of parallel plates creating a uniform electric field, which caused them to be deflected vertically downwards. Application of a magnetic field horizontally cancelled out the deflection due to the electric field and allowed the electron's charge-to-mass ratio e/m_e to be calculated (see Box 2.1).

The electron's mass and charge

While Thomson's apparatus enabled him to find the ratio e/m_e, it did not allow him to determine the values of e and m_e separately. In 1909 the American physicist Robert A. Millikan (1868–1953) devised an experiment based on the motion of charged oil drops in an electric field that enabled the value of the electron charge e to be determined. Figure 2.1 shows the essential features of Millikan's apparatus.

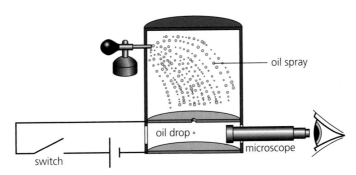

Figure 2.1 Millikan's oil drop apparatus for measuring the elementary charge e. When a charged oil drop drifted into the lower chamber through the hole in the upper plate, its motion was controlled by an electric field in the chamber

Box 2.1 Calculating the charge-to-mass ratio of the electron

Figure 2.2a is a diagram of the apparatus used by J. J. Thomson to measure the charge-to-mass ratio of the electron. A beam of electrons emerges from the cathode and is accelerated by the anode to a uniform velocity v_x. The beam then passes through a uniform electric field E provided by a pair of capacitor plates, while an electromagnet supplies a uniform magnetic field at right angles.

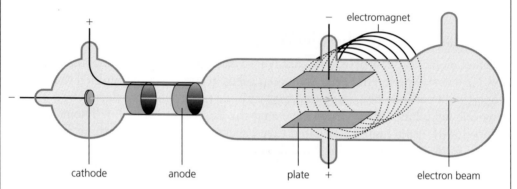

Figure 2.2(a) Thomson's apparatus for calculating e/m_e

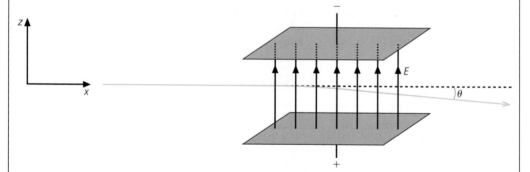

Figure 2.2(b) Deflection of the electron beam by the electric field

Now suppose that only the electric field is switched on. The electron beam will be deflected vertically towards the plate with positive polarity (Figure 2.2b). The force F_z exerted on an electron by the plates will be

$$F_z = -eE$$

and by Newton's 2nd Law the acceleration a_z will be

$$a_z = -\frac{eE}{m_e}$$

where m_e is the electron mass. The electron will now have a vertical velocity, while its horizontal velocity v_x remains constant. The initial vertical velocity u_z is zero, and the final vertical velocity v_z is $v_z = a_z t$. If the length of the plates is l, then the time t the electron spends in the electric field is $t = l/v_x$. We can therefore write v_z as

$$v_z = -\frac{eEl}{m_e v_x}$$

The angle of deflection θ of the electron beam is given by

$$\tan \theta = \frac{v_z}{v_x} = -\frac{eEl}{m_e v_x^2}$$

We now switch on the electromagnet and adjust the strength B of the magnetic field so that the force $-Bev_x$ on the electron due to the magnetic field exactly equals that due to the electric field. This brings the electron back to a horizontal path, with the net vertical force on the electron now being zero, that is

$$-eE = -Bev_x$$

so that

$$v_x = \frac{E}{B}$$

The ratio of the electric field to the magnetic field equals the unknown velocity of the electrons. Substituting for v_x in the expression for $\tan \theta$, we obtain

$$\tan \theta = \frac{elB^2}{m_e E}$$

or

$$\frac{e}{m_e} = \frac{E}{B^2 l} \tan \theta$$

This expresses the ratio of the electron charge to its mass, all in measurable quantities. The preferred modern value is

$$\frac{e}{m_e} = 1.758\,820\,174\,(71) \times 10^{11}\,\mathrm{C\,kg^{-1}}$$

A cloud of fine droplets of oil is sprayed into a dust-free chamber containing two plates across which a potential difference can be applied. As the oil drops enter the chamber, some of them become charged due to friction occurring during spraying, or from ions already present in the air. The drops are allowed to fall under gravity with *no* potential between the plates and are viewed through a low-power microscope. A single drop is observed until it is falling with terminal velocity v through the chamber in air of viscosity η. A viscous fluid will slow down the motion of an oil drop moving through it and the retarding force F_R is given by **Stokes' Law:**

$$F_R = 6\pi\eta r v$$

where r is the radius of the drop. At some terminal velocity v_1 there is no acceleration and the retarding force must equal the weight of the drop, so we can write

$$m_{\text{drop}} \times g = 6\pi\eta r v_1$$

where g is the acceleration due to gravity.

If we now turn the electric field on, the drop will accelerate upwards and attain a new terminal velocity v_2. The electric force produced is given by qE, where q is the charge on the drop and E is the electric field strength. The viscous force is now balanced by the electric force minus the weight, so that

$$qE - (m_{\text{drop}} \times g) = 6\pi\eta r v_2$$

If we divide the second of these equations by the first we get

$$\frac{qE - (m_{\text{drop}} \times g)}{(m_{\text{drop}} \times g)} = \frac{v_2}{v_1}$$

Rearranging, we can then write the charge on the drop as

$$q = \frac{m_{\text{drop}} \times g}{E} \times \left(\frac{v_1 + v_2}{v_1}\right)$$

The terminal velocities v_1 and v_2 are measured by timing the drops through a known distance on the microscope graticule. Several measurements of terminal velocities are made and the results averaged to give a value for q.

Millikan was careful to ensure that his apparatus gave accurate results. He used oil that was non-volatile to minimise evaporation, which would alter the mass of the drop. He maintained the chamber at a constant temperature to reduce convection currents between the plates and changes in the viscosity of the air. He also took into account inhomogeneities in the air composition that would cause departures from Stokes' Law and corrected for them.

By measuring the charges on a large number of drops, Millikan found that the charge on an oil drop was always an integral multiple of a lowest common value. His best value for the charge of a single electron was $-1.592 \times 10^{-19}\,\text{C}$. The modern preferred value is

$$e = -1.602\,176\,462(63) \times 10^{-19}\,\text{C}$$

The electron charge is an example of a **quantum number**. All particles that can be directly observed have integral multiples of this value: $0e$, $\pm 1e$, $\pm 2e$, $\ldots$, etc.

Using the known result for e/m_e, Millikan calculated the value for the electron mass. The preferred value for the mass of the electron is

$$m_e = 9.109\,381\,88(72) \times 10^{-31}\,\text{kg}$$

The electronvolt

The size of the electron charge can also be used as the basis for a unit of energy. We saw when considering the photoelectric effect (Box 1.1) that the energies of photons are given by the Planck Relation $E_{\text{photon}} = hf = hc/\lambda$. For example, a photon of visible red light with $\lambda = 630\,\text{nm}$ has an energy of

$$\frac{(6.67 \times 10^{-34}\,\text{J s}) \times (3.0 \times 10^8\,\text{m s}^{-1})}{(630 \times 10^{-9}\,\text{m})} = 3.2 \times 10^{-19}\,\text{J}$$

We can do a similar calculation for other electromagnetic waves. BBC Radio 4 broadcasts radio photons of frequency 91 MHz and hence energy $6 \times 10^{-26}\,\text{J}$, whereas in the gamma ray region of the electromagnetic spectrum a photon of frequency $10^{20}\,\text{Hz}$ would have an energy of about $7 \times 10^{-14}\,\text{J}$.

What you will notice is that if we measure the energies of photons in joules (J) then we get very small quantities. In order to make the numbers more manageable, physicists use a more convenient unit of energy – the **electronvolt**. One electronvolt (1 eV) is the energy gained by an electron when it is moved through a potential difference (p.d.) of one volt (1 V). Since the work done in moving a charge Q through a p.d. of V is given by $Q \times V$ then

$$1\,\text{eV} = \text{size of electron charge} \times 1\,\text{volt}$$
$$= 1.6 \times 10^{-19}\,\text{C} \times 1\,\text{V} = 1.6 \times 10^{-19}\,\text{J}$$

Converting energies in joules to electronvolts is easy – just divide by the value of the electron charge:

e.g. energy of a photon of red light $= \dfrac{3.2 \times 10^{-19}\,\text{J}}{1.6 \times 10^{-19}\,\text{C}} = 2\,\text{eV}$

Conversely, to convert eV to joules, multiply by the value of the electron charge:

e.g. energy of a 60 keV X-ray photon $= (60 \times 10^3\,\text{eV}) \times (1.6 \times 10^{-19}\,\text{C})$
$$= 9.6 \times 10^{-15}\,\text{J}$$

The electronvolt is not an SI unit, but it is widely used in particle physics. Some common multiples are:

kilo-electronvolts (keV)	$1\,\text{keV} = 1000\,\text{eV}$
mega-electronvolts (MeV)	$1\,\text{MeV} = 1\,000\,000\,\text{eV}$
giga-electronvolts (GeV)	$1\,\text{GeV} = 1\,000\,000\,000\,\text{eV}$
tera-electronvolts (TeV)	$1\,\text{TeV} = 1\,000\,000\,000\,000\,\text{eV}$

Essential features of an atomic model

Any model of the atom must account for its basic properties, which are the following:

1 *Atoms are very small.* They cannot be seen with even the most powerful light microscope. In fact atoms have sizes of the order of 10^{-10} m (Box 2.2), so that any attempt to image at the wavelengths of visible light will not work due to diffraction effects.

2 *Atoms are stable.* We do not generally observe matter breaking up spontaneously. This suggests that the forces that hold atoms together internally must be in balance, otherwise the atom would collapse.

3 *Atoms are electrically neutral.* We know from the photoelectric effect that electrons can be emitted from metals and this implies that atoms contain electrons. In addition electrons are emitted from atoms in some radioactive decay processes although, as we shall see later, the electron is 'manufactured' in the nucleus of an atom and is *not* present before the decay takes place. However, in general, large-scale matter is electrically neutral, which means that the atom must contain some distribution of positive charge to counteract the negative charge of the electrons it contains.

4 *Atoms emit and absorb electromagnetic radiation.* We know from the study of atomic spectra that a hot gas shows characteristic patterns of bright (emission) lines and that, when white light is shone through the gas, dark (absorption) lines can be seen. One example of this is the heating of sodium, which produces a distinctive bright yellow-orange light caused by the spectral emission lines in the yellow-orange part of the electromagnetic spectrum. This is what gives the yellow colour in common street lamps on motorways, where traces of sodium in the gas inside the lamps are excited to emission by the heating effect of an electric current. Any model of atomic structure must be able to account for the varied types of spectra observed from different elements.

Box 2.2 How big is an atom?

The properties of solids such as crystals can be modelled by assuming them to be made up of atoms arranged in a regular pattern called a **crystal lattice**. A technique called **Bragg diffraction** can be used to determine the atomic spacing. An X-ray beam of wavelength λ is incident on a plane of atoms where each atom scatters the X-rays and the result is that the plane reflects the beam rather like a mirror (Figure 2.3). By the principle of superposition, constructive interference of the X-rays reflected from adjacent planes occurs when the path difference is

$$AB + BC = 2d\sin\theta = m\lambda$$

where d is the spacing of adjacent planes, θ is the glancing angle and $m = 1, 2, 3, \ldots$ is the diffraction order. This is known as **Bragg's Law** after Sir William Henry Bragg (1868–1942), the English physicist who developed this X-ray diffraction technique.

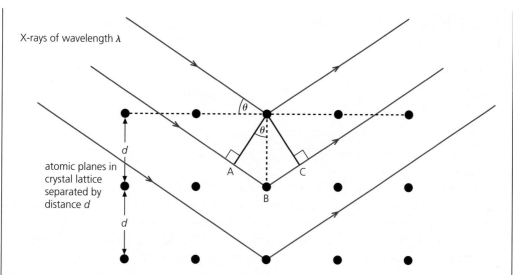

Figure 2.3 Bragg diffraction

If we assume that the atoms inside a solid are like spheres in contact, then we can find the radius of the spheres from the value of the spacing between the atomic planes. Bragg diffraction allows us to calculate the atomic spacing within molecules, and thereby atomic sizes.

WORKED EXAMPLE 2.1

A beam of X-rays of wavelength 0.250 nm is incident on a crystal of table salt (NaCl). The first-order diffraction maximum is observed at an angle of 26.3°. What is the atomic spacing of NaCl?

Using Bragg's Law,

$$d = \frac{m\lambda}{2\sin\theta} = \frac{1 \times (0.250 \times 10^{-9}\,\text{m})}{2 \times \sin 26.3°} = 2.8 \times 10^{-10}\,\text{m}$$

Working out the radius of the atoms depends on knowing how the atoms are packed in the solid. Figure 2.4 shows atoms packed in a solid in a simple cubic lattice structure. The radius r of each atomic 'sphere' is given by $r = a/2$, where a is the distance between atomic centres and is equal to the d in Bragg's Law.

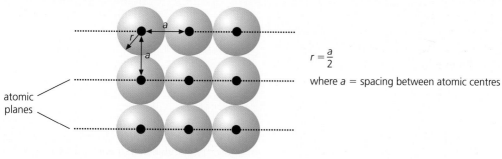

$r = \dfrac{a}{2}$

where a = spacing between atomic centres

atomic planes

Figure 2.4 Simple cubic lattice structure

The atomic radii of some atoms, as found from crystalline forms of the chemical elements, are shown in Table 2.1.

Table 2.1 Atomic radii

Element	Atomic radius/10^{-10} m
Lithium	1.52
Sodium	1.86
Potassium	2.31
Chlorine	0.99
Carbon	0.77
Bromine	1.14

Probing the nucleus

Thomson's 'plum pudding model'

In 1907 J. J. Thomson proposed an early model of the atom in which there were a number Z of electrons embedded in a uniform sphere of positive charge (Figure 2.5). This was commonly known as the 'plum pudding model' as the electrons could be thought of rather like raisins in a positively charged Christmas pudding!

The charge of the sphere was $+Ze$, which exactly neutralised the electrons' charge. In Thomson's model the atom, if undisturbed, would enable the electrons to rest in their equilibrium positions within the sphere, where the attraction of the positive charge balanced their mutual repulsion. If the atom became disturbed, for example by a collision with say a neighbouring atom, then the electrons would oscillate about their mean positions just like a mass on a spring and, by Maxwell's electromagnetic theory, emit light (recall that in Chapter 1 we stated Maxwell's assertion that oscillating charges emit electromagnetic radiation). However, while it can be shown that these oscillating electrons would emit electromagnetic radiation at visible wavelengths, they would not do so in a way that was consistent with the observed spectral pattern of lines characteristic of different chemical elements.

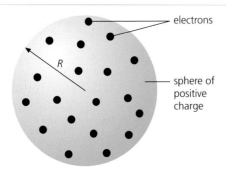

electrons

R

sphere of positive charge

Figure 2.5 The Thomson model of the atom: Z tiny electrons embedded in a uniform sphere of positive charge $+Ze$ and radius R. The sphere as a whole is electrically neutral

Figure 2.6 Ernest Rutherford (1871–1937)

Rutherford's experiments

The discovery of alpha and beta particles in radioactivity offered a new way to probe the internal structure of the atom. The method was to use the deflection or **scattering** of fast-moving alpha and beta particles as they travelled through matter. In 1911 two physicists, Hans Wilhelm Geiger (1882–1945) and Ernest Marsden (1889–1970), working under the direction of the New Zealand-born British experimental physicist Ernest Rutherford (1871–1937) (Figure 2.6), directed alpha particles at gold foil.

Geiger, Marsden and Rutherford made a remarkable discovery. While most of the particles were scattered through fairly small angles, some were scattered through extremely *large* angles, and very occasionally some almost travelled back along the way they had come. This startling observation caused Rutherford to make the remark at the start of this chapter.

To see why this led Rutherford to propose the nuclear model of the atom, see Figure 2.7, which shows the geometry of the scattering experiment. For each **impact parameter** b, which is the perpendicular distance from the particle's horizontal path to a line through the centre of the small, massive nucleus, there is a certain scattering angle θ. Upon deflection by the nucleus, the trajectory of the particle can be shown to follow a curve called a *hyperbola*, for which the relationship between the scattering angle and the impact parameter is given by the formula

$$b = \frac{q_\text{p} q_\text{nuc}}{4\pi\epsilon_0(\text{KE})} \cot\tfrac{1}{2}\theta$$

where q_p is the charge on the particle, q_nuc is the charge on the nucleus and KE is the particle's kinetic energy in joules.

An important assumption made here is that the nucleus is much more massive than the particle, so that it does not recoil when interacting with the particle. In this case the initial and final kinetic energies of the particle are the same.

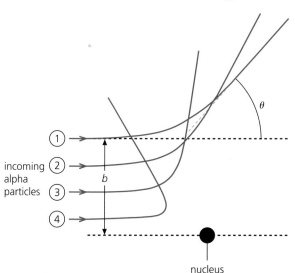

Figure 2.7 Alpha particles being scattered by a nucleus. The alpha particles can be shown to follow hyperbolic trajectories. The distance b is the impact parameter of particle ①, which is scattered through angle θ

WORKED EXAMPLE 2.2

An alpha particle of KE 7.7 MeV is incident on a gold nucleus ($Z = 79$) with an impact parameter of 1.7×10^{-12} m. What is the scattering angle?

We can rearrange

$$b = \frac{q_p q_{nuc}}{4\pi\epsilon_0 (KE)} \cot\tfrac{1}{2}\theta$$

to give

$$\cot\tfrac{1}{2}\theta = b \times \frac{4\pi\epsilon_0 (KE)}{q_p q_{nuc}} \quad \text{or} \quad \frac{1}{\tan\tfrac{1}{2}\theta} = b \times \frac{4\pi\epsilon_0 (KE)}{q_p q_{nuc}}$$

since for trigonometrical ratios the cotangent (cot) is the reciprocal of the tangent (tan).

$$\therefore \quad \tan\tfrac{1}{2}\theta = \frac{q_p q_{nuc}}{b \times 4\pi\epsilon_0 (KE)}$$

The Z number of gold is 79, so that $q_{nuc} = 79 \times (1.6 \times 10^{-19}) = 1.3 \times 10^{-17}$ C, and the charge on an alpha particle is $+2e$. We need to express the KE in joules; 7.7 MeV is $7.7 \times 10^6 \times 1.6 \times 10^{-19} = 1.2 \times 10^{-12}$ J. Putting in the values:

$$\tan\tfrac{1}{2}\theta = \frac{(2 \times 1.6 \times 10^{-19}\,C) \times (1.3 \times 10^{-17}\,C)}{(1.7 \times 10^{-12}\,m) \times 4 \times \pi \times (8.85 \times 10^{-12}\,F\,m^{-1}) \times (1.2 \times 10^{-12}\,J)} = 0.02$$

so

$$\tfrac{1}{2}\theta = \tan^{-1} 0.02 = 1.1°$$

and the scattering angle $\theta = 2 \times 1.1 = 2.2°$.

The main point to understand here is that, the smaller the impact parameter, then the greater the deflection of the alpha particles. The size of the impact parameter for large deflections is therefore a measure of the nuclear size. If we repeat the same calculation as in Worked Example 2.2 but using an impact parameter of 5.4×10^{-14} m, then we get a much larger deflection of 30°. For alpha particles of this energy to be deflected by more than about 2° the impact parameter must be less than 1.7×10^{-12} m, and for them to be deflected by more than 30° the impact parameter must be less than 5.4×10^{-14} m. These correspond to effective circular cross-sectional targets as seen by the alpha particles of $\pi \times (1.7 \times 10^{-12})^2\,m^2$ and $\pi \times (5.4 \times 10^{-14})^2\,m^2$ respectively (Figure 2.8).

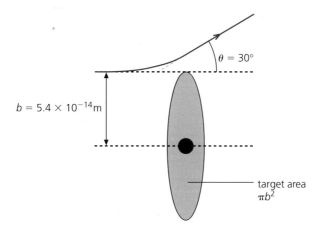

Figure 2.8 The impact parameter *b* is a measure of the nuclear size

Since the alpha particles are emitted from the source at random by radioactive decay, they strike the gold foil randomly and so the probabilities for deflections at these angles are also proportional to these target areas in the ratio

$$\frac{\text{scattering probability for } \theta \geqslant 1°}{\text{scattering probability for } \theta \geqslant 30°} = \frac{\pi \times (1.7 \times 10^{-12})^2 \, \text{m}^2}{\pi \times (5.4 \times 10^{-14})^2 \, \text{m}^2} = 991$$

This means that, for this energy, only about 1 in 1000 alpha particles are deflected through an angle greater than 30°. Most particles pass through with little deflection and large deflections are quite rare events, from which we can conclude – as Rutherford did – that the atom must be mainly empty space!

The Thomson model suggested that the probability for scattering at large angles was bound to be large, owing to the diffuse distribution of positive charge. This was contrary to the experimental evidence, and Rutherford showed that the low observed incidence of large θ could be accounted for if the atom were regarded as mainly empty space with a tiny core of positive charge. Rutherford determined that the atomic nucleus contains 99.9% of the mass of the atom, with the nucleus having a size of the order 10^{-14} m, and proposed that the electrons orbit the nucleus rather like planets around the Sun.

Rutherford scattering, as it is called, enables us not only to infer the size of the nucleus but also to deduce the distance of closest approach (Box 2.3). **Scattering experiments** are a standard technique used by particle physicists for exploring sub-atomic structure. While Rutherford only had projectiles of sufficient energy to touch the surface of the nucleus, particle physicists today use accelerators capable of producing very high-energy particles which can probe the interiors of the protons and neutrons within the nucleus. We will have much more to say about these later.

Box 2.3 Estimating the size of the nucleus

In their scattering experiments with alpha particles and metal foils, Geiger and Marsden found that on average 1 in 20000 particles were deflected through an angle of 90° or more using a metal foil 4×10^{-7} m thick (Figure 2.9).

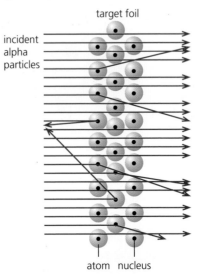

Figure 2.9 The angle through which an alpha particle is scattered depends on how close its incident path passes to an atomic nucleus. Large deflections, of 90° or more, are the result of very close encounters

Assuming that the atoms in the foil are separated by about 2.5×10^{-10} m then the number of layers of atoms in the foil will be

$$\frac{4 \times 10^{-7}}{2.5 \times 10^{-10}} = 1.6 \times 10^3 = 1600 \text{ layers}$$

So if 1600 layers can be expected to turn back 1/20000 particles then one layer can turn back

$$\frac{1}{20000 \times 1600} = \frac{1}{32 \times 10^6}$$

of them. This number represents the fraction of the surface area of the foil which turns back alpha particles through angles of 90° or more.

Assume each atom has a surface area of $\pi \times (0.5 \times 10^{-10})^2$ m^2 (taking the atomic diameter as 10^{-10} m) and let r be the radius of the nucleus. The area of each nucleus is therefore πr^2 and so the fraction of area is

$$\frac{\pi r^2}{\pi \times (0.5 \times 10^{-10} \text{ m})^2} = \frac{1}{32 \times 10^6}$$

$$\therefore \quad r = \sqrt{\frac{(0.5 \times 10^{-10}\,\text{m})^2}{32 \times 10^6}} \approx 10^{-14}\,\text{m}$$

A second method is to use Coulomb's Law and think about the changes in potential energy and kinetic energy as a positively charged particle approaches the nucleus. In this situation we consider a 'head-on' collision between the nucleus and an alpha particle (Figure 2.10).

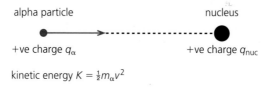

alpha particle

nucleus

+ve charge q_α

+ve charge q_{nuc}

kinetic energy $K = \frac{1}{2}m_\alpha v^2$

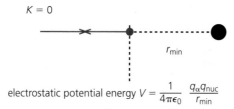

alpha particle has come momentarily to rest before reversing direction

$K = 0$

r_{min}

electrostatic potential energy $V = \dfrac{1}{4\pi\epsilon_0}\dfrac{q_\alpha q_{nuc}}{r_{min}}$

Figure 2.10 Changes in kinetic energy and potential energy as an alpha particle approaches a nucleus

Since the alpha particle has positive charge it will slow down as it approaches the positively charged nucleus due to their mutual electrostatic repulsion. The closer the alpha particle gets to the nucleus, the more of its initial kinetic energy it exchanges for electrostatic potential energy.

If the alpha particle has mass m_α and is travelling with a velocity v then it will have an initial kinetic energy K given by $K = \frac{1}{2}m_\alpha v^2$. The electrostatic potential energy V at a distance r from the nucleus is

$$V = \frac{1}{4\pi\epsilon_0}\frac{q_\alpha q_{nuc}}{r}$$

where q_α is the charge on the alpha particle and q_{nuc} the charge on the nucleus. As the alpha particle approaches the nucleus from infinity, K decreases and V increases from zero but $(K+V)$ remains constant. At the distance r_{min} the velocity is v_{min}, so that if the initial velocity of the alpha particle is v we can write

$$\tfrac{1}{2}m_\alpha v_{min}^2 + \frac{1}{4\pi\epsilon_0}\frac{q_\alpha q_{nuc}}{r_{min}} = \tfrac{1}{2}m_a v^2 = \text{constant}$$

In the Rutherford experiment, suppose an alpha particle collides head-on with a gold nucleus of atomic mass number $Z = 79$. Now at the instant of closest approach the alpha particle will momentarily come to rest and give up all its kinetic energy before being repelled by the nucleus in the opposite direction. At this point $v_{min} = 0$ so that

$$\frac{1}{4\pi\epsilon_0} \frac{q_\alpha q_{nuc}}{r_{min}} = \tfrac{1}{2}m_\alpha v^2$$

from which we can find r_{min}, which is an estimate for the size of the nucleus.

WORKED EXAMPLE 2.3

What is the distance of closest approach of an alpha particle of energy 5.0 MeV when it collides head-on with a gold nucleus?

The kinetic energy K of the alpha particle is $5\,\text{MeV} = 5 \times 10^6 \times 1.6 \times 10^{-19} = 8 \times 10^{-13}\,\text{J}$. At the distance of closest approach r_{min}, all this energy is converted into electrostatic potential energy, V. The charge of the gold nucleus is $79e$ and that of the alpha particle is $2e$, so that

$$r_{min} = \frac{1}{4\pi\epsilon_0} \frac{(2 \times 79)e^2}{V}$$

$$= (9 \times 10^9\,\text{Fm}^{-1}) \times \frac{158 \times (2.6 \times 10^{-38}\,\text{C}^2)}{8 \times 10^{-13}\,\text{J}} \approx 5 \times 10^{-14}\,\text{m}$$

This is of the same order of magnitude as obtained by the first method. The size of the nucleus is at least 10 000 times less than the size of the atom.

The discovery of the neutron

By the early 1920s Rutherford's model of the atom containing a nucleus made of positively charged protons was firmly established. However, there were still some things that did not fit. For the nucleus to contain most of the atomic mass as well as all the positive charge, then the protons should account for *all* the mass (the contributions from the mass of the electrons being negligible). If this were the case then a nucleus with twice as much charge as another should have twice as much mass, but this is *not* what we measure. Nuclei with twice as much charge have *at least double* the mass that would be expected from the number of protons they contain. To explain this, Rutherford suggested that the nucleus might contain another particle similar to the proton in mass but having no charge. This electrically neutral particle, called the **neutron**, was finally discovered in 1932 by the British physicist James Chadwick (1891–1974) (Figure 2.11).

Figure 2.11 James Chadwick (1891–1974)

One of the difficulties in detecting neutrons was that, being uncharged particles, they do not ionise molecules of air and cannot be deflected by electric and magnetic fields. Chadwick, however, was aware of experiments by other physicists which showed that, when certain light elements such as beryllium were bombarded by alpha particles, some kind of radiation was formed that had the ability to eject protons from paraffin wax. Chadwick repeated these experiments and showed that the best explanation for this phenomenon was to suppose that the alpha particles were knocking neutral particles out of the beryllium nucleus, which in turn were knocking protons out of the paraffin. By detecting the ejected protons, the neutron's presence could be inferred indirectly.

The neutron turned out to be a very important particle for initiating nuclear fission reactions and for his work Chadwick was awarded the Nobel Prize for Physics in 1935. The principle that Chadwick used for discovering the neutron by inferring its effect on another particle is an important one used frequently in particle physics, and we will see many examples of the 'inference method' later in this book.

The neutron is represented by the symbol $_0^1$n as it has an atomic mass number of 1 and zero charge. It is not a stable particle outside the nucleus. It has a half-life of approximately 15 minutes and will decay into a proton according to the nuclear decay process

$$_0^1\text{n} \rightarrow {}_1^1\text{p} + {}_{-1}^{\ 0}\beta + {}_0^0\bar{\nu}$$

where $_{-1}^{\ 0}\beta$ is a beta particle (electron) and $_0^0\bar{\nu}$ is a particle called an **antineutrino**.

By the end of the 1930s the basic model of the atom was complete. An atom consisted of a positively charged nucleus made up of protons and neutrons. Surrounding the nucleus were orbiting negatively charged electrons whose number was equal to the number of protons so that the overall electrical charge on the atom was zero. Atoms of any chemical element could be represented as $_Z^A$X, where A is the **atomic mass number**, equal to the number of protons and neutrons in the nucleus, and Z is the **proton number**, equal to the number of protons in the nucleus. It can be seen therefore that an electrically neutral atom will contain $N = A - Z$ neutrons and have Z electrons. N is called the **neutron number**. We refer to neutrons and protons collectively as **nucleons**, thus A is sometimes referred to as the **nucleon number**. X is the chemical symbol so that $_{79}^{197}$Au, for example, represents the element gold, which has 197 nucleons, 79 protons and $(197 - 79) = 118$ neutrons.

The Bohr model of atomic structure

Maxwell showed that charges that accelerate or decelerate, as in oscillation, will emit electromagnetic waves. A body moving with a uniform speed in a circle is subject to a centripetal acceleration, so an electron in orbit around the nucleus, as in Rutherford's model, must therefore be accelerating and emitting electromagnetic radiation continually. The electrons are thus losing energy in the form of electromagnetic waves and must soon spiral in to the nucleus, causing the atom to collapse. This clearly does not happen, so this 'classical' interpretation of the electron orbit must be wrong.

The Danish physicist Niels Bohr (1885–1962) (Figure 2.12a) provided the answer. In the **Bohr model** of the atom (Figure 2.12b), we obtain a useful picture that helps us to understand how the electron can remain in its orbit and also how electromagnetic radiation can be emitted and absorbed by the atom. According to Bohr, the electrons rotate around the nucleus in well defined orbits that correspond to specific **energy levels**. Bohr used classical mechanics to describe the orbit of the electron around the hydrogen nucleus but he added a number of important conditions or *postulates*.

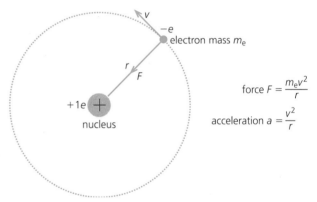

force $F = \dfrac{m_e v^2}{r}$

acceleration $a = \dfrac{v^2}{r}$

Figure 2.12(b) The Bohr model of a hydrogen atom ($Z = 1$). An electron of charge $-e$ and mass m_e orbits a positive nucleus of charge $+1e$, with a speed v and acceleration v^2/r, where r is the radius of the orbit

Figure 2.12(a) Niels Bohr (1885–1962)

Postulate 1 *The energy of an electron in its orbit is constant.*

If the energy were not constant, then the electron would radiate energy away and eventually spiral in to the nucleus. An electron in its orbit has both kinetic energy by virtue of its orbital motion and potential energy since it is attracted to the nucleus by electrostatic forces.

Postulate 2   *An electron can occupy only certain definite orbits and cannot exist 'in between' them.*

Postulate 2 is an important new concept. Here Bohr is bringing in the idea of the quantum. The orbits are 'quantised' and can only take discrete quantum states. This is a departure from the classical idea of continuous values. Since the electron in its orbit has a certain energy, we associate with it a specific energy level corresponding to the particular orbit it happens to be in at the time. Bohr stated that when the electrons are in these quantum states *the atom does not radiate*. The lowest energy level that an electron has in an atom is called its **ground state**. Energy levels are usually measured in electronvolts, and the ground state for the hydrogen atom has the value $-13.6\,\text{eV}$. The significance of the negative sign comes about due to the convention used by physicists in assigning differences in electron energies. A free electron, unattached to any atom, has positive or *greater* energy. We regard the interior of an atom as a potential energy well, with the potential energy at its 'surface' equal to zero. Inside the atom at some distance from the nucleus the electron therefore has negative or *less* potential energy. This is purely a convention but beware of confusion here! The further away from the nucleus the electron is, the greater is its electrical potential energy, since its potential energy is less negative.

Postulate 3   *The absorption or emission of electromagnetic radiation in an atom comes about due to a **transition** of an electron between orbits.*

When a photon is *absorbed* by an atom, the electron makes a transition from a lower energy orbit (or level) to a higher one. When an atom *emits* a photon, the electron makes a transition from a higher to a lower energy orbit. It is important to understand that, when atoms emit or absorb light, then the principle of energy conservation implies that they can *only* do so if the photon's energy is equal to the *difference* between two specific electron energy levels. The energy of the emitted/absorbed photon is said to be **quantised** and is equal to the difference in energy between the two orbits.

If an electron makes a transition from an orbit of higher energy E_2 to an orbit of lower energy E_1, then the emitted photon will have a frequency given by the Planck Relation:

$$E_2 - E_1 = hf \quad \text{or} \quad f = \frac{E_2 - E_1}{h}$$

When the electron is in the ground state it is in its lowest possible energy level, and the atom is then in its lowest quantum state. If energy is added to the atom either by heating it or by the absorption of photons, an electron can jump to a higher energy level and the atom is then said to be in an **excited state**. If an electron gains enough energy it can escape from the nucleus altogether and the atom is said to be **ionised**.

Notice that, in order to remove the electron completely from the hydrogen atom in its ground state, an energy of at least $13.6\,\text{eV}$ is required. This is called the **ionisation energy** for hydrogen. Once the electron has escaped it can have a continuous range of energies. Only while it is inside the atom is its energy quantised.

The Bohr model was able to explain many of the features observed in the spectrum of atomic hydrogen. The hydrogen spectrum has a sequence of spectral lines that start at 656.31 nm and end at 364.56 nm. These lines, which are at visible wavelengths, are called the **Balmer lines**, and correspond to electrons dropping down to the lowest excited state from higher energy levels; the entire sequence is called the **Balmer series**. Lines produced by electron transitions to the ground state are all emitted at ultraviolet wavelengths and are called the **Lyman series**; while those to the second excited state are emitted in the infrared and are called the **Paschen series**. The energy levels for hydrogen showing these electron transitions are given in Figure 2.13.

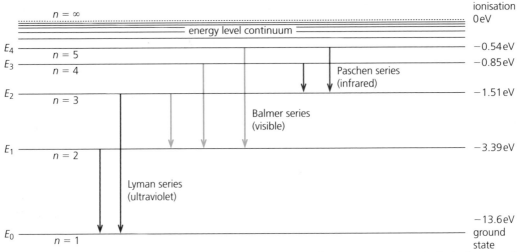

n is the quantum number of the particular energy level, $n = 1, 2, 3, \ldots$ ($n = \infty$ corresponds to the ionisation energy)

Figure 2.13 Energy levels and electron transitions giving rise to the spectrum of atomic hydrogen

It must be stressed that the Bohr model does not provide a complete picture of atomic spectra. While the model can explain the main features of the spectrum of hydrogen and its ionisation energy, it cannot account for the spectra of atoms with more than one electron. Using his model, Bohr was however able to propose a tentative explanation for Mendeleyev's Periodic Table. He suggested that the chemical behaviour of elements can be explained by the way the allowed orbits are filled by electrons in the atom; that is, the chemical properties of an element are determined by the arrangement of electrons in its energy level structure and the ability of the outermost electrons to interact with those of different elements. However, Bohr's most important contribution was to introduce quantum ideas into the model of the atom and to show how these could account for the observed phenomena.

Properties of the nucleus

Nuclear terminology

When physicists are considering the characteristics of atomic nuclei as specific nuclear species, they refer to the nuclei as **nuclides** and write them as $^A_Z X_N$. For example, nuclides for hydrogen, uranium and iron are written as $^1_1 H_0$, $^{238}_{92} U_{146}$ and $^{56}_{26} Fe_{30}$. When we measure the masses of nuclides we find that nuclei with a given proton number can have several different atomic mass numbers. Nuclei with the same Z number but different numbers of neutrons are called **isotopes** of each other. Many isotopes are unstable and change into stable nuclei by radioactive decay. Gold, for example, has 32 isotopes with atomic mass numbers ranging from $A = 173$ to $A = 204$, but only one, $^{197}_{79} Au_{118}$, is stable. Nuclides with the same N but different Z are called **isotones**, so that $^2_1 H_1$ and $^3_2 He_1$ are isotones. Nuclides that have the same A are called **isobars**, so that $^3_2 He_1$ (stable) and $^3_1 H_2$ (radioactive) are examples of isobars.

Nuclear sizes and shapes

Like the atom, the nucleus is not a solid object with a well defined boundary. Scattering experiments with beams of high-energy electrons have shown that most nuclei are approximately spherical in shape (although some are ellipsoidal) with an effective radius given by

$$R = R_0 A^{\frac{1}{3}}$$

where A is the atomic mass number and $R_0 = 1.2 \times 10^{-15}$ m is the average nucleon radius. On nuclear scales, physicists use a unit called the **femtometre** (also called the **fermi**) for measuring nuclear distances:

$$1 \text{ femtometre} = 1 \text{ fermi} = 1 \text{ fm} = 10^{-15} \text{ m}$$

R_0 in the nuclear radius equation above is therefore 1.2 fm.

Nuclear masses

The masses of atoms can be measured using a **mass spectrometer**, which is a device that can obtain a spectrum of masses of a beam of ions using specially arranged magnetic and electric fields; it works in a way not unlike the apparatus used to measure the charge-to-mass ratio of the electron (Box 2.1). Masses can also be found by nuclear reaction experiments where conservation of atomic mass and proton number are used to deduce the mass of reaction products. Atomic masses are measured using the **unified atomic mass constant** (u), also called the atomic mass unit (a.m.u.), which is referred to the atomic mass of $^{12}_6 C$ such that

$$1 \text{ a.m.u.} = 1 \text{ u} = \tfrac{1}{12} \times \text{mass of } ^{12}_6 C = 1.661 \times 10^{-27} \text{ kg}$$

The mass of a nuclide is expressed in terms of the unified atomic mass constant rounded off to the nearest integer, so that gold, actually 196.966573 u, is rounded to 197 u.

From Einstein's mass–energy relation $E = \Delta mc^2$, we see that $1 u = 931$ MeV (approximately) and we can use this relationship to find the energy equivalent (in MeV) of any mass or, conversely, the mass equivalent of an energy in atomic mass units.

Nuclear matter

What is nuclear matter like? The formula for the effective nuclear radius tells us that the radius is proportional to the cube root of the atomic mass number, which in turn suggests that the *volume* of the nucleus is proportional to A. The mass of the nucleus is also approximately proportional to A and so the densities of all the nuclei must be approximately equal. This is very similar to the density of a drop of liquid, which is constant and independent of its size. For this reason the nucleus is sometimes modelled as a liquid drop, and the **liquid drop model** (Figure 2.14) of the nucleus has been very successful in explaining a wide range of nuclear behaviour, particularly the fission of heavy nuclei.

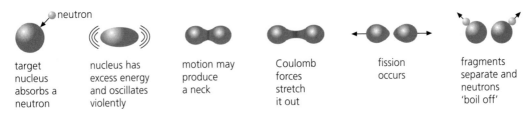

| target nucleus absorbs a neutron | nucleus has excess energy and oscillates violently | motion may produce a neck | Coulomb forces stretch it out | fission occurs | fragments separate and neutrons 'boil off' |

Figure 2.14 The liquid drop model of the nucleus may be used to explain nuclear fission

WORKED EXAMPLE 2.4

What is the density of nuclear matter?

The volume V of a sphere of radius R is given by $V = \frac{4}{3}\pi R^3$ and, since the effective nuclear radius is given by $R = R_0 A^{\frac{1}{3}}$, we can write the volume of a nucleus as

$$V = \frac{4}{3}\pi (R_0 A^{\frac{1}{3}})^3 = \frac{4}{3}\pi R_0^3 A$$

We also know that density = mass/volume, so that the nuclear density ρ_N expressed in nucleons per unit volume is

$$\rho_N = \frac{A}{V} = \frac{A}{\frac{4}{3}\pi R_0^3 A} = \frac{3}{(4\pi)(1.2\,\text{fm})^3} = 0.138\,\text{nucleon fm}^{-3}$$

You can now see why all nuclei can be considered to have a single density, since A cancels in the equation above. To express this density in terms of $kg\,m^{-3}$, we note that the mass of a proton (or a neutron) is about $1.67 \times 10^{-27}\,kg$ and $1\,fm = 10^{-15}\,m$, so that

$$0.138\,\text{nucleon fm}^{-3} = \frac{(0.138 \times 1.67 \times 10^{-27}\,\text{kg})}{(10^{-15}\,\text{m})^3} \approx 2 \times 10^{17}\,\text{kg m}^{-3}$$

Compare this with the density of water, which is $1000\,kg\,m^{-3}$: the density of nuclear matter is some 10^{14} times as great!

Nuclear binding energies

If we compare the mass of an atomic nucleus with the sum of the masses of its individual protons and neutrons, it is found that nuclei have a mass which is *less* than the total mass of their protons and neutrons (Figure 2.15). It is in fact this **mass defect** which explains why the nucleus should exist at all, since for nuclei containing more than one proton, the electrostatic Coulomb repulsion force of their like charges should cause the nucleus to fly apart. For nuclei to be stable, there must exist a force between the nucleons that is short-range, attractive and can overcome the Coulomb repulsion of the protons. Physicists call this force the **strong nuclear force** and we will have much more to say about it in later chapters, but for now we will simply regard it as a necessary feature of a nuclear model.

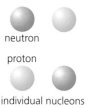

neutron

proton

individual nucleons   nucleus X

Figure 2.15 Mass defect $\Delta m =$ (mass of N neutrons + mass of Z protons) − (mass of nucleus $^{N+Z}_{Z}$X). By the mass–energy relation, binding energy $= \Delta m \times c^2$

Suppose we assemble a nucleus of N neutrons and Z protons. There will be an *increase* in the electric potential energy due to the electrostatic forces between the protons, trying to push the nucleus apart, but there is a *greater* decrease of potential energy due to the strong nuclear force acting between the nucleons, attracting them to one another. As a consequence, the nucleus has an overall net *decrease* in its potential energy. This decrease in potential energy is called the **nuclear binding energy** and the decrease per nucleon is called the **binding energy per nucleon**. This loss of energy is, by the mass–energy relation, equivalent to a loss of *mass*, hence the difference between the mass of a nucleus and the sum of the masses of its nucleons. It is the binding energy that holds the nucleus together, and its release is where the useful energy of the nucleus comes from.

WORKED EXAMPLE 2.5

a Given that the mass of a helium nucleus is 4.001 51 u, the mass of a proton is 1.007 28 u and the mass of a neutron is 1.008 66 u, show that the difference in mass between a helium nucleus and its constituent nucleons is 0.030 37 u.

b What is the binding energy of the helium nucleus $^{4}_{2}$He?

a Helium has two protons and two neutrons.
   Total mass of nucleons $= (2 \times 1.007\,28\,\text{u}) + (2 \times 1.008\,66\,\text{u}) = 4.031\,88\,\text{u}$
   ∴ difference in mass $= 4.031\,88\,\text{u} - 4.001\,51\,\text{u} = 0.030\,37\,\text{u}$

b The mass defect Δm between an assembled helium nucleus and its individual nucleons is 0.030 37 u, so $\Delta m = 0.030\,37 \times 1.66 \times 10^{-27}\,\text{kg} = 5.04 \times 10^{-29}\,\text{kg}$
   and binding energy $\Delta E = \Delta m c^2$
$$= 5.04 \times 10^{-29}\,\text{kg} \times (3.00 \times 10^{8}\,\text{m s}^{-1})^2\,\text{J}$$
$$= 4.5 \times 10^{-12}\,\text{J}$$
   Alternatively we can express binding energy in MeV by remembering that
   1 u = 931 MeV, so 0.030 37 u $= 0.030\,37 \times 931\,\text{MeV} = 28.27\,\text{MeV}$.

We can represent the binding energy per nucleon on a graph which shows how it varies with atomic mass number A (Figure 2.16). Notice that the shape of the curve indicates that the binding energy generally increases as the number of nucleons in the nucleus increases. Also, there are some nuclei, such as helium ($^4_2$He), carbon ($^{12}_6$C) and oxygen ($^{16}_8$O), which lie above the general trend of the curve and have binding energies greater than their neighbours. The curve reaches a maximum near iron ($^{56}_{26}$Fe). Iron, because of its high binding energy per nucleon, is a very stable nucleus. Beyond iron, the binding energy per nucleon tends to fall slightly as A increases towards the more massive nuclei. However, the most stable nuclide of all is that of $^{62}_{28}$Ni, which has the highest binding energy of 8.794 60 MeV/nucleon.

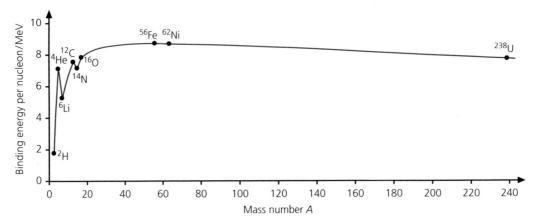

Figure 2.16 Variation of binding energy per nucleon with atomic mass number

Fission and fusion

Two processes can release energy from the nucleus of an atom: **nuclear fission** and **nuclear fusion**. To see how this can happen, we note that the binding energy of a nucleus is proportional to its mass defect.

In nuclear fission, a massive nucleus such as uranium splits into two lighter nuclei of approximately equal mass. From Figure 2.16 we see that this happens on the *falling* part of the curve, so that mass is lost and binding energy released when very heavy elements fission to nuclei of smaller mass number. Nuclear fission is responsible for the release of energy in nuclear reactors and atomic bombs.

In nuclear fusion, binding energy is released when two light nuclei are fused together to form a heavier nucleus. So if, for example, we form a helium nucleus by the fusion of two hydrogen nuclei, we find that the mass of the helium nucleus is *less* than that of the two hydrogen nuclei and the mass lost will be released as energy. The *rising* part of the curve in Figure 2.16 shows that elements with low mass number can produce energy by fusion. This is the principal source of energy in stars and thermonuclear explosions (hydrogen bombs). Since the nuclei of atoms carry positive charge due to

Figure 2.17 The power of an uncontrolled thermonuclear reaction

their protons, how can they fuse together if they are repelled by the Coulomb's Law force acting between them? The answer is that fusion can happen if each nucleus has sufficient kinetic energy to enable them to overcome their mutual repulsion, be captured by the strong nuclear force and stick together. In order to fuse two hydrogen nuclei (protons) together we need very high temperatures, of the order of 10^6 K, so that they are moving fast enough to overcome their electrostatic repulsion. (Remember that temperature is a measure of the average kinetic energy of particles.) It is because of the high temperatures needed to give the protons sufficient kinetic energy that these nuclear reactions are also known as **thermonuclear** fusion reactions. While it is possible to produce thermonuclear weapons in the form of hydrogen bombs, a controlled thermonuclear reaction in the form of a practical fusion reactor that can produce useful amounts of electric power is a formidable engineering challenge and has not as yet been achieved. Such a device would provide humankind with a virtually unlimited source of energy (Figure 2.17).

The Chart of the Nuclides

As we saw in Chapter 1, all the neutral atoms of the elements may be organised in terms of their chemical properties on the Periodic Table. However, the Periodic Table is of limited use to the nuclear physicist as it gives little information about the nuclear properties of an element. All known nuclides may be organised on a **Chart of the Nuclides** as shown in Figure 2.18. A nuclide is represented by plotting its neutron number N against its proton number Z. All the elements that are stable, that is, not radioactive (including stable isotopes), are marked by the black dots which form a **nuclear stability region**. The unstable nuclides are called **radionuclides** and lie either side of the stable ones. The chart shows that for proton numbers up to about $Z = 20$ the preference is to form nuclei with $N = Z$, while beyond this value of Z nuclei tend to have more neutrons than protons. The neutron excess for heavier nuclei may be considered to reduce the overall potential energy due to the electrostatic repulsive force between protons in the nucleus. The neutrons 'keep them apart', as it were, reducing the magnitude of the force on neighbouring protons, and help to keep the nucleus together. Very heavy nuclides have been made with atomic numbers as high as $Z = 112$ and $A = 278$. These unstable nuclides are not found naturally but are created in nuclear reactions using particle accelerators; they quickly decay to stable nuclei.

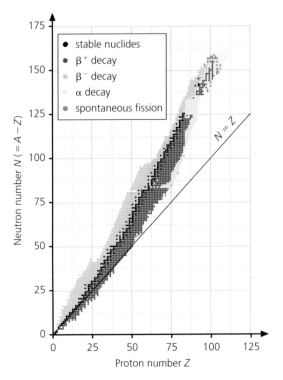

Figure 2.18 The Chart of the Nuclides

Nuclear stability and radioactive decay

Unstable nuclei will decay into other nuclear species by two different decay mechanisms which change the Z and N numbers of the nucleus. These are **alpha decay**, in which alpha particles are emitted, and **beta decay**, in which beta particles are emitted. In addition, an excited nucleus can also emit gamma rays.

In a similar way to the energy of an atom, the energy of a nucleus is quantised. Nuclei can exist only in discrete quantum states, each having a well defined energy. However, whereas atomic energy levels are measured in electronvolts (eV), those of nuclei are of the order of millions of electronvolts (MeV). When a nucleus makes a transition from a higher energy level to a lower one, a gamma ray is emitted. By studying the gamma rays emitted from excited nuclei, nuclear physicists can learn much about the distribution of nucleons inside the nucleus. Figure 2.19 shows the energy level structure for the palladium nucleus. **Gamma decay** *does not* change Z or N. The lifetimes of gamma decay events are very short, of the order of 10^{-11} s, and are usually observed because they follow either alpha or beta decay. These three processes – alpha, beta and gamma decay – are types of **radioactive decay** and originate from changes to the *nucleus*. They are in no way connected to the electron shell structure of the atom, although there will be a rearrangement of the orbital electrons subsequent to alpha and beta decay due to the change in Z.

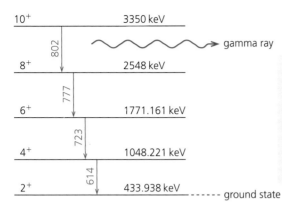

A nucleus has an energy level structure similar to that of an atom, with a nuclear 'ground state'. The numbers on the left-hand side are a measure of the total angular momentum of the nucleus in the particular energy state. The plus signs refer to the 'parity' of the state (see Chapter 6). When the nucleus descends from a higher energy state to a lower one, losing angular momentum, it emits a gamma ray of characteristic energy

Figure 2.19 The energy level structure of the palladium nucleus $^{108}_{46}Pd$

Why are some nuclei stable and others radioactive? Some clues are provided from the Chart of the Nuclides. Stability is favoured by an even number of protons and an even number of neutrons. 168 of the stable nuclei have even numbers of protons and neutrons ('**even–even nuclei**'), suggesting that two protons and two neutrons (which constitute an alpha particle) form a particularly stable combination of nucleons. On Earth, elements such as oxygen $^{16}_{8}O$, silicon $^{28}_{14}Si$ and iron $^{56}_{28}Fe$, all of which are even–even nuclei, make up some 75% of the crust – so fortunately for us the Earth's surface is not excessively radioactive!

As an example of instability, take the unstable nuclide $^{25}_{13}Al_{12}$, which lies just below the line of stability. This nuclide is unstable because its 12 neutrons are insufficient to form a stable nucleus with 13 protons. The nucleus can be made more stable if one of its protons transforms itself into a neutron. This may be done by the process of beta decay, so that

$$^{25}_{13}Al \rightarrow {}^{25}_{12}Mg + {}^{0}_{+1}\beta + {}^{0}_{0}\nu$$

Note that when we write nuclear equations, the A and Z numbers on each side must add up and be equal. The particle $^{0}_{+1}\beta$ is called a **positron** and we will have more to say about this in later chapters. All nuclei that fall below the stability region will decay by $^{0}_{+1}\beta$ emission.

Now consider the nuclide $^{29}_{13}Al_{16}$, which lies just *above* the region of stability. Such a nucleus has too many neutrons for the number of protons and will become more stable by converting a neutron into a proton and emitting an electron, $^{0}_{-1}\beta$. This can be done according to the nuclear decay process

$$^{29}_{13}Al \rightarrow {}^{29}_{14}Si + {}^{0}_{-1}\beta + {}^{0}_{0}\overline{\nu}$$

When beta decay occurs the new nucleus has the same A, but in the case of a $^{0}_{-1}\beta$ particle (electron emission) the Z number *increases* by one since the nucleus now has an extra proton, and for a $^{0}_{+1}\beta$ particle (positron emission) the Z number *decreases* by one since a proton has been transformed into a neutron.

Some radionuclides decay by the emission of an alpha particle, which is a helium nucleus ^4_2He. When this happens the ejected particle removes charge equal to two electronic charges ($2e$) and four atomic mass units ($4\,\text{u}$). The nucleus so formed has atomic mass number ($A - 4$) and proton number ($Z - 2$). Alpha decay occurs in very heavy nuclei ($Z > 83$) because the mass of the original radionuclide is greater than the sum of the masses of the decay products. An example is the decay of thorium $^{232}_{90}\text{Th}$ into radium $^{228}_{88}\text{Ra}$ via the emission of an alpha particle:

$$^{232}_{90}\text{Th} \rightarrow {}^{228}_{88}\text{Ra} + {}^4_2\alpha$$

$$\text{mass of } {}^{232}_{90}\text{Th} = 232.038\,124\,\text{u}$$
$$\text{mass of } {}^{228}_{88}\text{Ra} = 228.031\,139\,\text{u}$$
$$\text{mass of } {}^4_2\alpha = 4.002\,602\,\text{u}$$
$$\text{total mass of decay products} = (228.031\,139 + 4.002\,603)\,\text{u}$$
$$= 232.033\,742\,\text{u}$$

The total mass of the decay products is therefore less than the mass of $^{232}_{90}\text{Th}$ by $0.004\,382\,\text{u}$. Therefore $^{232}_{90}\text{Th}$ is unstable to alpha decay.

This illustrates a general point about nuclear stability. If the sum of the masses of the decay products is greater than the mass of the original nucleus, then that nucleus must be stable against such a possible break-up. We can also say that if the break-up of a nucleus into two nuclear species is energetically possible, then it will in general occur.

The decay product or **daughter** of a radioactive nuclide is often radioactive itself and decays by either alpha or beta decay, or both. For the naturally occurring radioactive substances, the only decay emissions are alpha particles ($Z = 2$, $A = 4$), negative beta particles (electrons) ($Z = -1$, $A = 0$) and gamma rays. It is because of this that for a parent nucleus with a mass number A, the daughter must differ by either 0 or 4 units in mass number. It follows that if the first member of a series of radioactive decays has a mass number A that is *exactly divisible by 4* then so does every successive member of the series. Three radioactive series occur in nature, with a fourth being produced artificially. The three natural ones are the uranium ($4n + 2$), thorium ($4n$) and actinium ($4n + 3$) series. The symbol n is an integer that decreases by 1 at each stage in the decay chain and the expression containing n gives the value of the mass number at each stage. The value of n is not the same for the first nucleus in each of the series, although all three finish up as an isotope of lead. The fourth series is called the ($4n + 1$) or neptunium series and was discovered after the isotope $^{241}_{94}\text{Pu}$ (plutonium) had been made artificially. The ($4n + 1$) series is not found naturally because its longest-lived member has a half-life of only 2×10^6 years, which is small compared to the accepted age of the Earth (4.5×10^9 years), and consequently this series has now disappeared.

The energies of alpha particles from natural radioactive sources range from about 4 to $7\,\text{MeV}$ and the half-lives from about $10^{-5}\,\text{s}$ to 10^{10} years and, in general, the smaller the energy, the longer the half-life. In Chapter 3 we will see how quantum mechanics can explain this enormous variation in half-lives and will take a more detailed look at the workings of both alpha and beta decay.

The nature of the strong nuclear force

We can now draw some general conclusions about the properties of the force inside the nucleus. As we have said earlier, the protons in the nucleus all carry positive charges and repel one another. If no other force were present, the nucleus would be unstable and fly apart. The fact that stable nuclei *do* exist and that everyday matter is stable means that there must be another force present in the nucleus that is attractive and is able to overcome the mutual electrostatic repulsive forces of the protons. The strong nuclear force binds both protons and neutrons so it must be *charge-independent*.

The graph of binding energy per nucleon (Figure 2.16) provides us with some clues. The average energy per nucleon is about 8.3 MeV. For values of $A > 50$ the curve is approximately flat and shows that the binding energy is approximately proportional to A, indicating that each nucleon is attracted only to its nearest neighbours and not to those further away. If, for example, each nucleon attracted every other, then for a nucleus containing A nucleons there would be $(A - 1)$ bonds with each nucleon and a total of $A \times (A - 1)$ bonds altogether. The binding energy would then be proportional to $A(A - 1)$ and the binding energy per nucleon would not be approximately constant as shown.

The steep rise in binding energy at low A is accounted for by the increase in the number of nearest neighbours and the number of bonds per nucleon. As A increases, the nuclear force becomes saturated among the nucleons and falls slightly due to the electrostatic repulsion of the protons (which can be shown to increase as Z^2) and thereby reduces the overall binding energy. At very large atomic mass numbers $(A > 300)$ the electrostatic repulsion becomes so large that the nucleus spontaneously fissions and splits in two. One way to think of it is to imagine yourself in a crowd of people, say at a rock concert. With a small number of people pushing together, the 'push' you feel will increase with the number of people, until a large enough crowd gathers and the 'push' everyone feels is then more or less the same – you only really feel a push from the person next to you.

The observation that the strong nuclear force cannot interact with all the nucleons suggests that it is *short-range*. Knowing the approximate size of the nucleus gives an upper limit of 10^{-14} m for the range, although experiments involving the interaction of neutrons with protons show that its range is less than 3 fm (3×10^{-15} m).

One final point to note from the above is that neutrons are able to change into protons and vice versa. This fact suggests that protons and neutrons are not fundamental particles but have an internal structure which enables them to do this.

In this chapter we have looked at the atomic and nuclear structure of matter. By the mid-1930s many physicists thought they were approaching a complete description of its fundamental constituents. However, the development of the quantum theory, the discovery of new energetic particles in cosmic rays and the construction of the first particle accelerators showed how wrong they were. In the next chapter we shall look at the modern theories of Special Relativity and quantum mechanics and see how they provide us with powerful conceptual tools for describing the nature of matter and energy at even more fundamental levels.

Summary

◆ The electron was the first elementary particle to be discovered. It has negative electric charge. Its **charge-to-mass ratio** was first determined by J. J. Thomson and its charge by Robert A. Millikan. All observable charged particles have values that are integral multiples of the electron's charge.

◆ Models of the atom must account for its small size, its neutral electric charge and its ability to emit and absorb electromagnetic radiation. J. J. Thomson's **plum pudding model** was one of the first atomic models but was unable to explain the characteristic spectral pattern of lines seen in various chemical elements.

◆ By scattering alpha particles off gold foil, Ernest Rutherford and his co-workers were able to show that the atom contained a tiny **nucleus** with a positive charge and that the atom's volume was mainly empty space. The particles in the nucleus containing the positive charge are called **protons**. Orbiting the nucleus are negatively charged electrons which exactly balance out the positively charged nucleus. An important measure of the size of any object that scatters incoming particles is the **impact parameter**. **Scattering experiments** are extensively used by particle physicists as a useful probe of sub-atomic structure.

◆ Shortcomings in this **Rutherford model** of the atom led to the discovery of the **neutron** by James Chadwick. This particle has a mass approximately the same as that of the proton but zero charge. The neutron can only be detected indirectly. The collective name for protons and neutrons is **nucleons**, and the number of nucleons in a nucleus is called the **nucleon number** or **atomic mass number**, A. Atoms of each chemical element contain a unique number of protons.

◆ The **Bohr model** of the atom was introduced to explain why electrons do not continually radiate away their energy as electromagnetic radiation. An important postulate of Niels Bohr's model was the concept of **quantised** energy – electron **energy levels**. The Bohr model could explain the periodic nature of the chemical behaviour of elements, but could not account for the spectra of atoms with more than one electron. The Bohr model introduced quantum ideas into models of the atom for the first time.

◆ Nuclei can be classified into different species called **nuclides**. Nuclei with the same **proton number** Z but different neutron numbers are called **isotopes**. Nuclei with the same N but different Z are called **isotones**. Nuclei that have the same mass number A are called **isobars**.

◆ The **fermi** is a unit of nuclear distances equal to 10^{-15} m. The radius of a nucleus is proportional to the cube root of its nucleon number. The masses of nuclei are expressed in terms of the **unified atomic mass constant** (u). Nuclear densities are very high, of the order of 10^{17} kg m^{-3}. The densities of all atomic nuclei are approximately equal and can be described in terms of a **liquid drop model**.

◆ The nucleus is bound together by a **strong nuclear force** which stops it flying apart. Every nucleus has a **mass defect**, which can be converted into nuclear binding energy to hold the nucleus together. Nuclear binding energy can be released from the atom by **fission** or **fusion**.

◆ All the nuclides, including stable and unstable nuclei, can be organised on a **Chart of the Nuclides** which plots their proton number against their neutron number. The chart includes a **nuclear stability region** containing all nuclei that are non-radioactive. Non-stable nuclei may disintegrate by **alpha, beta** and **gamma decay**, which originate from changes to the nucleus. There is a great variation in the half-lives of non-stable nuclides, which may decay in **radioactive series**. Three naturally occurring series are those of uranium, thorium and actinium.

◆ The strong nuclear force is **charge-independent** and **short-range**. Neutrons are able to change into protons and vice versa, suggesting that nucleons may have an internal structure of their own.

Questions

1 In a Millikan oil drop apparatus, two plane parallel conducting plates are separated by a distance of 15 mm in air. The upper plate is maintained at a potential of 1500 V while the other plate is earthed. What is the number of electrons that must be attached to a small oil drop of mass 4.90×10^{-15} kg for it to remain stationary between the plates? You may assume that the density of air is negligible in comparison with that of oil.

2 This question is about the scattering of a parallel beam of α-particles by gold nuclei in thin foil.

   a The figure below shows the path PQ of an α-particle as it passes close to a gold nucleus $^{197}_{79}$Au.

   On a copy of the diagram:

   i) draw an arrow to show precisely, the direction of the electrostatic force on the α-particle when it is at the point X.

   ii) draw carefully the possible path of another α-particle with the same initial energy which passes through point R.

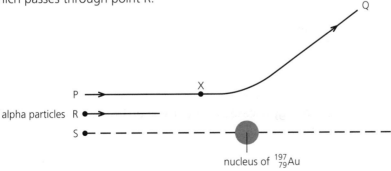

b By what factor would the **magnitude** of the electrostatic force on the α-particle change if the experiment was repeated using:

i) α-particles with **twice** the initial kinetic energy;

or

ii) silver foil containing $^{107}_{47}Ag$ nuclei instead of gold?

c i) Calculate the distance of closest approach to a gold nucleus of an α-particle with initial kinetic energy of 8.0×10^{-13} J. Start by considering the energy changes as the α-particle approaches the nucleus through point S.

ii) In carrying out this calculation what assumption did you make about the gold nucleus?

OCSEB, June 1991

3 Explain what is meant by *ground state*, *excited state* and *ionisation* when describing an atom.

Hydrogen has a ground state energy of -13.6 eV and the first and second excited states are -3.39 eV and -1.51 eV respectively.

Sketch an energy level diagram showing the ground state and the next two energy levels for hydrogen.

On your sketch show the electron transition that would result in the emission of radiation of the longest wavelength and calculate the value of this wavelength.

[$1 \text{ eV} = 1.6 \times 10^{-19}$ J, speed of light $= 3.0 \times 10^{8}$ m s^{-1}, Planck constant $= 6.63 \times 10^{-34}$ J s]

4 The hydrogen nucleus has a radius of 1.3×10^{-15} m. A teacher tells a class that no known nucleus has a radius seven or more times that of the hydrogen nucleus. Explain, with relevance to a relevant expression, whether the teacher is correct.

UODLE, June 1997

5 The radius of a lithium ($^{7}_{3}Li$) nucleus is 2.3×10^{-15} m and the radius of a proton is 1.2×10^{-15} m.

Assuming that the proton and the lithium nucleus act as point charges, calculate the electric potential energy of a proton when it is just in contact with a lithium nucleus. Express your answer in electronvolts.

With reference to your answer, why do you think that particle accelerators used for research into the structure of nuclei are often referred to as 'high-energy' accelerators?

6 In nuclear fusion, a nucleus of deuterium ($^{2}_{1}H$) fuses with a nucleus of tritium ($^{3}_{1}H$) to give a helium nucleus and a neutron.

a Write down a nuclear equation for this fusion reaction.

b How much energy is released in this reaction?

[Mass of deuterium $= 2.0141$ u, mass of tritium $= 3.0161$ u, mass of helium nucleus $= 4.0028$ u, mass of neutron $= 1.0087$ u, $1 \text{ u} = 931$ MeV]

Suppose it was possible to build a nuclear power station that worked using this fusion reaction. Calculate the energy released by the fusion of 1 kg of deuterium. If 50% of this energy were used to produce 1 MW of electrical power continuously, for how many days would the station be able to generate electricity?

7 The grid shown here enables different nuclei to be represented by plotting the number of neutrons *N* against the number of protons *Z* in a nucleus. The arrow shows a nucleus X decaying to a nucleus Y.

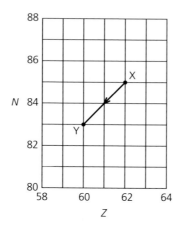

What type of radioactive decay is taking place?

Write a nuclear equation for this decay.

Copy the grid. Add another arrow to the grid to represent what happens if nucleus Y subsequently decays by β^- emission to nucleus W.

Mark a point on the grid that could represent the nucleus of an isotope of X.

Edexcel, January 2000

Special Relativity and quantum behaviour

3

' "I can't believe that," said Alice.

"Can't you?" the Queen said, in a pitying tone. "Try again: draw a long breath and shut your eyes."

Alice laughed. "There's no use trying," she said. "One can't believe impossible things."

"I daresay you haven't had much practice," said the Queen. "When I was your age, I always did it for half-an-hour a day. Why, sometimes I've believed as many as six impossible things before breakfast." '

Lewis Carroll (1832–1898), *Through the Looking Glass*

' "Quantum mechanics" is the description of the behavior of matter in all its details and, in particular, of the happenings on an atomic scale. Things on a very small scale behave like nothing that you have any direct experience about. They do not behave like waves, they do not behave like particles, they do not behave like clouds, or billiard balls, or weights on springs, or like anything that you have ever seen.'

R. P. Feynman, R. B. Leighton and M. Sands, 1964, *The Feynman Lectures on Physics*, Vol. I, p. 37-1, Addison-Wesley, Reading, MA

A new view of the world

Of the most significant theories that have influenced our understanding of the physical world in the 20th century, none have been more important than Einstein's Special Theory of Relativity and the modern theory of 'quantum behaviour', which is more commonly known as **quantum mechanics**. The Special Theory of Relativity leads to some very unusual predictions, particularly when describing the motion of particles travelling close to the speed of light. In Chapter 1, we discussed the origins of both Special Relativity and quantum theory, which caused us to rethink our view of nature from being continuous to being discrete or 'step-like'. Elementary particles of matter, such as electrons, photons, neutrinos, quarks and others, all exhibit quantum behaviour, and in this chapter we will take a closer look at what this means. While being two completely independent physical theories, Special Relativity and quantum mechanics are essential to understanding particle physics, and we will examine some of their key concepts and ideas. We will start first by looking at Einstein's Special Theory of Relativity (also known as the Principle of Special Relativity).

What is a relativity principle?

What does Einstein mean by a 'relativity principle'? A simple answer is that, in physics, a **relativity principle** is a statement that reconciles the observations that two observers (who may be in different places) make about the *same* physical event. Classical or Newtonian physics rests on the **Galilean Principle of Relativity**. Named after the Italian scientist and philosopher Galileo Galilei (1564–1642), this is the intuitive or 'common-sense' way of measuring things from different points of view.

However, at the very high velocities, close to the speed of light, at which elementary particles travel, the Galilean Principle of Relativity breaks down and modern physics requires this principle to be modified in a way that decidedly goes against our common-sense notions of space and time. Let's take a closer look at Galilean Relativity, and then see how, in the light of particular experimental evidence, Special Relativity is needed to reconcile the observations of physical events when very high velocities are involved.

Galilean Relativity

Galilean Relativity is so-called because it is based on the mechanics of Galileo, who was one of the first to study systematically the dynamics of moving bodies. Implicit in Galilean Relativity are two key assumptions:

1 Velocities are *additive*.
2 Time is *invariant*, that is, both observers agree about the time at which an event happens.

To see how Galilean Relativity works, consider the following example. Imagine two observers A and B watching the motion of a ball dropped from the mast of a moving boat (Figure 3.1). A, who is on the boat, sees the ball fall in a straight line towards the deck. However, observer B, who is standing on the quayside watching the boat go past, sees the ball move in a curved path called a *parabola*, because of the horizontal velocity v of the boat. Both A and B observe the event happen in the same interval of time.

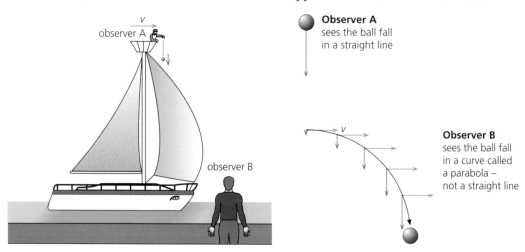

Figure 3.1 Observing a falling ball from two different frames of reference

Frames of reference

At this point it will help to explain in more detail some terminology that we introduced in Chapter 1. We say that A and B are observing the event of the falling ball from their respective **frames of reference**. A 'frame of reference' is simply the platform from which you observe an event. In physics it is convenient to specify a frame of reference as an *xyz* coordinate system to which all measurements can be referred (Figure 3.2). The frame of reference might be stationary with respect to the event (such as that of observer A on the deck of the ship), or it might be moving at some relative velocity (such as that of observer B, who sees the event of the falling ball moving at velocity *v*). Frames of reference that are either stationary or moving with *constant* velocity are called **inertial frames**. Frames of reference that are *accelerating* are called **non-inertial frames**. In Galilean and Special Relativity we concern ourselves only with measuring events from inertial frames of reference.

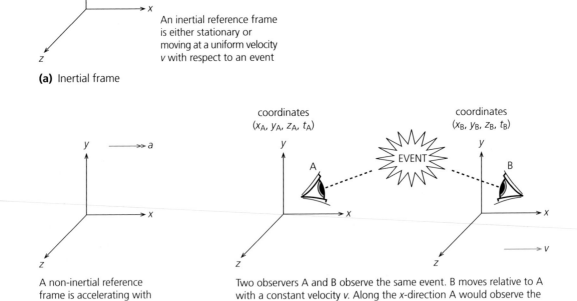

An inertial reference frame is either stationary or moving at a uniform velocity *v* with respect to an event

(a) Inertial frame

A non-inertial reference frame is accelerating with respect to an event

(b) Non-inertial frame

coordinates (x_A, y_A, z_A, t_A)

coordinates (x_B, y_B, z_B, t_B)

Two observers A and B observe the same event. B moves relative to A with a constant velocity *v*. Along the *x*-direction A would observe the event at $x_A = x_B - vt$ so that $x_B = x_A + vt$

(c) Galilean transformation

Figure 3.2 A frame of reference, often an *xyz* coordinate system, is a platform from which to observe an event

Galilean transformations

From their respective frames of reference, A and B are both able to write down equations of motion that describe the path of the ball using Newton's Laws. In the case of A this is the equation of a straight line, and for B it is that of a parabola. Both sets of equations are describing the same event and can be related by means of **transformations**. In relativity, a **Galilean transformation** is a mathematical expression that relates the measurements of position x_A, y_A, z_A and time t_A in A's reference frame with the corresponding measurements x_B, y_B, z_B and t_B in B's reference frame. For the inertial frames shown in Figure 3.2c, these Galilean transformation equations are:

$$x_A = x_B - vt$$
$$y_A = y_B$$
$$z_A = z_B$$
$$t_A = t_B = t$$

Galilean transformations assume that velocities are additive and that $t_A = t_B$ when the origins of A's and B's inertial frames coincide. So if the ship is travelling at $5\,\mathrm{m\,s^{-1}}$ and A at the top of the mast throws the ball forward at $10\,\mathrm{m\,s^{-1}}$, then B (who is in a moving frame of reference with respect to A) sees the initial horizontal velocity of the ball as $15\,\mathrm{m\,s^{-1}}$. Also, if the ship sails on a voyage from Southampton to New York with A aboard and B waiting on the quayside, both A and B will measure the same amount of elapsed time for the duration of the voyage.

All these things seem intuitively obvious. However, as we shall see next, velocities are *not* additive and time is *not* invariant with Special Relativity, and according to Einstein we have to abandon these 'common-sense' notions of velocity and time.

Special Relativity

In Chapter 1, we discussed the Michelson–Morley experiment, which showed that the ether did not exist and that the speed of light appeared to be constant to all observers. In this section we will summarise two key results to do with time and length that arise from the Special Theory of Relativity and (this is optional) explain how they are derived.

Lorentz and FitzGerald noticed that the results of Michelson and Morley could be explained if the Galilean transformations were adjusted on the basis that intervals of length when measured in one frame of reference appear *contracted* when observed from another. The adjustment factor depended on the relative velocity between the two frames. In effect, Lorentz and FitzGerald were saying that the constancy of the speed of light was due to compensating variations in the distance over which the light travelled. Suppose you measure the length x_0 of an object which is stationary in your inertial reference frame. If the object then moves along the direction of its length relative to you at speed v, you will measure its length to be x, given by the **length contraction transformation**:

$$x = x_0 \sqrt{1 - \frac{v^2}{c^2}}$$ where c is the speed of light

Since the factor $\sqrt{1 - v^2/c^2}$ is <1, the object is observed to have contracted in length.

Also, intervals of time are related by the **time dilation transformation**:

$$\Delta t = \frac{\Delta t_0}{\sqrt{1 - \frac{v^2}{c^2}}}$$

showing that time intervals Δt as measured from the moving frame are longer, or dilated.

What you should notice about these **Lorentz–FitzGerald transformations** is that they revert to the Galilean form when $v \ll c$ and that these effects of 'length contraction' and 'time dilation' only become apparent at speeds close to the speed of light. Another important result is that these transformations imply that nothing can travel faster than light, for if $v = c$ then the adjustment factor $\sqrt{1 - v^2/c^2}$ becomes zero, leading to division by zero, which is mathematically undefined. The factor $1/\sqrt{1 - v^2/c^2}$ is called the **Lorentz factor** and is sometimes given the symbol γ, and the term v/c in the Lorentz factor is given the symbol β, called the **speed parameter**, so that $\gamma = 1/\sqrt{1 - \beta^2}$ (don't confuse these with beta and gamma rays!).

These transformations by Lorentz and FitzGerald were proposed on an entirely speculative basis, and it was the genius of Einstein who combined them into a workable physical theory called the **Special Theory of Relativity**. Even more remarkable was that Einstein formulated the theory entirely in his head as a *Gedankenexperiment*. This is a German phrase meaning 'thought experiment', which is an idealised experiment whose outcome can be predicted by thought alone.

Einstein put forward two postulates, which we introduced in Chapter 1 and restate here in slightly different terms, that form the physical basis of the Special Theory of Relativity:

Postulate 1 *All inertial reference frames are equivalent for the observation and formulation of physical laws.*

Postulate 2 *The speed of light is constant to all observers.*

The first postulate is saying that there are no absolute or preferred inertial frames of reference in the universe. The laws of physics are the same in all reference frames and your frame of reference is just as good as mine for observing and recording measurements.

The second postulate is stating that the speed of light is independent of the motion of the source – an idea that violates the Galilean principle of addition of velocities and is far from obvious to our everyday experience. For example, suppose you were standing on an express train moving at $30\,\mathrm{m\,s^{-1}}$ and were shining a torch in the direction of motion. Intuitively we would think that an observer standing by the track would measure the speed of the photons emerging from the torch as the speed of light ($3 \times 10^8\,\mathrm{m\,s^{-1}}$) plus the speed of the train ($30\,\mathrm{m\,s^{-1}}$), that is, $300\,000\,030\,\mathrm{m\,s^{-1}}$. Not so says Einstein! Both you and the observer at the trackside would measure the photon speed to be the same, that is, $3 \times 10^8\,\mathrm{m\,s^{-1}}$!

Since all inertial frames are equivalent, measurements of space and time must vary between them for the speed of light to remain constant. In case you may be wondering whether this is just some fanciful speculation, these effects are real (see Box 3.1). Adding together combinations of speeds that are less than c can *never* be made to exceed c – something that seems utterly at odds with our everyday experience.

Box 3.1 The lifetime of muons – a test of Special Relativity

Muons (μ) are charged elementary particles heavier than an electron that are created by cosmic rays high in the atmosphere. A negatively charged muon (μ^-) can decay into an electron (e), a neutrino (ν) and an antineutrino $\bar{\nu}$ by the process

$$\mu^- \to e^- + \nu + \bar{\nu}$$

Because they decay, muons are like radioactive clocks. With a charged particle detector, we can detect the arrival of a muon and then, at a certain time later, the production of an electron, and so measure the decay time.

The lifetime of muons in a laboratory reference frame is found to be $1.5 \times 10^{-6}\,\mathrm{s}$. However, the muons travelling down towards the surface of the Earth to reach us have a lifetime in our frame of about $1.4 \times 10^{-5}\,\mathrm{s}$. The reason for this is *time dilation*.

The muons are travelling at speeds of about $0.994\,c$ and if we use the time dilation equation of Special Relativity

$$\Delta t = \frac{\Delta t_0}{\sqrt{1 - \dfrac{v^2}{c^2}}}$$

we find that, from our point of view, the muon lives for

$$\frac{1.5 \times 10^{-6}}{\sqrt{1 - (0.994)^2}} = 1.4 \times 10^{-5}\,\mathrm{s}$$

To us, the muon's decay time seems to have been slowed by a factor of almost 10, so that it has enough time to reach the Earth before decaying.

From the muon's point of view, it still has a lifetime of $1.5 \times 10^{-6}\,\mathrm{s}$ but the thickness of atmosphere through which it has to travel has been foreshortened by length contraction. Using

$$x = x_0 \sqrt{1 - \frac{v^2}{c^2}}$$

the muon finds that the distance it has to travel has contracted by the same factor, so that it has a shorter distance to go before it reaches the ground.

Another important relativistic concept is the **relativity of simultaneity**. To see what this means, imagine two observers Sally and Steve, who observe events from their respective frames of reference (Figure 3.3). Sally notes that two independent events, say a red light and a blue light coming on, happen at the same time, that is, simultaneously. Steve, who is moving with a constant velocity with respect to Sally, also records these same two events. Will Steve also observe that the lights come on at the same time? Strange as it may seem, the answer is that in general Steve will *not* agree that the red and blue lights come on simultaneously. This leads to the concept that, in Special Relativity, **simultaneity** is not an absolute concept but a relative one that depends on the speed of the observer. The reason for this is the constancy of the speed of light. The wavefronts from the red and blue lights travel outwards with a constant velocity c which is independent of the observer, and as a result Steve and Sally observe the same events at different times.

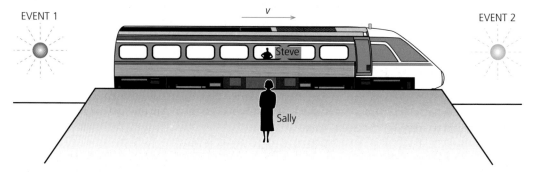

Figure 3.3(a) A red light and a blue light come on behind and ahead of a train travelling with velocity v past a station platform. Sally, who is on the platform and equidistant from the two events, sees the lights come on simultaneously

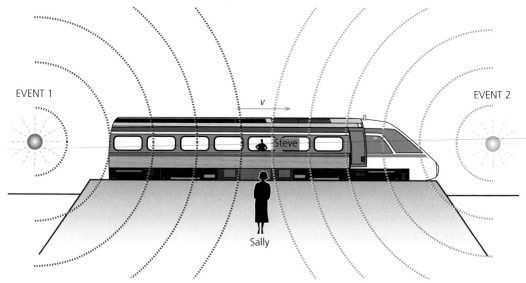

Figure 3.3(b) In Steve's reference frame, the moving train, the blue light from the front end of the train reaches him before the red light from the rear, so he does *not* see them light up simultaneously

The relativity of time

According to Einstein, if two observers who move relative to each other measure a time interval between two events, in general they will get different results. Why should this be? The answer is that the *separation in space* of the events can affect the time intervals measured by each observer. Let's consider a more specific example involving Sally and Steve.

In Figure 3.4 Sally is riding in a train moving at a constant velocity v relative to a station. Inside her train is a mirror (M) and a light source (A). Event 1 occurs when a pulse of light leaves the light source A; the pulse travels upwards a distance D, hits the mirror and is reflected vertically downwards, where it is detected at the source, which we will call event 2. Using a single clock on board the train, Sally times the interval between the two events as Δt_0. Using the formula time = distance/speed, Sally determines that $\Delta t_0 = 2D/c$. The two events happen at the *same location* in Sally's reference frame and she only needs one clock at that location to measure the time interval.

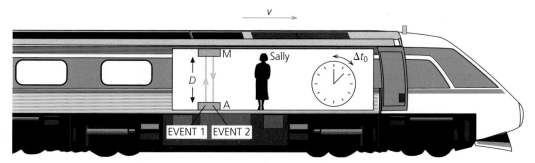

Figure 3.4 Sally is on the train moving at velocity v. She uses a single clock on the train, and measures the time interval between events 1 and 2 as $\Delta t_0 = 2D/c$

However, Steve, who is standing on the station platform watching the train go by, sees the two events somewhat differently. Since the mirror and light source move with the train between the two events, Steve sees the light travel in the path shown in Figure 3.5. For Steve, the two events occur at *different locations* in his reference frame. So he must use two clocks C_1 and C_2, which are synchronised and located at each event. Now according to Einstein the speed of light is constant in both reference frames. However, in Steve's frame, the light travels a distance $2L$ between events 1 and 2. The time interval that Steve measures between the two events Δt is given by $\Delta t = 2L/c$. Since the train is travelling at a velocity v, using Pythagoras' Theorem,

$$L = \sqrt{(\tfrac{1}{2}v\Delta t)^2 + D^2}$$

Substituting for D using the expression in Sally's frame, that is, $D = c\Delta t_0/2$, we get

$$L = \sqrt{(\tfrac{1}{2}v\Delta t)^2 + (\tfrac{1}{2}c\Delta t_0)^2}$$

Since in Steve's frame $L = c\Delta t/2$ we can substitute for L and write

$$\tfrac{1}{2}c\Delta t = \sqrt{(\tfrac{1}{2}v\Delta t)^2 + (\tfrac{1}{2}c\Delta t_0)^2}$$

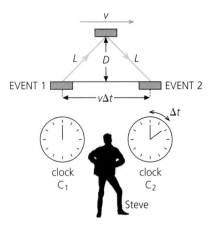

Figure 3.5 Steve is on the platform. He needs two *synchronised* clocks to measure events 1 and 2, and he finds that $\Delta t \neq \Delta t_0$ as measured by Sally on the train

Squaring both sides of the last equation gives

$$\tfrac{1}{4}c^2(\Delta t)^2 = \tfrac{1}{4}v^2(\Delta t)^2 + \tfrac{1}{4}c^2(\Delta t_0)^2$$

or, after multiplying each term by 4,

$$c^2(\Delta t)^2 = v^2(\Delta t)^2 + c^2(\Delta t_0)^2$$

Collecting like terms and factorising gives

$$(\Delta t)^2(c^2 - v^2) = c^2(\Delta t_0)^2$$

$$(\Delta t)^2 = \frac{c^2(\Delta t_0)^2}{c^2 - v^2}$$

Finally, dividing both sides by c^2 and taking the square root on both sides, we get

$$\Delta t = \frac{\Delta t_0}{\sqrt{1 - \dfrac{v^2}{c^2}}}$$

or, using the symbol γ for the Lorentz factor, as mentioned on page 60,

$$\Delta t = \gamma \Delta t_0$$

Don't worry too much if you find the algebraic derivation complicated. The important thing to understand about the above equation is that it tells us how the time interval between the two events as seen in Steve's frame compares with the interval between the same two events as seen from Sally's frame. They are *not* the same. Since v must always be less than c, the denominator must always be less than unity. This means that the time interval between the two events in Steve's frame is *longer* than that in Sally's. This effect is called **time dilation** (to 'dilate' something means to expand or stretch it).

What you should see now is that time is a *relative* concept rather than an absolute one. When two events take place at the *same* location in an inertial reference frame (such as Sally's), we say that the interval between them is called the **proper time**. When the same two events are measured from another inertial frame in a different location, the time interval measured from that frame is always greater. Be assured that there is nothing 'improper' about any other time interval, it's just that 'proper' is the term that physicists use to emphasise that the time interval is that between two events in one frame as observed in that frame, rather than the time interval measured from another frame, which is dilated due to the relative motion of the observer. This effect is real and, as we will see in Chapter 4, particle accelerators have to be designed to take it into account.

The relativity of length

Length contraction is a direct consequence of time dilation. Let's go back to Sally and Steve again. Recall that Sally is on a train passing through a station and Steve is on the platform. Using a tape measure, Steve measures the length of the platform to be L_0. The length L_0 is the length of the platform when it is stationary with respect to Steve, meaning that the platform and Steve are in the same reference frame, and L_0 is thus called the **proper length** (or rest length). Using his synchronised clocks, Steve notes that Sally travels the length of the platform in an interval of time $\Delta t = L_0/v$ so that $L_0 = v\Delta t$. However, as we have seen before, this is *not* a proper time interval since his observations of the two events, that is, Sally passing the beginning of the platform (event 1) and Sally passing the end of the platform (event 2), take place at *different locations* and so Steve needs two synchronised clocks to measure Δt.

However, from Sally's frame of reference, she sees the platform moving, and the two events that Steve observes happen for her at the *same location* in her frame. She can measure the interval with a single clock, so that her time interval Δt_0 is a proper time interval. Sally measures the length of the platform as $L = v\Delta t_0$. We can eliminate V between these two equations by dividing one into the other,

$$\frac{L}{L_0} = \frac{v\Delta t_0}{v\Delta t} = \frac{\Delta t_0}{\Delta t}$$

and since (from the previous section)

$$\Delta t = \frac{\Delta t_0}{\sqrt{1 - \dfrac{v^2}{c^2}}}$$

we have

$$\frac{\Delta t_0}{\Delta t} = \sqrt{1 - \frac{v^2}{c^2}}$$

and we can write

$$\frac{L}{L_0} = \frac{\Delta t_0}{\Delta t} = \sqrt{1 - \frac{v^2}{c^2}}$$

so

$$L = L_0\sqrt{1 - \frac{v^2}{c^2}} \quad \text{or} \quad L = \frac{L_0}{\gamma}$$

Since γ is always greater than unity, L is always less than L_0, and the relative motion between Sally and Steve has brought about a **length contraction**. This is another odd consequence of Special Relativity. If an object's length is measured from any reference frame other than the rest frame, that is, from a frame moving with uniform relative motion parallel to that length, then we will find that the length we measure is always less than its proper length measured in its rest frame.

The effects of time dilation and length contraction only become significant at speeds that are an appreciable fraction of the speed of light. Figure 3.6 is a plot of the Lorentz factor γ as a function of the speed parameter β. The difference is not significant unless $v > 0.1c$, so 'everyday' Newtonian kinematics (Galilean Relativity) works well enough at speeds below $0.1c$. For $v > 0.1c$, physicists use Special Relativity to get the correct answers. Note that the length contraction takes place only along the length in the *direction* of relative motion and does not necessarily have to be the length of a rigid object such as a ruler. Length contraction would also be observed from a moving reference frame of two objects whose *distance apart* was constant in the same rest frame, like two cars on a motorway for example (although the length contraction in this case would be very small!).

Experiments have in fact been carried out by placing portable atomic clocks in commercial airliners and flying them around the world. Before and after the flight, the clocks were compared with identical atomic clocks kept on the ground. The flying clocks were found to have run slow by as much as 10^{-7} s.

But is time really getting longer? Do objects actually get shorter as they travel faster? What we see is what we can measure and observe. It is more accurate to say that *what is measured* appears to differ, and since that is our reality, then to us this is indeed what happens.

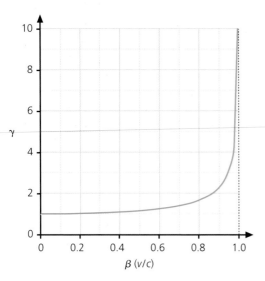

Figure 3.6 Lorentz factor γ as a function of β (v/c)

Relativistic mass, energy and momentum

The relativistic behaviour of time and length means that we have to adopt new definitions of momentum and energy and new relativistic equations of motion. The differences between the Newtonian equations of motion and the relativistic ones are small at speeds that are small compared with the speed of light. But, as we shall see, the differences become highly significant in the world of sub-atomic particles, which are routinely accelerated to near light speeds. The Lorentz–FitzGerald transformations tell us how to relate one inertial reference frame to another at speeds close to the speed of light. Earlier, we stated the Galilean transformation equations for objects travelling at speeds much less than the speed of light between two inertial reference frames A and B as

$$x_A = x_B - vt$$
$$y_A = y_B$$
$$z_A = z_B$$
$$t_A = t_B = t$$

It can be shown that the equivalent **Lorentz transformation** equations (which we shall not prove) may be written as

$$x_A = \gamma(x_B - vt)$$
$$y_A = y_B$$
$$z_A = z_B$$
$$t_A = \gamma\left(t_B - \frac{vx_A}{c^2}\right)$$

These are valid for all speeds up to the speed of light.

In the Galilean transformations it is automatically assumed that $t_A = t_B = 0$ when the origins of A and B coincide. However, the relativity of simultaneity tells us that we can no longer assume that this true. Notice how the Lorentz transformations revert to the Galilean form when $v \ll c$, and for objects moving at everyday velocities the Galilean transformations are perfectly adequate for our purposes.

So what happens if we give an object sufficient energy to try to make it go faster than light? In order to answer this question, Special Relativity requires that we invent a new definition of momentum. Let's get Sally and Steve to make some momentum measurements from their respective inertial reference frames. Consider a particle of mass m_0 moving with constant speed v in the x-direction. From a classical point of view its momentum is equal to the familiar

$$p = m_0 v = m_0 \frac{\Delta x}{\Delta t}$$

where Δx is the distance moved by the particle in time Δt, that is, the distance covered by a moving particle *as seen by* Steve, who is *watching* that particle go by. However, the momentum measured by Sally, who is *moving with* the particle, is

$$p = m_0 \frac{\Delta x}{\Delta t_0}$$

where Δt_0 is the time taken to cover the same distance Δx from Sally's reference frame. The particle is at rest with respect to Sally, with the result that the time Sally measures is the *proper time*, Δt_0. Using the time dilation formula we can then write

$$p = m_0 \frac{\Delta x}{\Delta t_0} = m_0 \frac{\Delta x}{\Delta t} \frac{\Delta t}{\Delta t_0} = m_0 \frac{\Delta x}{\Delta t} \frac{1}{\sqrt{1 - \left(\dfrac{v}{c}\right)^2}} = \gamma m_0 v$$

Notice that the expression for **relativistic momentum**

$$p = \frac{m_0 v}{\sqrt{1 - \left(\dfrac{v}{c}\right)^2}}$$

becomes our everyday non-relativistic definition of momentum when $v \ll c$. We call the mass m_0 the **rest mass**, which is the mass of the particle when measured from an inertial frame that is at rest with respect to the particle. Figure 3.7 shows a graph of relativistic momentum versus v/c. You can see that, as the particle approaches the velocity of light, its momentum becomes infinite.

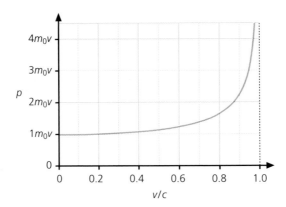

Figure 3.7 A plot of relativistic momentum $\dfrac{m_0 v}{\sqrt{1 - \left(\dfrac{v}{c}\right)^2}}$ versus $\dfrac{v}{c}$

Relativistic mass

If length and time measurements change at relativistic speeds, then mass must change too. Look at the relativistic equation for momentum

$$p = \frac{m_0 v}{\sqrt{1 - \left(\dfrac{v}{c}\right)^2}}$$

When compared with the non-relativistic definition of momentum $p = mv$ with which you are familiar, then these two expressions for momentum differ only by the Lorentz factor γ, that is, $p = \gamma m_0 v$. This implies that the two masses in these expressions differ by the Lorentz factor, to give the **relativistic mass**

$$m = \gamma m_0 = \frac{m_0}{\sqrt{1 - \left(\dfrac{v}{c}\right)^2}}$$

In fact, if we try to make particles in accelerators go faster than light by giving them more kinetic energy, what happens is that they become more massive. Figure 3.8 is a graph of relativistic mass versus v/c. We see that m_0 is the mass of the particle when it is at rest in the inertial reference frame of the observer (the mass you would measure if it was sitting on a table in front of you), and m is the mass you would measure when it was travelling past you at high speed.

From our discussion of relativistic momentum and Einstein's postulate that the laws of physics are the same in all reference frames, we arrive at the startling conclusion that, for the law of conservation of momentum to remain valid at all speeds, the mass of a particle can no longer be regarded as a fixed quantity, and we have to distinguish between its rest mass and relativistic mass.

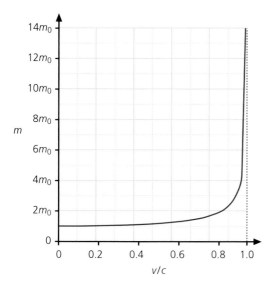

Figure 3.8 A plot of relativistic mass $m = \dfrac{m_0}{\sqrt{1 - \left(\dfrac{v}{c}\right)^2}}$ versus $\dfrac{v}{c}$

Relativistic energy

Since mass increases with velocity, then you might wonder whether the familiar equation for kinetic energy $KE = \frac{1}{2}mv^2$ has a relativistic equivalent at high speeds. As with momentum, it turns out that we need to introduce a new definition of relativistic energy if the law of conservation of energy is to remain true. We need a new formula for kinetic energy that is conserved for all velocities and reduces to the form $KE = \frac{1}{2}m_0 v^2$ for small values of v.

If an object has a rest mass of m_0, then by Einstein's mass–energy relation, its rest energy is $m_0 c^2$. If the same mass is moving with a velocity v, then its mass will increase with velocity to

$$m = \frac{m_0}{\sqrt{1 - \left(\dfrac{v}{c}\right)^2}}$$

and its total energy will increase to mc^2.

When a force is applied to accelerate a given mass, then the work done = force × distance. This work represents an increase in the kinetic energy of the particle E_k so that

$$\text{work done} = \text{force} \times \text{distance} = E_k$$

By the law of conservation of energy, the total energy E of the particle is therefore

$$E = \text{rest-mass energy} + \text{kinetic energy}$$

or

$$mc^2 = m_0c^2 + E_k$$
$$\therefore \quad E_k = mc^2 - m_0c^2$$

or

$$E_k = \frac{m_0c^2}{\sqrt{1 - \left(\dfrac{v}{c}\right)^2}} - m_0c^2$$

On factorising we get

$$E_k = m_0c^2 \left(\frac{1}{\sqrt{1 - \left(\dfrac{v}{c}\right)^2}} - 1 \right)$$

and using the symbol γ we can write this as

$$E_k = m_0c^2(\gamma - 1)$$

which is the relativistic formula for kinetic energy. Therefore

$$E = m_0c^2 + m_0c^2(\gamma - 1)$$
$$= m_0c^2 + m_0c^2\gamma - m_0c^2 = m_0c^2\gamma$$

or

$$E = \frac{m_0c^2}{\sqrt{1 - \left(\dfrac{v}{c}\right)^2}}$$

which is the expression for the **relativistic total energy**.

WORKED EXAMPLE 3.1

Electrons with rest mass 9.1×10^{-31} kg have been accelerated at the Stanford Linear Accelerator Center (SLAC) to speeds of $0.999\,999\,999\,67\,c$. Calculate

a the rest energy of an electron,
b the relativistic mass increase at this speed,
c the relativistic kinetic energy of an electron moving at this speed,
d the total energy of such an electron at SLAC.

Note: For v/c close to 1, $\sqrt{1 - (v/c)^2}$ may be approximated by $\sqrt{2} \times \sqrt{1 - v/c}$. This results from using 'the difference of two squares', that is, $a^2 - b^2 = (a + b)(a - b)$. In this case a^2 is 1^2 and b^2 is $(v/c)^2$, and we get

$$1 - \left(\frac{v}{c}\right)^2 = \left(1 + \frac{v}{c}\right)\left(1 - \frac{v}{c}\right)$$

and since $v/c \approx 1$, $1 + (v/c) \approx 2$, and so

$$\sqrt{1 - \left(\frac{v}{c}\right)^2} = \sqrt{1 + \frac{v}{c}} \times \sqrt{1 - \frac{v}{c}} \approx \sqrt{2}\sqrt{1 - \frac{v}{c}}$$

a The rest energy is given by

$$m_0 c^2 = 9.1 \times 10^{-31}\,\text{kg} \times (3.0 \times 10^8\,\text{m s}^{-1})^2 = 8.2 \times 10^{-14}\,\text{J}$$

b The relativistic mass is given by

$$m = \frac{m_0}{\sqrt{1 - \left(\frac{v}{c}\right)^2}}$$

which, using the approximation, becomes

$$m = \frac{9.1 \times 10^{-31}\,\text{kg}}{\sqrt{2} \times \sqrt{\left(1 - \frac{0.999\,999\,999\,67\,c}{c}\right)}} = \frac{9.1 \times 10^{-31}\,\text{kg}}{\sqrt{2} \times \sqrt{3.3 \times 10^{-10}}} = 3.5 \times 10^{-26}\,\text{kg}$$

The mass increase $= m/m_0 = (3.5 \times 10^{-26})/(9.1 \times 10^{-31}) \approx 39\,000$

c The relativistic kinetic energy of the electron is given by

$$E_k = m_0 c^2 \left(\frac{1}{\sqrt{1 - \left(\frac{v}{c}\right)^2}} - 1 \right)$$

$$= 9.1 \times 10^{-31}\,\text{kg} \times (3.0 \times 10^8\,\text{m s}^{-1})^2 \times \left(\frac{1}{\sqrt{2} \times \sqrt{3.3 \times 10^{-10}}} - 1 \right) = 3.2 \times 10^{-9}\,\text{J}$$

d The total energy E is given by

$$E = \frac{m_0 c^2}{\sqrt{1 - \left(\dfrac{v}{c}\right)^2}}$$

$$= \frac{9.1 \times 10^{-31}\,\text{kg} \times (3.0 \times 10^8\,\text{m s}^{-1})^2}{\sqrt{2} \times \sqrt{3.3 \times 10^{-10}}} = 3.2 \times 10^{-9}\,\text{J}$$

Note that the expression for the total energy is almost the same as that for the relativistic kinetic energy, and to the number of significant figures given here the values are the same. This is because the rest energy is a much smaller fraction of the total energy. We now have a new definition of conservation of energy based on Special Relativity which states:

In an isolated system of particles, the relativistic total energy is conserved.

(Note that by an 'isolated system of particles' we mean a system in which no extra energy is introduced to the particles by some external means. The energy of the particles in the system consists entirely of their individual rest and kinetic energies.)

Relativistic energy and momentum

Two other expressions follow from the expression for relativistic mass

$$m = \frac{m_0}{\sqrt{1 - \dfrac{v^2}{c^2}}}$$

Squaring both sides, we get

$$m^2 = \frac{m_0^2}{\left(1 - \dfrac{v^2}{c^2}\right)} = \frac{m_0^2 c^2}{c^2 - v^2}$$

Rearranging,

$$m_0^2 c^2 = m^2(c^2 - v^2)$$

Now $p = mv$, so that $v = p/m$ and we can write

$$m_0^2 c^2 = m^2\left(c^2 - \frac{p^2}{m^2}\right) = m^2 c^2 - p^2$$

Multiplying throughout by c^2,

$$m_0^2 c^4 = m^2 c^4 - p^2 c^2$$

or

$$m_0^2 c^4 = E^2 - p^2 c^2$$

which shows that the relativistic energy E can be expressed in terms of the relativistic momentum p as

$$E^2 = p^2c^2 + m_0^2c^4$$

This is a very important and useful result in particle physics, and an easy way to remember this formula is to express it as

$$E^2 = (pc)^2 + (m_0c^2)^2$$

and compare it with Pythagoras' Theorem for a right angled triangle (Figure 3.9).

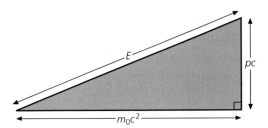

Figure 3.9 Right angled triangle for remembering the relationship between the relativistic energy E and relativistic momentum p, that is, $E^2 = (pc)^2 + (m_0c^2)^2$

For a particle moving close to the speed of light, the first term $(pc)^2$ is much greater than the second term $(m_0c^2)^2$ and the relativistic energy E can be approximated as

$$E \approx pc$$

This shows that the momentum and energy of an ultra-relativistic particle are proportional to each other.

The 'mass' of a photon

Earlier we referred to particles of electromagnetic radiation called photons which travel at exactly the speed of light. What mass do they have? Since

$$m = \frac{m_0}{\sqrt{1 - \left(\dfrac{v}{c}\right)^2}}$$

then

$$m_0 = m\sqrt{1 - \left(\frac{v}{c}\right)^2}$$

When $v = c$ then the rest mass m_0 becomes zero. We therefore think of a photon having a finite mass–energy but a zero rest mass, and for particles that travel at the speed of light c, $E \approx pc$ is an exact relation ($E = pc$).

Units of energy, mass and momentum in particle physics

Energies in atomic and nuclear physics are usually expressed in terms of electronvolts (eV), mega-electronvolts (MeV) or giga-electronvolts (GeV), where $1\,eV = 1.6 \times 10^{-19}\,J$. In particle physics, convenient units of *mass* are eV/c^2, MeV/c^2 or GeV/c^2, which from Einstein's mass–energy relation is just the rest energy of the particle divided by c^2.

Similarly, because particle physicists measure such small values of momentum, they do not generally use units of $kg\,m\,s^{-1}$. We saw earlier that $E \approx pc$ for $E \gg m_0 c^2$ so that units of momentum can be expressed in terms of eV/c, MeV/c or GeV/c.

WORKED EXAMPLE 3.2

a A particle has a rest mass of $2\,MeV/c^2$. What is its mass in kg?
b Express $1000\,MeV/c$ in units of $kg\,m\,s^{-1}$.

a $2\,MeV/c^2 = \dfrac{2 \times 10^6 \times 1.6 \times 10^{-19}\,J}{(3 \times 10^8\,m\,s^{-1})^2} = 3.6 \times 10^{-30}\,kg$

b $\dfrac{1000\,MeV}{c} = \dfrac{1000 \times 10^6 \times 1.6 \times 10^{-19}\,J}{3.0 \times 10^8\,m\,s^{-1}} = 5.3 \times 10^{-19}\,kg\,m\,s^{-1}$

If you find the conceptual ideas of Special Relativity hard to swallow, don't worry! They astounded most physicists when Einstein first presented them in 1905. Do not be overly concerned about the mathematical derivations. They are there to show you where the equations come from and how they are connected. This section's main purpose is to give you a flavour of what Special Relativity is like. The main thing to understand is that space and time are not absolute concepts. Measurements depend on how you look at them, and the speed at which you are travelling. The important results are summarised for you at the end of this chapter. Later, we will see that Special Relativity plays a very necessary and vital role in particle physics, including the design of particle accelerators that can accelerate particles to near light speeds.

Quantum mechanics

De Broglie's electron waves

In 1924 the French physicist Prince Louis Victor de Broglie (1892–1987) (Figure 3.10) put forward the extraordinary idea that, just as light waves could behave like particles (photons), so could particles act like waves. In particular, he proposed that electrons could exhibit diffraction and interference effects. These ideas were put to the test in 1927 the two American physicists Clinton Joseph Davisson (1881–1958) and Lester Germer (1896–1971) and also independently by George Paget Thomson (1892–1975), the son of J. J. Thomson who discovered the electron. Davisson, Germer and Thomson found that if electrons were passed through metallic foils they showed interference fringes just like light passing through a diffraction grating.

De Broglie (pronounced broh-lee) stated that the wavelength of a particle is inversely proportional to its momentum p, with the constant of proportionality being the Planck constant h, that is,

$$\lambda_p = \frac{h}{p} = \frac{h}{mv}$$

where λ_p is called the **de Broglie wavelength**. The higher the momentum of the particle, the smaller the wavelength. Physicists refer to these 'matter waves' as **de Broglie waves**. Every particle, and not just electrons or elementary particles, has its own de Broglie wavelength. Even you and I have one!

Figure 3.10 Louis Victor de Broglie (1892–1987)

WORKED EXAMPLE 3.3

What is the de Broglie wavelength of

a a student of mass 70 kg moving with a velocity of $2\,\mathrm{m\,s^{-1}}$,
b an electron with a kinetic energy of 30 keV? (Take $h = 6.63 \times 10^{-34}\,\mathrm{J\,s}$.)

a The momentum of the student is

$$p = m \times v = 70\,\mathrm{kg} \times 2\,\mathrm{m\,s^{-1}} = 140\,\mathrm{kg\,m\,s^{-1}}$$

The de Broglie wavelength is

$$\frac{h}{p} = \frac{6.63 \times 10^{-34}\,\mathrm{J\,s}}{140\,\mathrm{kg\,m\,s^{-1}}} = 4.7 \times 10^{-36}\,\mathrm{m}$$

b $\mathrm{KE}_{\mathrm{electron}} = \frac{1}{2}m_e v^2$. We can rewrite this as

$$\mathrm{KE}_{\mathrm{electron}} = \frac{1}{2m_e}(m_e v)^2$$

so that the momentum of the electron is

$$m_e v = \sqrt{2 \times m_e \times \mathrm{KE}_{\mathrm{electron}}}$$

We have $\mathrm{KE}_{\mathrm{electron}} = 30\,\mathrm{keV}$ from the question, so

$$\mathrm{KE}_{\mathrm{electron}} = 30 \times 1000\,\mathrm{V} \times 1.6 \times 10^{-19}\,\mathrm{C} = 4.8 \times 10^{-15}\,\mathrm{J}$$

$$\therefore \quad p = m_e v = \sqrt{2 \times (9.1 \times 10^{-31}\,\mathrm{kg}) \times (4.8 \times 10^{-15}\,\mathrm{J})}$$

$$= 9.3 \times 10^{-23}\,\mathrm{kg\,m\,s^{-1}}$$

Therefore the de Broglie wavelength is $\dfrac{6.63 \times 10^{-34}\,\mathrm{J\,s}}{9.3 \times 10^{-23}\,\mathrm{kg\,m\,s^{-1}}} = 7.1 \times 10^{-12}\,\mathrm{m}$

For diffraction effects to be easily visible, the wavelength of an object must be of the same order of magnitude as or *commensurate with* the aperture through which it diffracts. This explains why you and I aren't 'smeared out' as we walk through, say, a doorway. Our de Broglie wavelength is some 10^{36} times smaller than the width of a door. However, the wavelength of electrons, photons and other sub-atomic particles can easily be of the same order of magnitude as the atomic and nuclear spacing in a material, and diffraction effects for them become important. It is the size of the Planck constant that determines the importance of de Broglie effects, and there is no experiment that can be done which would reveal the wave nature of macroscopic or everyday sized objects.

The meaning of wave–particle duality

De Broglie's ideas marked a completely new way of looking at nature. All matter can be regarded as *both* particles and waves. It really depends on the kind of experiment that we are doing. In the photoelectric effect (Box 1.1, page 15) light, previously thought of as a wave, behaves like a particle, but light can also be shown to be diffracted. Similarly, the electron, whose mass can be measured as a particle, can also be diffracted as a wave. Both philosophers and physicists have struggled to reconcile these apparently contradictory descriptions of matter. We have to conclude that neither the wave nor the particle picture is entirely correct all of the time. The two explanations are *complementary* to each other. This is rather like describing a work of art such as the Sistine Chapel either in terms of Michelangelo's artistic use of form or as a chemical description of the pigments he used in the paint to create the blends of colour!

When light is diffracted through a double slit such as Young's slits, we might say that atoms of the light source are emitting individual photons. When the photons hit the film after diffracting through the slits then they are still regarded as quantised. In between, we can think of electromagnetic energy propagating as a wave as it passes through the slits and spreads out into an interference pattern (Figure 3.11). Where the wave has a large *intensity*, that is, there are a large number of photons hitting the film per second, then we observe that the film is strongly exposed, indicating the presence of many photons. Where it is dark, few photons are seen. The chance of seeing photons is therefore proportional to the intensity of the light at a particular point on the film, and we will shortly see that this is an essential relationship that links the particle and wave descriptions of matter together.

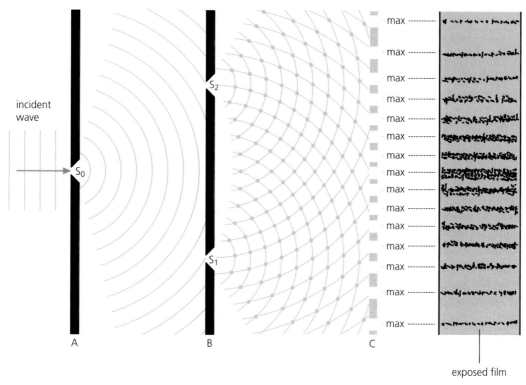

incident wave

max
max
max
max
max
max
max
max
max
max
max
max
max
max

A B C

exposed film

Figure 3.11 The interference of light at the slits in the Young's double-slit experiment can be explained by the wave theory of light, yet the exposure of the film is explained by individual photons activating tiny photographic 'grains' in the film to give an overall intensity distribution

The Heisenberg Uncertainty Principle

Uncertainty in the 'everyday world'

Whenever we measure something, there is always a certain amount of error in the measurement. Sometimes the error can be very small. For example, the standard unit of mass is the kilogram (kg), and is the mass of a piece of platinum–iridium kept in a carefully controlled environment at Sèvres, near Paris. Other masses can be compared with this standard to an accuracy of one part per hundred million. In contrast, the error in measuring the age of the universe can be very large, mainly because of the large uncertainty in the value of the *Hubble constant* (the Hubble constant and how it is related to the age of the universe is described in Chapter 8).

If we measure a length of, say, 10 m to the nearest metre, then what we mean is that the actual measurement is greater than or equal to 9.5 m and not more than or equal to 10.5 m. We can be certain about these limits – but we are uncertain as to what the exact value of the length is *between* these limits. So our uncertainty in the measurement of 10 m is 1 m. It is usual to denote the uncertainty in a measurement with the Greek letter Δ. Thus the uncertainty in a length x is written as Δx (for example, for our 10 m length, $\Delta x = 1$ m).

77

Now we clearly have measuring devices that can measure a length of 10 m far more accurately than this. We could use, say, a laser, and reflect the light from a mirror 10 m distant, timing the interval between the transmitted and received pulses with an atomic standard clock to calculate the distance and reduce the uncertainty in the measurement. However, in the quantum world, measuring things is rather different.

Uncertainty in the 'quantum world'

In 1926, the German physicist Werner Heisenberg (1901–1976) (Figure 3.12) argued that the only way to describe a physical system was to think about the *way* we observe it. Heisenberg realised that the very *act* of observing a physical system (including one involving quantum states) will disturb it, so denying a perfectly precise knowledge of the system to the observer.

To understand why, consider what happens when we try to measure the position of an object with electromagnetic radiation such as light. When we shine light on an object, we scatter light from it and determine its position by the direction of the scattered light. Suppose we try to observe the position of an electron in an atom by scattering a photon off it. Then the photon's wavelength is related to its momentum by the de Broglie relation

Figure 3.12 Werner Heisenberg (1901–1976)

$$\lambda_{\text{photon}} = \frac{h}{p}$$

But the photon is a particle, so how do we visualise it as a wave? The answer is that we describe a particle as a wavepacket. To see how we do this, imagine a sinusoidal wave that is infinitely long (Figure 3.13a). An infinite wave like this cannot describe any particle because one of the features a particle must have is that it must be confined to a relatively small region of space such as, for example, the size of an atom or a nucleus. In other words, for a wave to describe any kind of finite object, it must be confined to a region of space, or be **localised**.

Now suppose we add other waves of slightly different wavelengths. By the Principle of Superposition, at certain positions along the x-axis, the wave amplitude starts to become more obvious at some regions and less so at others. If we added enough waves, then we would eventually get a wave in which the amplitude outside a certain narrow region of space was very small. The region between A and B is not precisely defined, but is merely the distance over which the wave has a recognisably large amplitude (Figure 3.13b). Such a localised wave disturbance is called a **wavepacket**. A group of

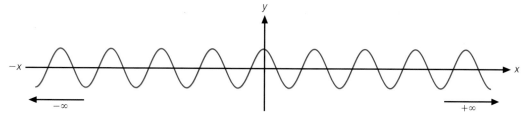

Figure 3.13(a) An infinitely long sine wave

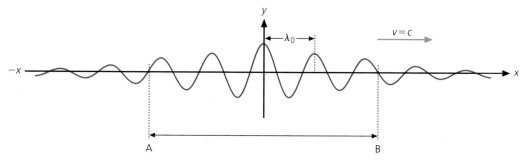

Figure 3.13(b) A wavepacket representing a photon. The average wavelength is λ_0

waves superimposed in this way will have a group velocity v with which they all (and so the wavepacket) will move. A range of wavelengths having a mean value λ_0 is associated with the wavepacket. In this way, we can regard a photon as being a localised wave disturbance with a wavelength λ_0 and travelling at $v = c$.

In order to 'see' the electron, the wavelength of the photon must at least be more or less the size of the atom. A long-wavelength (and by the de Broglie relation, low-momentum) photon can only give a rough estimate of the position of the electron, whereas a short-wavelength (high-momentum) photon can localise the electron more accurately (Figure 3.14a and b). In the first case, the photon's wavelength is too large a 'measuring stick', while in the second, the wavelength is more commensurate in determining the electron's position.

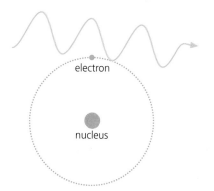

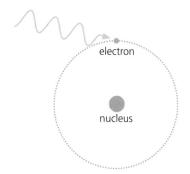

Figure 3.14(a) A long-wavelength photon can only give a rough estimate of the position of the electron

Figure 3.14(b) A short-wavelength photon can determine the position of the electron more accurately

The 'resolving power' of the light used to see the electron is determined by the wavelength of the light used, so the uncertainty in the electron's position is given by $\Delta x \approx \lambda_0$. However, when viewed as a particle, the photon when striking the electron could give up some, or all, of its momentum p to the electron. Since we do not measure the properties of the photon, we don't know how much momentum it gave up, so there is an uncertainty Δp in the momentum of the electron given by

$$\Delta p = \frac{h}{\lambda_0}$$

If we multiply Δx and Δp together we get

$$\Delta x \times \Delta p \approx \lambda_0 \times \frac{h}{\lambda_0} = h$$

This is saying that the product of the uncertainties in locating the position of an electron and its momentum is at least equal to the Planck constant.

This statement is a formulation of the **Heisenberg Uncertainty Principle** (often just referred to as the Uncertainty Principle), which states that it is impossible to measure simultaneously the position and the momentum of a particle with unlimited precision.

The Uncertainty Principle also applies to other pairs of quantities, such as energy and time. It can be shown that the precision in measurement of a particle's energy ΔE is at the expense of the uncertainty in time Δt such that

$$\Delta E \times \Delta t \approx h$$

A more precise formulation of the Uncertainty Principle relations, defining the degrees of uncertainty to be the standard deviations in the mean values of the measurements of the relevant physical quantities, leads to the following expressions:

$$\Delta x \times \Delta p \geq \hbar$$

and

$$\Delta E \times \Delta t \geq \hbar$$

These are called the **Heisenberg uncertainty relations**. The first states that:

We cannot know precisely both the position and the momentum of a particle simultaneously.

and the second that:

We cannot know precisely both the energy and the time coordinate of a particle simultaneously.

The quantity $\hbar$ (pronounced 'aitch-bar') occurs frequently in quantum theory and is equal to $h/(2\pi)$, and thus $\hbar = 1.05 \times 10^{-34}\,\mathrm{J\,s}$.

Don't worry if you find these ideas mathematically complicated. Intuitively, you should think of the Uncertainty Principle as a consequence of the fact that any act of observation disturbs the system under study, thus destroying perfect knowledge of it. Accuracy in one of these measurements is at the expense of its corresponding partner. It's rather like trying to measure, say, the position of a finely balanced top spinning on a table with a metre rule. As soon as the rule touches the top, the top spins away in an unpredictable direction, and the certainty of its original position is lost since the very act of measurement has disturbed the system. For macroscopic phenomena, this effect is negligible, but on atomic and nuclear scales it becomes very significant. You should note that the Heisenberg Uncertainty Principle is *not* saying that 'everything is uncertain'. It is simply a statement that tells us where the limits of uncertainty lie when we make measurements of sub-atomic events. In practice, the experimental uncertainties are usually much greater than the fundamental limit that results from the Uncertainty Principle.

WORKED EXAMPLE 3.4

What is the uncertainty in the position of a cricket ball of mass 0.145 kg moving at a speed of $9\,\mathrm{m\,s^{-1}}$ towards a wicket? Assume that the speed of the ball can be measured to a precision of 1%.

The momentum p of the cricket ball is $(0.145\,\mathrm{kg}) \times (9\,\mathrm{m\,s^{-1}}) = 1.3\,\mathrm{kg\,m\,s^{-1}}$.

The uncertainty Δp is 1% of this value, or $0.01 \times 1.3\,\mathrm{kg\,m\,s^{-1}} = 1.3 \times 10^{-2}\,\mathrm{kg\,m\,s^{-1}}$.

The uncertainty in the position of the ball is therefore

$$\Delta x = \frac{\hbar}{\Delta p} = \frac{1.05 \times 10^{-34}\,\mathrm{J\,s}}{1.3 \times 10^{-2}\,\mathrm{kg\,m\,s^{-1}}} = 8.1 \times 10^{-33}\,\mathrm{m!!}$$

WORKED EXAMPLE 3.5

What is the uncertainty in the position of an electron moving at a speed of $4.0 \times 10^{6}\,\mathrm{m\,s^{-1}}$ if we can only measure its speed to a precision of 1%?

Note that the electron isn't travelling fast enough for relativistic effects to be significant. The electron's momentum is

$$p = mv = (9.11 \times 10^{-31}\,\mathrm{kg}) \times (4.0 \times 10^{6}\,\mathrm{m\,s^{-1}}) = 3.6 \times 10^{-24}\,\mathrm{kg\,m\,s^{-1}}$$

The uncertainty in speed is 1%, so the uncertainty in the momentum is

$$\Delta p = 3.6 \times 10^{-24}\,\mathrm{kg\,m\,s^{-1}} \times 1\% = 3.6 \times 10^{-26}\,\mathrm{kg\,m\,s^{-1}}$$

The uncertainty in the position is therefore

$$\Delta x \approx \frac{\hbar}{\Delta p} = \frac{1.05 \times 10^{-34}\,\mathrm{J\,s}}{3.6 \times 10^{-26}\,\mathrm{kg\,m\,s^{-1}}} = 2.9 \times 10^{-9}\,\mathrm{m} = 2.9\,\mathrm{nm}\ \text{(roughly 30 atomic diameters)}$$

The modern theory of quantum mechanics

Figure 3.15 Erwin Schrödinger (1887–1961)

Louis de Broglie's wave–particle ideas were developed by the Austrian physicist Erwin Schrödinger (1887–1961) (Figure 3.15), who used the mathematics of differential calculus to formulate them into a theory of **wave mechanics** describing the behaviour of matter in terms of a wave equation or **wavefunction**. The wavefunction is a mathematical function derived from a **differential equation** called the **Schrödinger equation** that shows how a particle's behaviour evolves in time and space under a specific set of physical conditions. A mathematical description of the Schrödinger equation is beyond the scope of this book. However, it is essentially a statement about the total energy of a quantum particle, made up of the sum of its kinetic and potential energies, that is,

total energy of particle = kinetic energy of particle + potential energy of particle

Schrödinger published his equation in 1926 and won the Nobel Prize for Physics in 1933 for his work.

Now just like an ordinary linear equation, where the solution is a particular value of x say, the solution of a differential equation is a *mathematical function* that describes the behaviour of an object subject to a set of initial constraints or **boundary conditions**. The boundary conditions are a set of physical conditions that confine the motion of a particle to within certain limits.

The wavefunction is usually denoted by the Greek letter ψ (psi). If we turn the mathematical 'crank' and generate a solution for the Schrödinger equation, we get a wavefunction with a solution $\psi(x, t)$ for a particular set of boundary conditions, and this has to have certain mathematical properties. The first is that it must be 'well behaved'. What mathematicians mean by this is that the solution is a **continuous function**; in other words, it doesn't have any 'sharp edges' or discontinuities (Figure 3.16a). If it did, then this would be tantamount to a particle suddenly disappearing at one point in space and suddenly reappearing at another! The solution must also be *linear*. This has to be the case since de Broglie waves obey the **Principle of Linear Superposition**. For two or more particles interacting, their wavefunction amplitudes must be combined together to form a composite wavefunction that describes the quantum state of the system (Figure 3.16b).

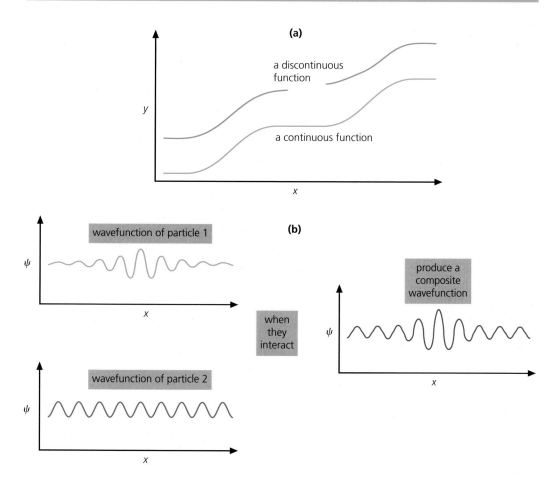

Figure 3.16 In order to represent a particle, the solution of the Schrödinger equation must be **(a)** *continuous*, that is, wavefunctions are 'smooth' and have no sharp edges, and **(b)** *linear*, that is, wavefunctions obey the Principle of Linear Superposition when they interact

Interpreting the wavefunction

What does the wavefunction mean in physical terms? In 1926 the German physicist Max Born (1882–1970) proposed that

The square of the amplitude of the wavefunction at any point in space is directly proportional to the probability of locating the particle at that point.

Born's idea was that the wavefunction had no physical interpretation other than to indicate the chance of finding the particle at the point of measurement (Figure 3.17a, b).

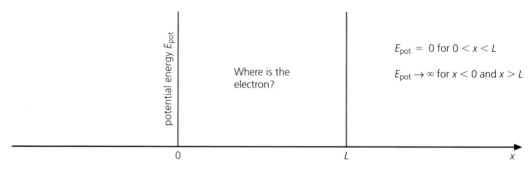

Figure 3.17(a) A quantum particle like an electron can be trapped in a 'potential well'. It is called a 'well' because an electron placed inside it cannot escape, since it cannot have enough energy. To simplify things, it is useful to visualise an infinitely deep potential energy well in the shape of a cylinder of diameter L

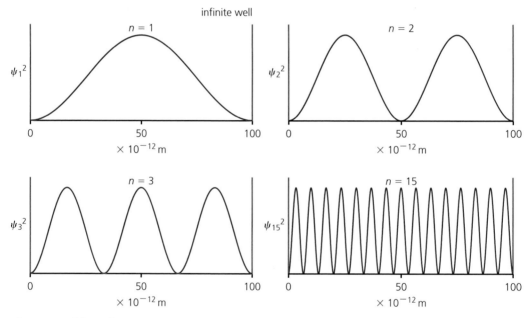

Figure 3.17(b) Finding an electron using the wavefunction. The probability for finding an electron in an infinite well is shown for four quantum numbers $n = 1, 2, 3$ and 15. The square of the wavefunction ψ for these states represents the probability of finding the electron at a particular location in the well. The electron is most likely to be found where ψ_n^2 is high and least likely to be found where ψ_n^2 is low

This idea can be applied to the motion of the electron in a hydrogen atom. The wavefunction of the electron tells us that it could be found anywhere within the atom, but the highest probability is for finding it in the allowed orbits of the Bohr atomic model. However, unlike the Bohr model, the wavefunction also indicates that there is a very small probability of finding it in the nucleus of the hydrogen atom! Figure 3.18 shows a representation of an electron's location in the hydrogen atom (a) from the Bohr perspective and (b) by a probability wavefunction.

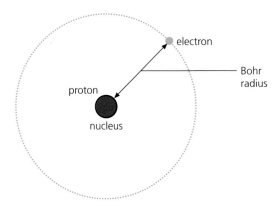

Figure 3.18(a) The Bohr interpretation of the electron in a hydrogen atom in the ground state. The electron is at a fixed distance (the Bohr radius) from the nucleus

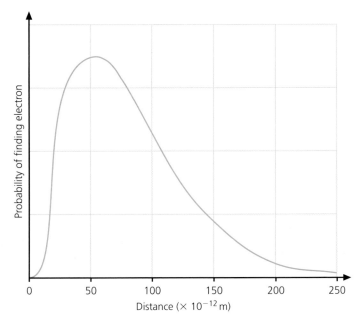

Figure 3.18(b) The probability of finding the electron in a hydrogen atom in the ground state. The probability reaches a maximum at a little over 50×10^{-12} m from the origin, which represents the centre of the atom

We mentioned earlier that we can localise a particle by means of a wavepacket by adding a number of different waves together. However, according to de Broglie, a particle has a single wavelength and must be a single wave. Or is it? If it is merely a single wave, then how can it interfere with itself to become localised? This is a very interesting question, and to answer it we need to take a closer look at some of the very strange features of the quantum world.

Re-interpreting quantum mechanics

Things that behave in a quantum way do so in a manner that is quite unlike our common-sense ideas of everyday life. The wavefunction can tell us the probability of locating a particle in space and time, and is somewhat mathematically involved. However, there is another way of looking at quantum behaviour, which was formulated by Richard Phillips Feynman (see page 185) in the 1940s. Feynman's approach is a more useful way of understanding quantum behaviour, and so it is to Feynman's ideas that we will turn next.

Getting from A to B

You may think that if you want to travel from one point to another then all you have to do is take the most direct path. However, if you are a quantum particle then life isn't quite as simple as that.

To understand the motion of quantum objects, let's look at a very simple case – the problem of a quantum particle travelling from A to B. In classical Newtonian physics, in the absence of any external force, we may consider a particle travelling in a straight line. However, in the quantum world, its behaviour is quite different. A quantum particle will consider all possible paths between A and B, and then add them together to produce the most likely way of getting from A to B! (See Figure 3.19.)

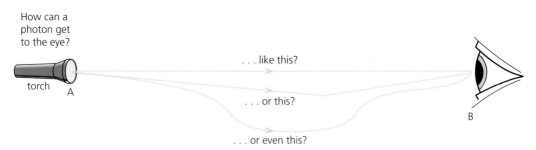

Figure 3.19 Getting from A to B, from a quantum particle's point of view. The quantum rule is to 'add up' all the possible ways

Young's slits

You may be familiar with the Young's double-slit experiment. In this, light from a source passes through two slits to give an interference pattern of light and dark fringes on a screen. However, light is made of photons, so what does a photon do when confronted by two slits? Which one should it go through?

Let us take a closer look at the quantum behaviour of the photon, which is intimately connected with electromagnetic waves. A photon does not have charge or mass, which would complicate our explanation unnecessarily. In Chapter 1 we saw that light can be regarded as quanta of energy $E = hf$ called photons, and the particle nature of light can be seen in the photoelectric effect as described in Box 1.1 (page 15). However, light also demonstrates interference effects, such as in the Young's double-slit experiment and diffraction effects in gratings. If light behaves as a quantum particle, then it is reasonable to suppose that the Principle of Linear Superposition must in some way be an essential feature of quantum behaviour. However, if a photon is a particle, how does it decide which one of the slits in the Young's experiment to go through? Can it be in two places at once? Quantum mechanics says to the photon – *do both* – explore both possible paths through the slits! This may seem odd if you are used to thinking of photons as little packets of energy travelling in space. Quantised energy cannot mean that space is discrete. However, waves can be spread out in space – they don't concentrate their energy in one place like particles – so a photon cannot be simply a particle. Photons have a wave–particle duality, a bit of both the qualities of a wave and those of a particle, but not behaving exclusively as one or the other.

The wavefunction can tell us where the photon should be for the bright and dark fringes to be in the right positions. However, we can use some ideas from the wavefunction to explain what is going on without buying the idea that the photon is merely a wave. This is what Feynman did. In the Young's double-slit experiment, what determines whether there is a bright or dark fringe is the difference in the lengths of the paths, corresponding to a difference in **phase** for light coming along each path. If the phase difference is 360°, then we will see a maximum. If it is 180°, then we see a minimum.

Feynman's idea was to use a simple mathematical concept called a **phasor** to describe the quantum behaviour of a particle. A phasor is simply an arrow of a certain length or amplitude that points in a specific direction. One way is to imagine it like a circular wheel on which there is an arrow drawn as a single spoke (Figure 3.20a). Possible paths from the light source to the screen are traced out by rolling the wheel along each path and noting the direction of the arrow. How fast should the wheel turn? The answer is that it should rotate at the frequency of the radiation in accordance with the Planck relation $f = E/h$. If we do this for two possible paths, we will end up with two phasors pointing in different directions. We then combine them from tip to tail to give a **resultant phasor** pointing in a specific direction (Figure 3.20b). What can this tell us?

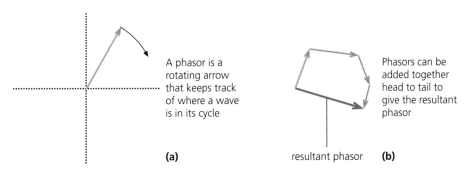

A phasor is a rotating arrow that keeps track of where a wave is in its cycle

(a)

Phasors can be added together head to tail to give the resultant phasor

resultant phasor **(b)**

Figure 3.20 Phasors

Well, the length of the resultant phasor arrow, or its *amplitude*, is used to calculate the *probability* that a photon will arrive at the end of the path. The probability must be proportional to how bright the fringe is, that is, its *intensity*. In Born's interpretation of the wavefunction, the amplitude of the wave at any point gives the probability of finding the quantum particle at that point. We apply this probability idea to the phasor description in the following way. The *intensity* of a wave (the rate at which energy arrives) is proportional to the square of its amplitude. So in order to calculate the quantum probability of arrival, we *square* the length of the resultant phasor arrow. So, for example, if the resultant phasor at one point (A) is three times the length of the resultant phasor at another point (B), then the probability of arrival of photons is *nine* times greater at point A.

Figure 3.21 shows the phasor idea applied to the two-slit experiment. The great advantage of the phasor idea is that we are not compelled to think exclusively in terms of wave or particle behaviour. We can just define a quantum behaviour that borrows from both these concepts but depends on neither. You need to remember that quantum behaviour combines phasors from all possible paths. Figure 3.22 shows how to do quantum calculations in general. In all cases we have a source of photons and some means of detecting them. In the case of Young's double slits, the source can be a lamp and the detector can be photographic film. We are not worried too much about the details of the source and detector, only that we can produce photons and detect them.

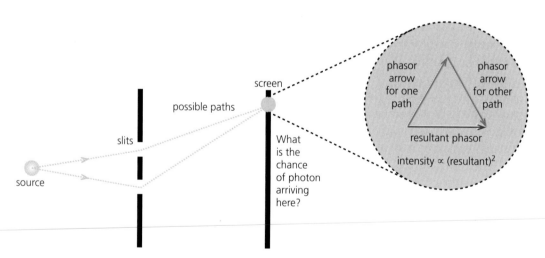

Figure 3.21 Phasor diagram for the Young's double-slit experiment with photons

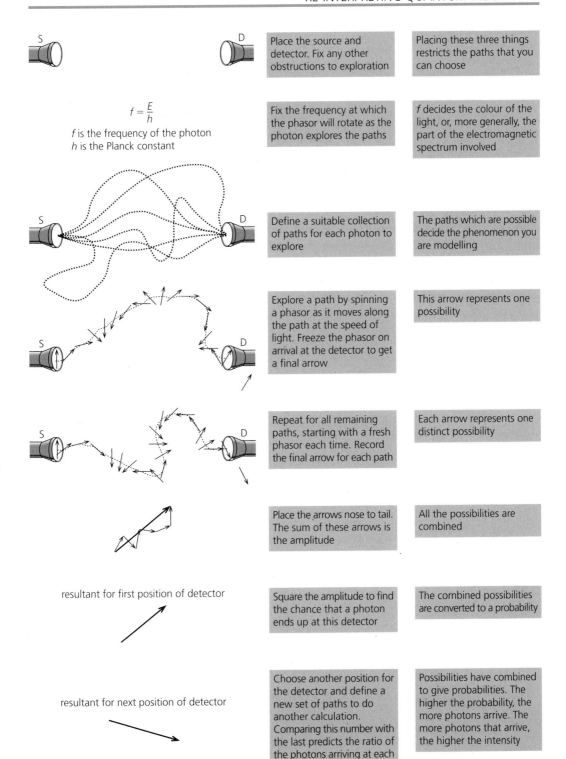

$$f = \frac{E}{h}$$

f is the frequency of the photon
h is the Planck constant

Place the source and detector. Fix any other obstructions to exploration

Placing these three things restricts the paths that you can choose

Fix the frequency at which the phasor will rotate as the photon explores the paths

f decides the colour of the light, or, more generally, the part of the electromagnetic spectrum involved

Define a suitable collection of paths for each photon to explore

The paths which are possible decide the phenomenon you are modelling

Explore a path by spinning a phasor as it moves along the path at the speed of light. Freeze the phasor on arrival at the detector to get a final arrow

This arrow represents one possibility

Repeat for all remaining paths, starting with a fresh phasor each time. Record the final arrow for each path

Each arrow represents one distinct possibility

Place the arrows nose to tail. The sum of these arrows is the amplitude

All the possibilities are combined

resultant for first position of detector

Square the amplitude to find the chance that a photon ends up at this detector

The combined possibilities are converted to a probability

resultant for next position of detector

Choose another position for the detector and define a new set of paths to do another calculation. Comparing this number with the last predicts the ratio of the photons arriving at each

Possibilities have combined to give probabilities. The higher the probability, the more photons arrive. The more photons that arrive, the higher the intensity

Figure 3.22 How to do quantum calculations

Quantum behaviour of electrons

Quantum behaviour is common to all the fundamental particles that we will meet in particle physics. Electrons are not like photons, as they have mass, which means that they can travel at different speeds (recall that, in our discussion on relativity, massless particles travel at the speed of light). However, electrons follow the same quantum rules – they explore all possible paths and a phasor maps out the direction along every path and superimposes them to give a resultant, which represents the probability of the electron arriving at a particular location.

Quantum mechanics in action

You will probably have gathered by now that quantum phenomena are quite unlike anything you have directly experienced. The way we have described quantum behaviour is just one way of looking at things. The 'explore-many-paths' idea was worked out by Richard P. Feynman (1918–1988) and his fellow American physicist Julian Schwinger (1918–1984) as well as the Japanese physicist Sin-Itiro Tomonaga (1906–1979) shortly after the Second World War. We add up phasors (amplitude and phase) in a spinning arrow to give a resultant from which we calculate a probability, and this is a useful way of helping us to do quantum calculations. You shouldn't think of quantum particles trying paths one after another, but imagine them trying them all at once. The paths in most cases add up to nothing at all. In general, only one path adds up to a probability where we have a reasonable chance of observing the arrival of the particle, and this is what we record in our experiments such as the Young's double slits and electron diffraction.

Quantum particles manage to go everywhere but can always be detected somewhere. In this book we will not be doing any advanced quantum calculations, but the important thing you should understand is how the quantum mechanical description of matter arises, and what it can tell us about the nature of matter. Before we leave our discussion on quantum mechanics, we will look at some examples of how it can be used to explain some properties of matter that can only be understood in terms of quantum behaviour.

Alpha decay

Some atoms decay by alpha emission. Alpha particles are helium nuclei, which move at speeds of about $1600 \, \text{km s}^{-1}$ but can be stopped by a sheet of paper. An example is the uranium nucleus, which decays by the process

$$^{238}_{92}\text{U} \rightarrow ^{234}_{90}\text{Th} + ^{4}_{2}\text{He} \qquad Q = 4.25 \, \text{MeV}$$

Here Q is the amount of energy released in the process, and the half-life of this decay is $4.47 \times 10^9 \, \text{yr}$.

We imagine the alpha particle to be already formed inside the nucleus before it escapes. Figure 3.23 shows the potential energy $V(r)$ of the alpha particle and the daughter thorium nucleus as a function of their separation r. Energy $V(r)$ is the sum of the potential well associated with the strong nuclear force that acts in the nuclear interior, and the **Coulomb potential**, which is associated in this case with repulsive electric force between the nucleus and the alpha particle before and after the decay has occurred. The nuclear potential of the strong force is constant, whereas that of the Coulomb potential follows a $1/r$ law.

The dotted line marked X is the escape energy (Q-value) for the disintegration. If this is the total energy E of the alpha particle, then it is well below the height U of the potential energy barrier that represents the boundary of the nucleus. The alpha particle is well and truly trapped inside the potential well of the nucleus. So how *does* it get out? Quantum mechanics explains how. The wavefunction that describes the alpha particle has amplitudes *outside* of the barrier, indicating that there is a finite probability of it existing there, and once outside it can immediately escape. This **quantum mechanical tunnelling**, as it is called, has no classical explanation. It is a consequence of the de Broglie matter-wave explanation of particles and the Heisenberg Uncertainty Principle. The alpha particle can be thought of as briefly 'disobeying' the law of conservation of energy by borrowing an amount ΔE which it pays back in a time Δt so that the uncertainty relation $\Delta E \times \Delta t \approx \hbar$ is satisfied.

Notice that an alpha particle 'higher up' in the potential well of the nucleus will escape sooner and with higher energy, because it has a shorter distance to tunnel through the potential barrier. It is for this reason that alpha particles with the highest Q-values come from nuclides with the *shortest* half-lives. In the case of $^{238}_{92}U$, the very long half-life is due to the low probability for such a tunnelling process.

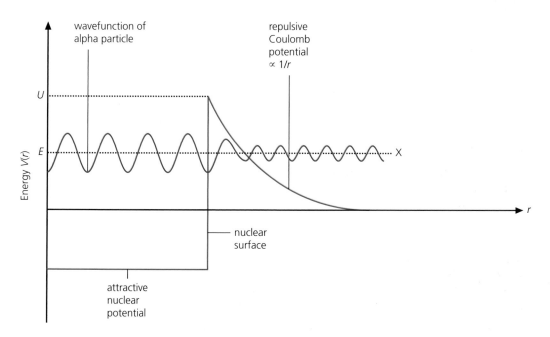

Figure 3.23 Alpha decay can be explained by quantum mechanical tunnelling

The scanning tunnelling microscope

To understand how scanning tunnelling microscopy works, we need to look at the behaviour of electrons in metals. In the classical picture, even the most weakly bound electrons cannot leave the metal unless they are given enough energy to go over the potential energy barrier represented by the work function ϕ. We have seen that this happens when electrons escape from the metal through the photoelectric effect. However, because the electron is a quantum particle, there is another way that electrons can leave a metal.

If we bring two metals very close together so that they are just a few nanometres apart, then a small electric current can be measured. The current gets larger the closer the metals are brought together, until it reaches a maximum when the metals are touching. We can explain this in terms of the electron behaving like a quantum particle with a wavefunction. Since the amplitude of a wavefunction represents the probability of finding the electron at any particular point, quantum calculations show that the wavefunctions of electrons have small but finite probabilities beyond the potential barrier, and thus beyond the metal surface. The electron 'tunnels' from one metal to the other, even though this contradicts classical physics, which says that an electron would have insufficient energy to escape the potential barrier that exists on the surface of the metal and so must remain within it. This tunnelling creates the electrical current that can be measured and analysed, and this is the principle behind the scanning tunnelling microscope (STM).

A practical STM has a small metal probe that is very carefully manipulated by piezoelectric actuators when it is brought up against a sample of metal that is to be analysed. A small voltage is applied between the probe and the sample, and electron tunnelling occurs. The probe is scanned along X and Y directions on the surface of the sample, and the variation in electron current allows an image of the atomic structure to be formed when processed by a computer. Figure 3.24 shows a typical STM image. The STM was developed by Gerd Karl Binnig (1947–) and Heinrich Rohrer (1933–), who were awarded the Nobel Prize for Physics in 1986 for their work, and it is a tribute to the power of quantum theory that a practical instrument based entirely on quantum principles has been developed in this way.

Figure 3.24 An STM image of gold atoms on a graphite substrate

Summary

◆ Two important physical theories in particle physics are Einstein's **Special Theory of Relativity** and **quantum theory**. These theories are independent of each other.

◆ A **relativity principle** is a statement that reconciles the observations that two or more observers make about the same physical event. Observers view an event from their own **frame of reference**, each of which is a convenient coordinate system to which all measurements of the event can be referred. Frames of reference that are either stationary or moving with constant velocity are called **inertial frames**. Frames of reference that are accelerating are called **non-inertial frames**.

◆ **Special Relativity** is concerned with objects moving at high velocities. Two important postulates are (i) the laws of physics are the same in all inertial frames and (ii) the speed of light is constant to all observers. Measurements in one inertial frame may be related to those in another by the **Lorentz–FitzGerald transformations**. These are

length contraction $$x = x_0 \sqrt{1 - \left(\frac{v}{c}\right)^2}$$

time dilation $$\Delta t = \frac{\Delta t_0}{\sqrt{1 - \left(\frac{v}{c}\right)^2}}$$

mass increase $$m = \frac{m_0}{\sqrt{1 - \left(\frac{v}{c}\right)^2}}$$

which are all functions of an object's velocity v and become significant at speeds approaching the speed of light c. A body in an inertial frame that is stationary is said to have a **rest mass** m_0.

◆ Einstein showed that mass and energy are equivalent. A body will have a **rest energy** $E_0 = m_0 c^2$ and

relativistic total energy $$E = \frac{m_0 c^2}{\sqrt{1 - \left(\frac{v}{c}\right)^2}}$$

For an isolated system of particles, the total relativistic energy is constant.

◆ Since time is not invariant between reference frames, a new definition of momentum is needed:

relativistic momentum $$p = \frac{m_0 v}{\sqrt{1 - \left(\dfrac{v}{c}\right)^2}}$$

Relativistic momentum is always conserved in particle collisions. The relativistic total energy is related to the relativistic momentum by the relation

$$E^2 = m_0^2 c^4 + p^2 c^2$$

For particles moving at near light speeds and for ones with zero rest mass, the total energy can be written as

$$E \approx pc$$

◆ In particle physics, convenient units of *mass* are eV/c^2, MeV/c^2 or GeV/c^2. From Einstein's mass–energy relation this is just the rest energy of the particle divided by c^2. Similarly, because particle physicists measure such small values of momentum, they do not generally use units of $kg\,m\,s^{-1}$. Since $E \approx pc$ for $E \gg m_0 c^2$, units of momentum can be expressed in terms of eV/c, MeV/c or GeV/c.

◆ Matter can behave as particles or as waves, a phenomenon called **wave–particle duality**. Each particle has a

de Broglie wavelength $$\lambda_p = \frac{h}{p}$$

which depends inversely on its momentum. Particles are described in terms of a **wavepacket**, which is a **de Broglie wave** that is **localised** over some region of space.

◆ The **Heisenberg Uncertainty Principle** is a statement that there is always an unavoidable amount of uncertainty in the simultaneous measurement of the energy and time, or of the position and momentum of a particle. Precise measurement of one of these physical quantities is always at the expense of knowledge of the other.

◆ A key mathematical relationship in the modern theory of **quantum mechanics** is the **Schrödinger equation**, which is a **differential equation** that describes the total energy of a particle. Solving the Schrödinger equation subject to boundary conditions leads to **wavefunctions** that describe how a particle behaves in terms of its wave properties. The wavefunctions of a system of particles may be combined together by the **Principle of Linear Superposition** to form a single composite wavefunction that describes the overall state of a quantum system. The physical meaning of the wavefunction is understood by interpreting the square of its amplitude at any point in space as being proportional to the **probability** of locating the particle at that point.

◆ Another way of doing quantum calculations involves the use of **phasors**, which represent the probability of a quantum particle arriving at the end of its path. Quantum behaviour combines phasors from all possible paths between two points, and the **resultant phasor** represents the most likely outcome. The phasor approach to quantum calculations means that we do not have to worry about whether we are dealing exclusively with a wave or a particle.

◆ Quantum mechanics can explain phenomena that classical physics cannot explain. Some examples are **radioactive alpha decay** in the form of **quantum mechanical tunnelling**, and the invention of the **scanning tunnelling microscope**.

Questions

1 a State the two postulates of Special Relativity. Comment briefly on the significance of the effects they predict.

b Describe a 'thought experiment' to illustrate length contraction.

c An Unidentified Flying Object 300 m long is observed to pass the Earth travelling at $2.650 \times 10^8 \, \text{m s}^{-1}$. What is its apparent length as seen from the Earth?

d A beam of alpha particles each with mass $6.65 \times 10^{-27} \, \text{kg}$ has a velocity of $2.450 \times 10^8 \, \text{m s}^{-1}$. What is their mass as seen by an observer at rest in a laboratory?

e A proton has a rest mass of $1.673 \times 10^{-27} \, \text{kg}$. What is its total energy when it is:
i) at rest,
ii) moving with a velocity of $2.50 \times 10^8 \, \text{m s}^{-1}$?
[Proton rest mass $= 1.66 \times 10^{-27} \, \text{kg}$]

2 a A particle called the *positive K meson* (K$^+$) has a lifetime of 0.1237 μs when stationary or measured in a laboratory rest frame. If positive K mesons emerge from an accelerator with a speed of 0.990c relative to a laboratory rest frame, how far can they travel during their lifetime?

b Suppose you have a starship that can travel at speeds of 0.9990c relative to an observer on Earth. You leave the Earth at this speed and travel to a distant star system. The clock on board your starship tells you that it has taken 10.0 years to get there. After travelling through the star system you turn your starship about and travel back to Earth with the same relative speed. Your clock tells you that it takes another 10 years to get back to Earth. Your clock records that you have been travelling in space for 20 years.
What does mission control on Earth measure as to how long your voyage has taken? You may neglect any effects due to accelerations in stopping and turning the starship about.
[Hint: You need to think carefully about what is meant by the 'relativity of time' and 'proper time'.]

3 Show that the units for mass, that is, eV/c^2, MeV/c^2, and for momentum, that is, eV/c, MeV/c, are dimensionally correct.

An electron with rest energy 0.511 MeV is moving with a speed of 0.8c.

a What is its total energy?

b What is its kinetic energy?

c Using the relativistic equation for total energy $E^2 = p^2c^2 + (m_0c^2)^2$, find the magnitude of its momentum.

4 a What is meant by *wave–particle duality*?

b Calculate the momentum p of an electron travelling in a vacuum at 5% of the speed of light.

What is the de Broglie wavelength of electrons travelling at this speed?

Why are electrons of this wavelength useful for studying the structure of molecules?

Part (b) only: ULEAC, June 1995 (part)

5 a A particle called the *charged π meson* has a rest energy of $E = 140$ MeV and a lifetime of 26 ns. Using the Uncertainty Relation for energy

$$\Delta E = \frac{\hbar}{\Delta t}$$

calculate the uncertainty ΔE and the quotient $\Delta E/E$, which is a measure of the precision of our knowledge of the rest energy of the particle.

Do the same calculations for:

b a *neutral π meson* with a rest energy of $E = 135$ MeV and a lifetime of 8.3×10^{-17} s,

c another particle called the *rho meson*, which has a rest energy of 765 MeV and a lifetime of 4.4×10^{-24} s.

Compare your values of $\Delta E/E$ for **a**, **b** and **c**. If the precision of our particle detectors is such that we can measure particle energies to about 1 part in 10^6, comment on the effect of the Uncertainty Principle in obtaining accurate measurements of energy for the *charged π meson*, the *neutral π meson* and the *rho meson*.

6 Suppose we can imagine the hydrogen atom as being like a box with an electron represented by a wave that is oscillating in the fundamental mode.

a If the size of the atom is 2×10^{-10} m, what is wavelength L of the fundamental oscillation of the electron?

b If the de Broglie wavelength of the electron is given by

$$\lambda_p = \frac{h}{m_e v}$$

what is the momentum of an electron having this wavelength?

c Calculate the kinetic energy of the electron. Compare this with the ionisation energy needed to remove an electron from the hydrogen atom (13.6 eV). Can an electron of the kinetic energy you have calculated remain in the hydrogen atom?

d For a string oscillating between points separated by a distance L, the wavelength is given by $2L/n$, where n is the number of 'loops' (in the case of the fundamental mode, the wave has just one loop). Using the fact that the kinetic energy of the electron $= \frac{1}{2}mv^2$, the momentum

$$mv = \frac{h}{\lambda_p}$$

and the wavelength $= 2L/n$, combine these three expressions to show that the kinetic energy of an electron in a hydrogen atom can be expressed as

$$\frac{h^2 n^2}{8mL^2}$$

Using your value for L calculated in **a**, find the kinetic energy of the electron in the hydrogen atom for $n = 1$, 2 and 3.

Particle accelerators and detectors

'In a previous paper we have described a method of producing high velocity positive ions with energies up to 700,000 electron volts. We first used this method to determine the range of high-speed protons in air and hydrogen and the results will be described in a subsequent paper. In the present communication we describe experiments which show that protons having [accelerating voltages] above 150,000 volts are capable of disintegrating a considerable number of elements.'

J. D. Cockcroft and E. T. S. Walton, 1932, *Proceedings of the Royal Society*, A, **137**, 229–242

Radioactive isotopes as particle sources

In this chapter we are going to look at some of the tools that particle physicists use to probe the structure of matter. The earliest nuclear physics experiments such as Rutherford's scattering experiment were carried out using alpha particles, which are emitted by a range of different radioactive isotopes when their nuclei disintegrate. However, difficulties are encountered when using sources that have to be of high radioactivity to be useful.

Nuclear radiations are damaging to biological tissues and, since the higher the activity, the greater the radiation hazard, it is important to minimise a source of radiation by **shielding** it from the operator. The radioactive sources used in school laboratories are relatively weak and pose a negligible hazard, as they have activities of only a few 10 to 100 kBq [one becquerel (1 Bq) = one nuclear disintegration per second]. Sources that emit alpha particles and are used in nuclear physics experiments commonly have activities ranging from a few tens to several thousands of megabecquerels (MBq). These are very strong sources, and special shielding and safety precautions need to be taken when using them.

Particles from a radioactive source are emitted uniformly in all directions. Physicists usually position the source so that the particles emerge in a **collimated beam** from a small aperture, with the rest of the source being shielded. It is because of this that, for a source of given activity, only a small fraction, say about 0.25%, of the emitted particles may emerge to strike a target, so this method of producing beams is relatively inefficient. A source of 1000 MBq, when properly shielded and collimated, might only deliver approximately

$$\frac{0.25}{100} \times 1000 \times 10^6 = 2.5 \times 10^6 \text{ particles per second at the target}$$

Atoms are mainly empty space. Consequently, the probability of a nuclear reaction taking place between particles of the beam and the target is not very high. It also depends on the depth of the target material, so one might only obtain a few reactions per second from a strongly radioactive source. In addition, radioactive sources cannot be 'turned off', so they always require careful storage even when they are not being used for experiments. A more important consideration is that the type and energy of the particles emitted are limited by the type of radiation available. The highest alpha particle energies available are of the order of a few MeV, and these have a limited range even in a vacuum.

To overcome these limitations, **particle accelerators** have been designed that can produce a range of sub-atomic particles such as electrons and protons as well as high-speed nuclei of many different atoms, together with their antiparticle counterparts. Such machines, which are popularly known as 'atom smashers', are able to accelerate charged particles from a few keV up to energies exceeding $1\,\mathrm{TeV}$ ($1\,\mathrm{TeV} = 1000\,\mathrm{GeV}$ or $10^6\,\mathrm{MeV}$), and there are a number of different designs.

In this chapter we will look at different kinds of particle accelerators and also at the various instruments that have been developed to detect particles. We will see that higher and higher energies have meant that a number of novel techniques have had to be developed in order to measure the properties of the many kinds of particles produced in accelerators. These range from simple fluorescent screens right up to sophisticated computer-controlled detection systems. Finally, we will examine the ways in which particle accelerators and detectors can be used in order to yield the maximum amount of information from a particle physics experiment.

Electrostatic accelerators

The first particle accelerator was built by the two physicists Ernest Walton (1903–1995) and John Cockcroft (1897–1967) and used the principle of **electrostatic acceleration**. A picture of their apparatus is shown in Figure 4.1.

Electrostatic accelerators work by accelerating a charged particle through a constant potential difference. If a particle has a charge q and mass m and moves through a potential difference of V, then it will gain a kinetic energy of

$$\tfrac{1}{2}mv^2 = qV$$

WORKED EXAMPLE 4.1

The 'electron gun' in the cathode ray tube (CRT) of a television set accelerates electrons through a potential difference of $3\,\mathrm{kV}$.

a What is the kinetic energy of the electrons?
b What is the speed at which they hit the screen?
   (Note that, at this accelerating voltage, the electron is accelerated to non-relativistic velocities.)

a The kinetic energy (KE) of the electrons is

$$3\,\mathrm{keV} = 3 \times 10^3\,\mathrm{V} \times 1.6 \times 10^{-19}\,\mathrm{C} = 4.8 \times 10^{-16}\,\mathrm{J}$$

b This is equal to $\tfrac{1}{2}m_\mathrm{e}v^2$, where m_e is the mass of the electron. Therefore

$$v = \sqrt{\frac{2 \times \mathrm{KE}}{m_\mathrm{e}}} = \sqrt{\frac{2 \times (4.8 \times 10^{-16}\,\mathrm{J})}{9.1 \times 10^{-31}\,\mathrm{kg}}} = 3.2 \times 10^7\,\mathrm{m\,s^{-1}}$$

In the Cockcroft–Walton accelerator, a network of capacitors are charged in parallel to achieve a potential of up to 800 kV by means of a special capacitor–diode voltage multiplication circuit. In theory, there is no limit to the potential that can be obtained with the Cockcroft–Walton multiplier but, in practice, the resistances of the diodes and the leakage of charge from the capacitors mean that accelerating potentials up to one to two million volts only can be generated.

The Cockcroft–Walton accelerator performed the first nuclear disintegration by artificial means, which was that of lithium. Cockcroft and Walton accelerated protons up to a voltage of 700 kV and bombarded them onto a target of lithium, producing the reaction

$$^1_1p + ^7_3Li \rightarrow ^4_2He + ^4_2He$$

This was the first experiment to show that one element (lithium) could be artificially transformed or **transmuted** into another element (helium).

The Van de Graaff accelerator

Figure 4.1 The Cockcroft–Walton accelerator (the first particle accelerator)

The accelerating voltage achievable by a Cockcroft–Walton accelerator is limited by the physical properties of its capacitors and diodes. To obtain higher accelerating voltages, a **Van de Graaff accelerator** is used. You may have seen a Van de Graaff machine in your school laboratory. These are the devices with shiny metal spheres on columns which make your hair stand on end if you touch them!

The Van de Graaff machine was designed by the American physicist Robert Jemison Van de Graaff (1901–1967). It works as follows. A continually moving belt of insulating material runs between two pulleys A and B (Figure 4.2), which are separated by an insulated column C. The lower pulley A is earthed. Located near to pulley A is a sharp metallic comb D, which is maintained at a potential difference of a few kilovolts between itself and the pulley. A high electric field is produced at the tip of the comb and an electrical discharge occurs through the belt from the comb to the pulley, removing electrons from the belt and so making it positively charged. The belt carries the charge up to the top pulley B, which is inside a large hollow ball-shaped metal electrode E. Electrons from a second comb at F flow off E and neutralise the positive charge on the belt. The electrode E thus acquires an increasing positive charge with a correspondingly high electric potential. The voltage on E rises rapidly until an equilibrium is established where the rate of loss of positive charge from E balances the positive charge current carried by the moving belt. In this way, the Van de Graaff machine can reach very high electric potentials of a few million volts.

In a Van de Graaff accelerator the high electric potential is coupled to an acceleration tube, at the top of which is an **ion source** that produces the particles to be accelerated. In the case of positive ions, the ion source will be at the end of the tube, which has the highest positive potential, with the experimental target being connected to earth. The positive ions are repelled away from the positive electrode and travel down an **acceleration tube** (see Figure 4.2) containing cylindrical electrodes, where they gain kinetic energy. In an evacuated tube, it is not possible to maintain electric potentials between electrodes much above 200 kV, and special multi-stage electrode tubes are used, with each pair of electrodes maintaining a potential of 50–100 kV. Surrounding each electrode are metal **corona rings**, which are outside the insulating wall of the tube. The purpose of these is to ensure that the electric field within the tube is parallel to its axis and that the particles do not hit the tube wall. The tube is evacuated of air down to a pressure of about 1.3×10^{-3} Pa, and this reduces collisions between the accelerated particles and gas molecules; these would produce gas ions and electrons, which would themselves be accelerated.

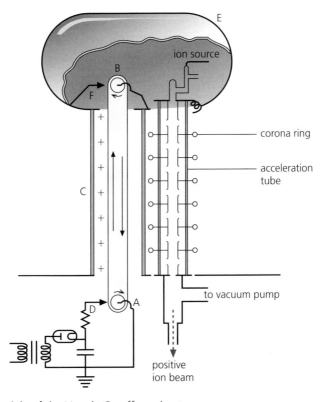

Figure 4.2 The principle of the Van de Graaff accelerator

By using cylindrical electrodes, a built-in focusing effect is achieved which serves to confine the particles to a narrow beam. Figure 4.3 shows the lines of force due to the electric field between two pairs of cylindrical electrodes at opposite potentials. If an incoming particle is slightly off-centre, then it will be deflected towards the centre of the axis on its way to the next pair of electrodes.

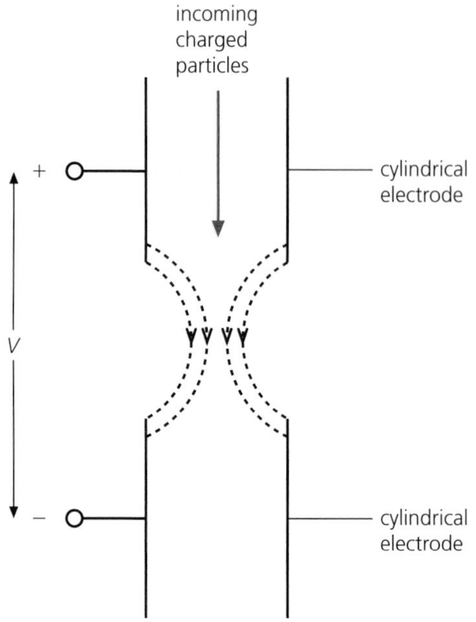

In an accelerator tube, such as that employed in a Van de Graaff accelerator, there is usually only one source of potential between the first and last electrodes, and the electrodes in between are held at intermediate potential by regarding them as a number of capacitors in series.

The energy required to carry the charge to the top electrode comes from an electric motor that drives the pulley at A. The maximum accelerating potential is limited by breakdown across the acceleration tube and charge leakage from the high-voltage electrode E. However, with careful design, Van de Graaff accelerators operating at accelerating potentials in excess of 25 MV have been built. It is also possible to obtain higher electrostatic potentials by connecting Van de Graaff accelerators in a tandem fashion, and this arrangement can accelerate protons to energies in excess of 40 MeV.

Figure 4.3 Cylindrical electrodes in the acceleration tube focus the charged particles into a narrow beam

Linear accelerators (linacs)

In a **linear accelerator** particles are accelerated horizontally in an acceleration tube to high energies by means of a series of hollow cylindrical electrodes of increasing length, called **drift tubes**, placed in a long line. Linear accelerators or **linacs** use the principle of **synchronous acceleration**. Figure 4.4 shows how it works.

Alternate electrodes electrically linked together are connected to an alternating source of potential difference $V = V_0 \sin \omega t$, which varies sinusoidally at radio frequencies. Now suppose we consider a particle, a positive ion of charge q and mass m, which enters the accelerator through a collimator A that is earthed and at zero volts. Let this occur at an instant in the alternating cycle where the electrode B is at its maximum negative value $-V_0$. The particle will then be accelerated across the gap between A and B and enter the electrode B with a velocity v_1 and gain kinetic energy given by

$$\tfrac{1}{2}mv_1^2 = qV_0$$

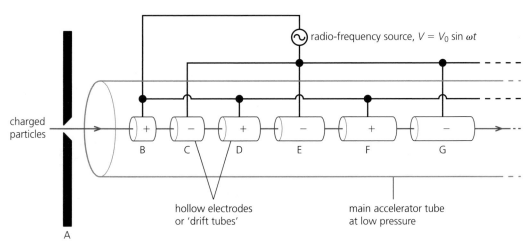

Figure 4.4 Linear accelerators (linacs) use the principle of synchronous acceleration. The length of the cylindrical electrodes or drift tubes must become greater as the particles are accelerated

There is no net electric field inside the electrode and so the particle travels at uniform velocity, or 'drifts', taking a time $t_1 = l_1/v_1$ to travel the length l_1 of the cylinder. Now if we make t_1 equal *exactly one-half* of the period of the alternating voltage, then the particle will emerge from B, into the gap between B and the next electrode C, at the instant when the alternating voltage is at $+V_0$. The electrode C is now negative with respect to B, and the particle will accelerate across the gap BC through the potential difference $+V_0$. It therefore gains another amount of energy qV_0. If its velocity in C is v_2, then it now has total kinetic energy given by

$$\tfrac{1}{2}mv_2^2 = 2qV_0$$

You can see that the action of the electrodes is first to 'pull' the particle in and then to 'push it out' as the potentials on the electrodes change synchronously. If the frequency f of the alternating voltage is $\omega/(2\pi)$, then the period will be $1/f$ and the second acceleration of the particle will take place when

$$t_1 = l_1/v_1 = \tfrac{1}{2}f$$

The same thing happens at each successive electrode, and synchronous acceleration will occur at each successive gap provided that the lengths of the cylinders are correctly related to the velocity of the particle as it travels through.

In general, for a linear accelerator containing n electrodes, the particle energy at the nth electrode is nqV_0. Figure 4.5 shows the Stanford Linear Accelerator at Stanford, USA, which is 3 km long and contains 100 000 cylindrical electrodes. It can accelerate electrons to energies of 30 GeV. The main disadvantage of linear accelerators is that the maximum particle energies they can produce depend on their length, and very long linacs would be impracticable to build. Smaller linear accelerators are employed in hospitals for the treatment of tumours. Figure 4.6 shows the linear accelerator at the Royal Surrey County Hospital.

Figure 4.5 The Stanford Linear Accelerator at the Stanford Linear Accelerator Center (SLAC)

Figure 4.6 The linear accelerator therapy machine at Royal Surrey County Hospital, Guildford. The accelerated beam is bent through 90° by magnets before it is directed at the patient

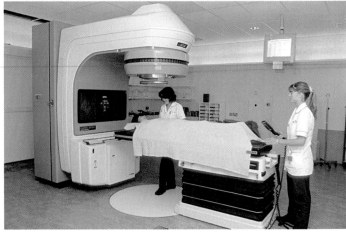

Cyclotrons

In 1930 the American physicist Ernest Orlando Lawrence (1901–1967) invented a particle accelerator called the **cyclotron**. The cyclotron uses the principle of synchronous acceleration, but in addition uses a magnetic field to make the charged particles move in a 'spiral' path (Box 4.1).

In a cyclotron, two hollow D-shaped electrodes called *dees* with a gap between them are arranged as in Figure 4.9a (page 108). These are placed inside a vacuum and the entire assembly positioned so that it lies perpendicular between the two poles of an electromagnet of magnetic flux density B. The dees are connected to a radio-frequency source of alternating voltage so that an alternating electric field is maintained across the gap, and the regions inside the hollow dees are free of electric fields.

At the centre of the gap between the dees is an ion source. Suppose that a positive ion is introduced into the gap at A. If dee X is at a negative potential with respect to dee Y, then the ion is accelerated into X, and it travels in a semi-circle inside the dee because it is moving perpendicular to the applied magnetic field. The frequency of the alternating voltage is such that, when the ion emerges into the gap again, dee Y is negative with respect to dee X and the particle is accelerated again across the gap. As it enters dee Y, it again moves in a semi-circle, but because of its increased kinetic energy, the radius is larger. In this way the particle is repeatedly accelerated and moves in a 'spiral' path towards the perimeter of the dees.

Box 4.1 The dynamics of charged particles in electric and magnetic fields

Many particles carry electric charge. This property means that they are affected by electric and magnetic fields. Here we review some essential principles of charged particle dynamics using Newton's Laws of Motion involving electric and magnetic forces.

(a) Speed of charged particles

In particle accelerators, charged particles are usually accelerated by means of an electric field. A particle of charge q will experience a force due to the electric field, and work is done on it.

In the case of electrons, these are usually emitted from a **hot cathode** (an electrode of negative potential) by **thermionic emission** and accelerated towards a **positive electrode** or **anode**. If the potential difference between anode and cathode is V, and an electron is initially at rest after leaving the cathode, then the work done in moving the electron from the cathode to the anode is $q \times V$, where q here is the value of the electronic charge. This is equal to the gain in kinetic energy by the electron, so that we can write

$$qV = \tfrac{1}{2}mv^2$$

where m is the mass of the electron and v is its velocity. Rearranging gives

$$v = \sqrt{\frac{2qV}{m}}$$

which is valid for all charged particles accelerated by a potential V.

(Note that this expression is not valid, however, for charged particles moving at appreciable fractions of the speed of light. For this latter case a relativistic correction needs to be made.)

(b) The deflection of charged particles by an electric field

If a charged particle enters an electric field acting at right angles to its direction of motion, it will be deflected from its original path. Suppose a uniform field of strength E exists between two parallel plates X and Y of length L as shown in Figure 4.7. A negatively charged particle of mass m and charge q travelling at a velocity v enters the plates from left to right. If plate X is at a positive potential, then the particle (if it has a negative charge) will experience an upwards force of qE and, by Newton's 2nd Law, a uniform upwards acceleration given by

$$a_y = \frac{qE}{m}$$

Since the force due to the electric field is vertical, the horizontal component of velocity is unaffected. Now the vertical displacement of the charged particle in a time t is given by

$$y = \tfrac{1}{2}a_y t^2 = \tfrac{1}{2} \times \frac{qE}{m} \times t^2$$

and in the same time period the horizontal displacement x is $x = vt$, so $t = x/v$. Substituting this expression for t in the equation for the vertical displacement, we obtain

$$y = \left(\frac{qE}{2mv^2} \right) \times x^2$$

This has the form $y = kx^2$, which is the equation of a curve called a *parabola*. We therefore conclude that a charged particle moving in a uniform electric field at right angles to its motion will move in a parabolic trajectory.

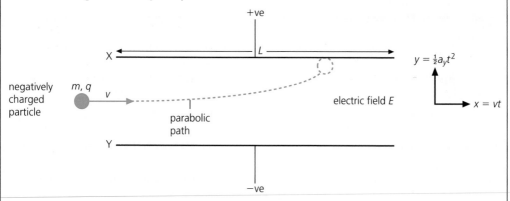

Figure 4.7 The deflection of charged particles by an electric field

(c) The deflection of charged particles by a magnetic field

A charged particle of charge q travelling with velocity v in a magnetic field of strength B perpendicular to its motion will experience a force F given by

$$F = Bqv$$

The direction of the force can be found using Fleming's Left-Hand Rule (remember to use the direction of *conventional* current flow – that is, a negative charge moving in one direction is equivalent to conventional current in the *opposite* direction).

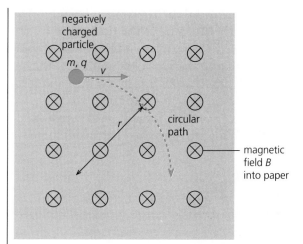

Figure 4.8 The deflection of charged particles by a magnetic field

Figure 4.8 shows the path of a negatively charged particle (q) moving in a perpendicular magnetic field (the direction of the field is into the paper). If the speed of the particle remains the same, it will move in a circular arc of radius r. The force due to the magnetic field is Bqv and this is the centripetal force, so that we can write

$$Bqv = \frac{mv^2}{r} \quad \text{so that} \quad r = \frac{mv}{Bq}$$

and since the momentum p is mv, we can write this as

$$p = Bqr$$

This shows that for any particle with charge q moving in a constant perpendicular magnetic field, its momentum $p = Bqr$ and the *radius of its circular path is proportional to its momentum*. This equation is used by particle physicists to determine the momentum of particles by measuring their radius of curvature as they move through a perpendicular magnetic field inside a particle detector.

WORKED EXAMPLE 4.2

A proton (mass 1.67×10^{-27} kg) moves in a perpendicular magnetic field of 0.5 T and has a radius of curvature of 5 cm.

a What is its momentum?
b What is its velocity?

a Using $p = Bqr$, the momentum is

$$(0.5\,\text{T}) \times (1.6 \times 10^{-19}\,\text{C}) \times (5 \times 10^{-2}\,\text{m}) = 4.0 \times 10^{-21}\,\text{kg m s}^{-1}$$

b Its velocity is

$$\frac{p}{m} = \frac{4.0 \times 10^{-21}\,\text{kg m s}^{-1}}{1.67 \times 10^{-27}\,\text{kg}} = 2.4 \times 10^{6}\,\text{m s}^{-1}$$

For synchronous acceleration to occur in the cyclotron, the time taken for the particle to travel half a revolution must be exactly the same as the time taken for the alternating voltage to go through one half-cycle. This means that the period of the alternating voltage must be the same as the period of the particle's motion. If the particle is travelling with a non-relativistic velocity v while it is in the dee, and it travels in a curve of radius R, then the time T taken to make one complete revolution is $T = 2\pi R/v$. If the ion has a charge q, then the force F_B it experiences due to the magnetic field is given by $F_B = Bqv$. This provides the centripetal force $(mv^2)/R$, so that

$$Bqv = \frac{mv^2}{R} \quad \text{and} \quad \frac{R}{v} = \frac{m}{Bq}$$

Now, since $T = 2\pi R/v$, we can write

$$T = \frac{2\pi m}{Bq}$$

and the frequency of the alternating voltage or the **cyclotron frequency** is

$$f_{\text{cyc}} = \frac{1}{T} = \frac{Bq}{2\pi m}$$

Note that the radius R of the path of the particle can be written as

$$R = \frac{mv}{Bq} = \frac{p}{Bq}$$

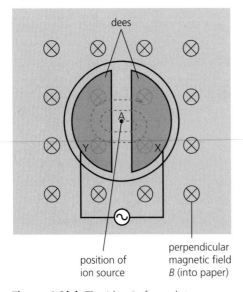

Figure 4.9(a) The 'dees' of a cyclotron

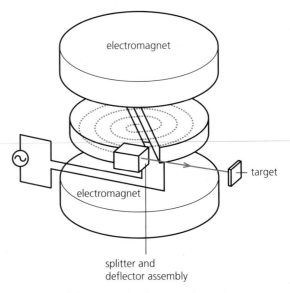

Figure 4.9(b) Schematic diagram of a cyclotron

Each time the particle crosses the gap between the dees, it gains kinetic energy and the radius of its path inside the dees increases. The kinetic energy E of the particle for a given radius R is

$$E = \tfrac{1}{2}mv^2 = \tfrac{1}{2} \times \left(\frac{q^2 B^2 R^2}{m} \right)$$

since, from the expression for centripetal force, $mv = BqR$. This expression tells us that the maximum particle energy obtainable in a cyclotron depends on the effective radius of the dees and the maximum available magnetic field strength.

The accelerated particles are extracted from the cyclotron by means of a special channel made up of a **splitter** and a **deflector** (Figure 4.9b). The deflector is held at a negative potential of about 50 kV with respect to the splitter, and the particles experience a radial force due to the electric field between these electrodes. This force draws the particles clear of the magnetic field, where they are directed to an experimental target. Particle energies in excess of 100 MeV are achievable using cyclotrons. Figure 4.10 shows the CIME cyclotron accelerator at Grand Accélérateur National d'Ions Lourds (GANIL) in France. This machine is used to study nuclear reactions.

Figure 4.10 The CIME cyclotron at GANIL, France. It is operated in the range 9.6 to 14.5 MHz and produces beams of unstable nuclei with energy in the range 1.7 to 25 MeV per nucleon, delivering a maximum beam energy of about 1 GeV

The synchrocyclotron

The cyclotron relies on the constant orbital frequency of a charged particle moving in a circular path perpendicular to a magnetic field. We have seen that the cyclotron frequency is given by

$$f_{cyc} = \frac{1}{T} = \frac{Bq}{2\pi m}$$

This is the orbital frequency of the particle, but at relativistic speeds ($\approx 0.2c$) this equation for a constant orbital frequency breaks down. Using the Special Theory of Relativity it can be shown that the expression for the cyclotron frequency must be replaced by

$$f_{cyc} = \frac{Bq}{2\pi m_0} \sqrt{1 - \frac{v^2}{c^2}}$$

Because of the relativistic mass increase as the particles spiral out from the centre, a lower value of frequency is required in order to maintain synchronism. As the speed v of the particles increases, the frequency decreases. So in order to accelerate particles to relativistic speeds, we can either adjust the frequency of the alternating electric field between the dees to compensate for this or change the perpendicular magnetic field strength B.

Cyclotrons that adjust the frequency of the electric field are called **synchrocyclotrons**. Their successful operation is practical confirmation that the Special Theory of Relativity must be true. Early synchrocyclotrons were modifications of existing cyclotrons; modern versions are able to produce beams of particle with energies of several hundred MeV.

WORKED EXAMPLE 4.3

A cyclotron has a magnetic field of 1.5 T and a maximum radius of 0.5 m. It is used for accelerating protons.

a What is the cyclotron frequency?
b What is the kinetic energy of the protons as they emerge from the machine?

a
$$f_{cyc} = \frac{Bq}{2\pi m} = \frac{(1.5\,\text{T}) \times (1.6 \times 10^{-19}\,\text{C})}{2\pi \times (1.67 \times 10^{-27}\,\text{kg})} = 2.3 \times 10^7\,\text{Hz or 23 MHz}$$

b
$$E = \tfrac{1}{2}mv^2$$

$$= \tfrac{1}{2} \times \left(\frac{q^2 B^2 R^2}{m} \right)$$

$$= \tfrac{1}{2} \times \left[\frac{(1.6 \times 10^{-19}\,\text{C})^2 (1.5\,\text{T})^2 (0.5\,\text{m})^2}{(1.67 \times 10^{-27}\,\text{kg})} \right] = 4.3 \times 10^{-12}\,\text{J}$$

$$= 4.3 \times 10^{-12}\,\text{J} \times \frac{1\,\text{eV}}{1.6 \times 10^{-19}\,\text{J}} = 26.9\,\text{MeV}$$

Synchrotrons

The most widely used type of particle accelerator is the **synchrotron**. As we mentioned earlier, relativistic effects need to be taken into account when particles reach velocities close to the speed of light. As we have seen with the cyclotron, the particles get out of step with the cyclotron frequency and eventually their energy stops increasing. We saw in the previous subsection that this difficulty can be overcome by the synchrocyclotron, which varies the frequency accordingly, but there is another problem.

If we want to go to even higher energies, then the size of the cyclotron becomes impossibly big. To see why, consider a proton of energy $0.5\,\text{GeV}$ ($1\,\text{GeV} = 10^9\,\text{eV}$), moving at relativistic speeds in a magnetic field strength of $1.5\,\text{T}$. The energy E of the proton is given by $E \approx cp$, so its momentum is

$$p = \frac{E}{c} = \frac{0.5 \times 10^9 \times 1.6 \times 10^{-19}\,\text{J}}{3 \times 10^8\,\text{m s}^{-1}} = 2.7 \times 10^{-16}\,\text{kg m s}^{-1}$$

The path radius of the proton is therefore

$$R = \frac{p}{Bq} = \frac{(2.7 \times 10^{-16}\,\text{kg m s}^{-1})}{(1.5\,\text{T}) \times (1.6 \times 10^{-19}\,\text{C})} = 1.1\,\text{km}$$

where we use the same equation as in the non-relativistic case, but this time the momentum is the relativistic momentum. Despite using a synchrocyclotron, this would require a magnet whose pole faces had an area of $\pi \times (1.1 \times 10^3\,\text{m})^2 \approx 4 \times 10^6\,\text{m}^2$!

Higher energies are realised using an accelerator of a different design, called a **synchrotron**. Instead of a single large magnet, a synchrotron uses many individual magnets along the circumference of a large circle, consisting of an evacuated circular beam-pipe or 'ring' of a fixed radius which acts as a 'racetrack' for the high-velocity particles. Each magnet assembly consists of two magnets, a **quadrupole magnet**, which keeps the particle beam tightly focused, and a **bending magnet**, which ensures that they move in a curved path (Figure 4.11).

The particles are first accelerated by means of an **injector**. This is a low-energy linear or electrostatic accelerator that injects particles into the ring at just the right instant, so that the magnetic field of the ring magnets has the right strength to hold them in the beam-pipe of fixed radius. The particles are then accelerated by special **radio-frequency cavities** containing electrodes, which cause them to be accelerated several times per revolution. In a synchrotron (as also with a synchrocyclotron), charged particles are accelerated in 'bunches' and the machine goes through a cycle of magnetic field and frequency adjustment and then initialises itself for the next bunch.

Once the particle bunch is in the ring, the accelerating potential applied at radio frequencies to the electrodes is smoothly increased, and the strength of the magnetic field of the bending magnets is varied as the particles are accelerated, thereby

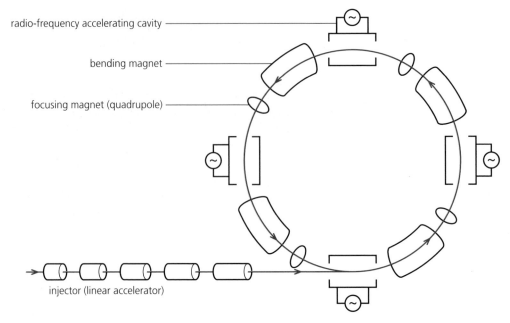

radio-frequency accelerating cavity

bending magnet

focusing magnet (quadrupole)

injector (linear accelerator)

Figure 4.11 The main components of a synchrotron. The particles are accelerated in radio-frequency accelerating cavities and kept in a circular path by the bending magnets. The focusing magnets keep the beam particles close together as they are accelerated in the ring

keeping the orbital radius constant. During this period of acceleration the particles complete many millions of revolutions and may travel hundreds or thousands of kilometres inside the ring before reaching their maximum energy. To reach high energies, particle physicists sometimes use a succession of increasingly larger synchrotrons, each one injecting particles into the next one.

The first synchrotron to come into operation was the Cosmotron in the USA, and in 1952 it accelerated protons to energies of 900 MeV. At the time of writing (2002), the highest-energy synchrotron in the world is the Tevatron at the Fermi National Accelerator Laboratory (Fermilab) in the USA (Figure 4.12). This machine is capable of producing collisions between beams of protons and antiprotons at energies of about 1 TeV per beam ($1\,\text{TeV} = 10^{12}\,\text{eV}$) with speeds of up to 99.999 95% of the speed of light, and is buried underground in a circular tunnel of circumference 6.3 km. In the Tevatron, protons and antiprotons circulate in opposite directions and repeatedly experience electric fields. After about 400 000 circuits, they reach their final energies of 1 TeV. The beams cross each other at the centres of two detectors, the CDF (see Box 4.2, page 129) and the DZero, where they collide to produce a large variety of new particles by the conversion of kinetic energy into mass.

At Fermilab, the acceleration of the protons and antiprotons is accomplished in a number of stages. An ion source produces negatively ionised hydrogen gas, consisting of two electrons and one proton. The ions are accelerated in a Cockcroft–Walton accelerator to 750 keV. Next, the hydrogen ions enter a linear accelerator, where

Figure 4.12 The circular tunnel containing the Tevatron at Fermilab, illuminated by the lights of a car travelling around the service road above the tunnel

they are accelerated to 400 MeV, before passing through a carbon foil, which removes the electrons, leaving only the positively charged proton. The protons then enter the Fermilab Booster, which is a smaller synchrotron that circulates the protons about 20 000 times, reaching an energy of 8 GeV. An important component of the Fermilab accelerator complex is the Main Injector, which accelerates particles and transfers beams into the Tevatron. The Main Injector accelerates the 8 GeV protons to 150 GeV and injects them directly into the Tevatron. In addition, it also generates beams of 120 GeV protons, which are sent to the Antiproton Source, where they are made to collide with a nickel target. The collisions produce a wide range of secondary particles, including many antiprotons. The antiprotons are collected and stored in a special accumulator ring. When a sufficient number of antiprotons have been produced, they are sent to the Main Injector for acceleration and injection into the Tevatron.

Electrons are also charged particles and can be accelerated by synchrotrons. They are lighter than protons, so for a given energy, the radius of the ring has to be larger. This is because of **synchrotron radiation** (Figure 4.13),

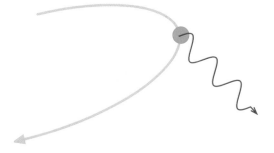

Figure 4.13 Synchrotron radiation consists of the electromagnetic radiations emitted by charged particles circulating in a synchrotron

which is the name given to the electromagnetic radiation emitted by charged particles circulating in a synchrotron. Any accelerated charged particle produces some electromagnetic radiation, and charged particles are being accelerated in synchrotrons since they move in curved paths inside the magnets in the ring. The intensity of synchrotron radiation is inversely proportional both to the fourth power of the particle's mass ($\sim 1/m^4$) and also to the radius of its path ($\sim 1/r$). Synchrotron radiation reduces the kinetic energy of the particles circulating in the ring, and this must be replaced by accelerating them around again. For this reason synchrotron radiation is more significant when accelerating electrons, which because of their smaller mass produce more synchrotron radiation than do protons.

Several electron synchrotrons have been built, and the largest one in the world is the **Large Electron–Positron collider (LEP)** at CERN in Switzerland. This monster accelerator is contained in an underground tunnel 27 km in circumference and up to 150 m deep. It runs under France and Switzerland, and first became operational in 1989, accelerating electrons and positrons up to 55 GeV in opposite directions. LEP contains 1312 quadrupole focusing magnets and 3304 bending magnets spaced around its ring, and was upgraded to LEPII with an increase in energy to over 100 GeV per beam. LEP consumed about 160 MW of power during peak operation, which is the entire output of a medium-sized power station! LEP is now being dismantled to make way for the **Large Hadron Collider (LHC)**, a synchrotron that will be able to accelerate protons and antiprotons in the existing tunnel.

Synchrotron radiation used to be considered a nuisance by particle physicists. However, it has opened up an abundance of research possibilities in all areas of science. The radiation is polarised and emerges from the accelerator in a high-intensity pulsed narrow beam. By varying the energy of the accelerated particles, the radiation can be 'tuned' from infrared to X-ray wavelengths, and it can be used as a probe to study the atomic and molecular structure of matter (including biological structures) as well as aiding in medical diagnosis. The SPEAR machine at SLAC (see below) is now used exclusively to study the properties and applications of synchrotron radiation.

Some synchrotrons are used as **storage rings**. These are accelerators that are able to store particles until they are needed in collisions. The particles are accelerated to high energies and kept circulating for long periods of time at a constant energy. The particles must pass through at least one accelerating cavity each time they circle the ring just to compensate for the energy they lose due to synchrotron radiation. LEP acts in this mode, circulating positrons and electrons in opposite directions until they are brought together in a collision.

A **beam collider** is a synchrotron that has two rings which overlap at several points where particles can collide with antiparticles so that much more of the collision energy can be converted into mass, thus creating new particles. The **SPEAR** machine or the **Stanford Positron–Electron Asymmetric Rings** at the Stanford Linear Accelerator Center (SLAC) was one of the first colliders to be built and came into operation in 1972. It forms part of the Stanford Synchrotron Radiation Laboratory (SSRL). SPEAR is able to store positrons and electrons with

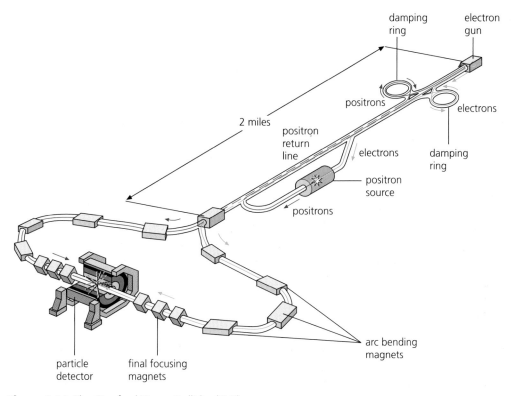

Figure 4.14 The Stanford Linear Collider (SLC)

energies of up to 2.4 GeV per beam and collide them together. LEP is also a collider and can bring beams of electrons and positrons together at various points in the ring with a collision energy of 180 GeV.

At SLAC, the **Stanford Linear Collider (SLC)** is a modification of the original SLAC linac in which electrons and positrons are brought round in a tightly curving oval to produce a single collision at an energy of 100 GeV (Figure 4.14).

Table 4.1 overleaf shows a list of particle accelerators currently in use and the energies they are capable of producing. Do not think that, because these machines are so big, the particle energy must be enormous. Remember that an electronvolt is a tiny amount of energy ($1 \, eV = 1.6 \times 10^{-19} J$). A single particle circulating in LEP has less kinetic energy than a mosquito! In fact, a pencil falling off a table has an energy of about 1.5 J or $10^{19} \, eV$ or over 1 000 000 TeV. Consider the amount of heat poured into the sea by the Sun. It is enormous, but since it is diffused in a huge quantity of water, its concentration is very low. In an accelerator, what makes the difference is that particles are accelerated almost to the speed of light, and since the kinetic energy they have is applied to such a tiny object, it is tremendously *concentrated*. When one particle meets another, then this kinetic energy becomes apparent at the moment of impact and transforms itself into a number of new particles that did not exist a moment before.

Table 4.1 Some particle accelerators and their energies*

Accelerator	Location	Start of operation	Type	Energy
Proton Synchrotron Complex (PS)	CERN, Geneva, Switzerland	1959	Synchrotron – used to accelerate particles to initial energies ready for further acceleration	500 MeV (electrons) 50 MeV (protons)
Super Proton Synchrotron (SPS)	CERN	1976	Synchrotron collider	Collides protons and antiprotons at 900 GeV
Large Electron–Positron Collider (LEP), upgraded to LEPII (now being dismantled to make way for the LHC)	CERN	1989	Synchrotron collider	Collides electrons and positrons at 180 GeV
Large Hadron Collider (LHC)	CERN	2005	Synchrotron collider	Planned to collide 7 TeV protons and antiprotons at 14 000 GeV using existing LEP tunnel
Stanford Linear Collider (SLC)	Stanford Linear Accelerator Center (SLAC), Stanford, USA	1989	Linac collider	Collides electrons and positrons to 100 GeV using special curving magnets
SPEAR	SLAC	1972	Synchrotron	In 1970s used to collide electrons and positrons at energies up to 4 GeV. Now used for synchrotron radiation research
PEP II	SLAC	1998	Synchrotron storage ring with linac injector	PEP II is an *asymmetric* storage ring. Electrons are accelerated to 9 GeV and positrons to 3.1 GeV
Tevatron	Fermi National Accelerator Laboratory (Fermilab), Chicago, USA	1984	Synchrotron collider	Collides protons and antiprotons at 2.0 TeV
Hadron–Electron Ring Accelerator (HERA)	Deutsches Elektronen-Synchrotron (DESY), Hamburg, Germany	1992	Synchrotron collider	Collides 30 GeV electrons with 920 GeV protons

*It should be remembered that particle accelerators are often modified and 'tweaked' to improve their energy performance.

Passive particle detectors

Since they are so small, elementary particles cannot be directly observed, and so particle physicists make use of a number of different kinds of particle detectors in order to detect and measure their properties. Most particle detectors depend on *ionisation* for their operation, and the type of detector used depends on the properties of the particle and the objectives of the experiment. The detectors used in modern particle physics experiments can be divided into two broad groups: **counters**, which simply register the passage of a charged particle, and **track recording devices**, which provide images of the paths of particles. Often these two types of detectors act in tandem, with one providing an input or **trigger** for the other. As particle physicists have probed deeper and deeper into the structure of matter, many different kinds of detection techniques have been developed, and we will now look at some of the methods employed to detect particles. The detectors we discuss in this section are what we might term *passive*. They do not require any form of electronic amplification for their operation.

Photographic emulsion

We saw in Chapter 1 that Becquerel first discovered radioactivity by accidentally leaving a photographic plate next to some uranium salts, causing the plates to become fogged. This was the first example of photographic material being used to detect ionising radiations. The emulsions from which photographic film is made consist of chemical grains. In 'normal' use, they react on exposure to light. But they also react due to the ionisation of their atoms by the passage of charged particles. Development of a film that has been exposed to charged particles causes the particle tracks to show up as dark lines. In normal photographic film, the particle tracks are not very well defined, and in the 1930s special **nuclear emulsions** were invented containing higher densities of grains that are sensitive to all kinds of charged particles. Photographic film provides a permanent record of charged particle events and has been extensively used in studying the charged particles produced by cosmic ray events. Particle tracks produced in nuclear emulsions are often very short due to their high photographic grain densities, which quickly bring the particle to a halt. For this reason it is usually necessary to view them under a microscope. Figure 5.6 (page 153) shows an example of a cosmic ray event captured on nuclear emulsion.

Cloud chambers

Charged particle tracks can be made visible by means of a **cloud chamber**. This device, invented by Charles T. R. Wilson (1869–1959) in 1911, is shown in Figure 4.15a and was an early invention for studying radioactivity, enabling photographs to be taken of alpha and beta particle tracks and gamma ray tracks. A chamber is filled with air containing saturated water or alcohol vapour. The vapour is rapidly cooled by expanding the gas by means of a piston. Then, when a charged particle passes through the chamber, it will ionise the air molecules. This causes condensation on the ions, which show up as a track when the chamber is suitably illuminated. Note that it is the *track* we see and not the radiation itself.

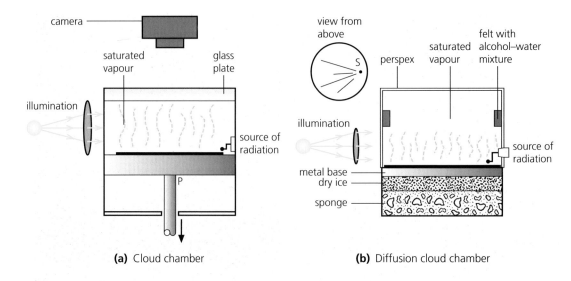

Figure 4.15 Cloud chambers and diffusion cloud chambers make the paths of ionising particles visible

A variation on Wilson's chamber is the **diffusion cloud chamber** (Figure 4.15b). This is a chamber made up of two compartments. The lower compartment contains frozen carbon dioxide (dry ice), at −78°C. The upper compartment contains air, which is at room temperature at the top and about −78°C at the bottom, the air at the bottom being cooled by the dry ice. A temperature gradient is thus maintained across the air in the upper chamber. A felt ring soaked in alcohol is fixed near to the top of the chamber and the alcohol vaporises in the warmer upper compartment. The alcohol diffuses downwards, cooling as it does so. The air on the floor of the upper compartment then forms a layer saturated with alcohol vapour. If a charged particle passes through this layer, then it will ionise air molecules, causing condensation of the alcohol vapour on to the air ions. This shows up as a track.

Alpha particles give short straight tracks (Figure 4.16a) whereas beta particles leave heavy tortuous tracks (Figure 4.16b). Gamma rays cause electrons to be ejected from air molecules, giving tracks with a wispy straggly appearance, similar to those of high-energy X-rays in Figure 4.16c.

Bubble chambers

Liquids are better than gases for detecting particles because their much greater density means that they contain many more nuclei with which particles can interact. The **bubble chamber** was invented in 1952 by the American physicist Donald Arthur Glaser (1926–), and consists of a tank filled with a liquid such as liquid hydrogen that is maintained at a temperature just below its boiling point. If the pressure of the liquid is suddenly lowered, then the liquid begins to boil. This effect is familiar to mountaineers – you can boil a kettle on top of a mountain at a lower temperature than that at sea-level, due to the decreased atmospheric pressure.

Figure 4.16(a) Alpha particle tracks

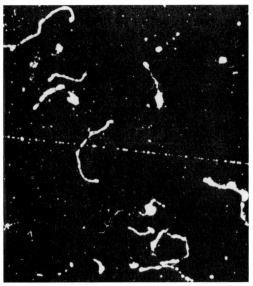

Figure 4.16(b) Beta particle tracks

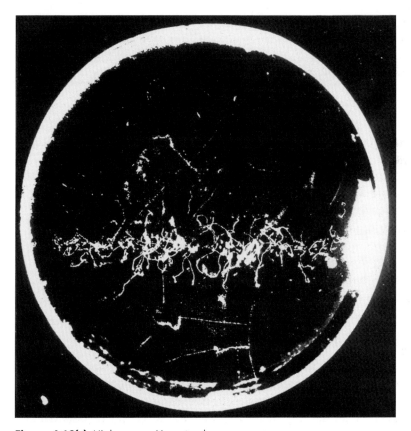

Figure 4.16(c) High-energy X-ray tracks

Figure 4.17 The Gargamelle bubble chamber at CERN

When liquid hydrogen changes state in this way, it is said to be in a **superheated** state. It will start to boil, but not immediately. During this initial superheated condition, if a high-energy charged particle passes through the chamber, then it will ionise atoms of the liquid along its path, ejecting electrons in the process. The electrons deposit all their energy in the liquid and trigger the formation of bubbles. High-speed cameras photograph the tracks. Then, within a few hundredths of a second, the chamber is recompressed, which raises the boiling point and quenches the bubbles. The cycle is repeated for the next passage of charged particles. Figure 4.17 shows a very famous bubble chamber at CERN called Gargamelle, which was used to provide evidence of an interaction mediated by the Z particle (see page 217).

Electronic particle detectors

Passive detectors have their disadvantages. The spatial resolution of photographic grains and bubble size mean that it is not easy to resolve tracks which are very short and close together. The information they provide is not instant and needs to be processed. The bubble chamber has a 'dead time' while it is recycling, during which it is unable to record any particle events or decays. A typical bubble chamber may perform 30 expansion cycles a second, but this is not fast enough to cope with the number of particle reactions per second that modern particle physics experiments routinely generate. Despite this, passive detectors act as both target and detector, where the actual interaction point can directly be seen. To overcome these limitations, particle physicists have devised detectors in which the position information of a particle can be detected by electronic amplification and then reconstructed with the aid of a computer.

Scintillation detectors

These are detectors in which a flash of light or **scintillation** is produced when charged particles or photons pass through certain materials. It was this kind of detector that Rutherford and his co-workers used in their alpha particle scattering experiments. At that time, physicists counted the flashes on a fluorescent screen with the naked eye, a tedious and error-prone process. The most common scintillator material is sodium iodide (NaI), but some plastics and even liquids can be used as scintillation materials. A NaI crystal is attached to a device called a **photomultiplier tube (PMT)**, which makes use of the photoelectric effect (see Box 1.1, page 15). A photomultiplier consists of a series of dynodes arranged in an evacuated glass tube (Figure 4.18, overleaf). When a particle passes through the crystal, the flash of light emitted falls on the first dynode, called the **photocathode**, and ejects electrons by the photoelectric effect. These are then accelerated through a potential difference to the next dynode, where more electrons are released by **secondary emission**. As a result, an electron amplification process occurs at each dynode stage right through to the final electrode, where a useful current is produced. The size of the current pulse is proportional to the energy deposited in the crystal by the particle.

A variation of the scintillator is the **Cerenkov detector**. This makes use of the fact that charged particles moving through a transparent medium at a speed *greater than* the speed of light in that medium emit electromagnetic radiation called *Cerenkov radiation* (Figure 4.19a). (Note that, while no particle can exceed the speed of light *in a vacuum*, the speed of light is reduced in a medium by a factor η, where η is the refractive index of the medium.) Cerenkov radiation takes the form of a cone around the direction in which the particle is travelling. The Cerenkov effect is similar to the sonic boom produced by a supersonic aircraft when it is moving through the air at a speed faster than the speed of sound through the air. It can be shown that the angle of the cone is related to the speed of the particle. By using a special optical system, light from a given angle can be intercepted via a photomultiplier whose angular position enables the velocity of the particle to be determined. Cerenkov radiation can be seen as a bluish glow in water surrounding the core of a nuclear reactor. Energetic charged particles emitted by the radioactive material travel faster than the speed of light in the water (Figure 4.19b).

A particle entering the scintillator crystal causes a flash of light to be emitted which strikes the photocathode and releases an electron.

The dynodes are shaped and positioned so that the electrons are channelled towards the next dynode. The photocathode is held at a negative potential of some 100 V with respect to earth. Each successive dynode is more positive than the photocathode by 100 V or so. The chain of dynodes produces an avalanche of electrons by secondary emission which are collected to produce an output current proportional to the energy of the incoming particles

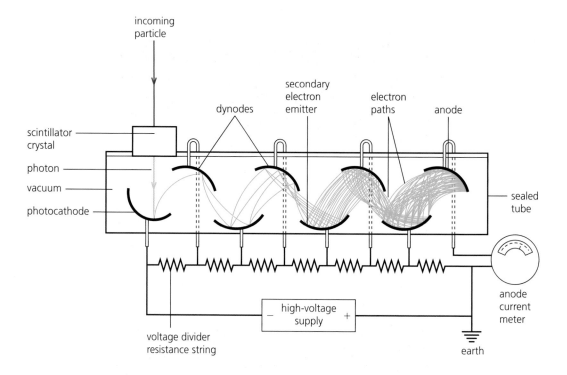

Figure 4.18 The principle of a photomultiplier tube

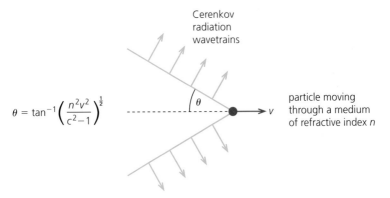

$$\theta = \tan^{-1}\left(\frac{n^2v^2}{c^2-1}\right)^{\frac{1}{2}}$$

Cerenkov radiation wavetrains

θ

particle moving through a medium of refractive index n

v

Figure 4.19(a) Cerenkov radiation is produced by charged particles moving through a transparent medium at a speed greater than the speed of light in that medium

Figure 4.19(b) The blue glow that can be seen in many nuclear reactors is Cerenkov radiation. The example shown here is from the advanced TRIGA reactor at the University of Illinois, USA

Spark chambers

The spark chamber was developed at CERN and other laboratories in the 1960s and consists of a series of parallel metal plates separated by a few millimetres and immersed in an inert gas. When a charged particle passes through such a detector, it ionises the gas atoms, leaving a trail of ions. If a high voltage is applied to every other plate immediately after the ion trail has been formed, then sparks like miniature lightning bolts form along the trail, revealing the path of the particle

123

(Figure 4.20). The sparks can be photographed to reveal the path of the particle track. Spark chambers can be used in conjunction with scintillation counters placed outside the chamber, which note the arrival of charged particles and trigger the electric discharges between the plates.

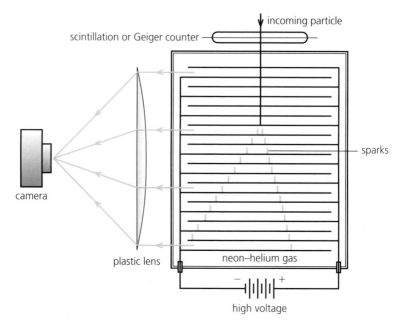

Figure 4.20 Spark chambers render the tracks of particles visible by producing sparks where the gas has been ionised. They have now been largely replaced by detectors with better time and spatial resolution

Multi-wire proportional chambers and drift chambers

A development of the spark chamber is the **wire chamber**. Here the metal plates are replaced by sheets of fine wires a few millimetres apart. When a charged particle passes through the chamber, an ion trail is formed. The pulse of current is sensed by the wires nearest to the spark, and an electrical pulse is sent down the wire and is then amplified. The current pulses provide all the information about the particle's trajectory in a form that can be readily processed by a computer. Wire spark chambers can be operated 1000 times faster than bubble chambers, allowing fast decay events to be imaged.

The speed and precision of particle detectors were greatly improved by the **multi-wire proportional counter (MWPC)** and the **drift chamber**. These were developed at CERN in the 1970s by the Frenchman Georges Charpak (1925–). The MWPC consists of a 'sandwich' of fine parallel wires such that a middle layer is held at a positive potential of 3–5 kV with respect to two outer layers (Figure 4.21). The sandwich is filled with an inert gas, and when a charged particle passes through it, an avalanche of ionisation electrons is produced in the vicinity of the nearest wire in the central layer. The position of the wire that generates the current pulse locates a point along the particle track, and by placing several MWPC sandwiches together, the path of a particle can be reconstructed. MWPCs can be built in many shapes and sizes ranging from a few square centimetres up to several square metres, containing thousands of wires. In colliding beam experiments (see page 136), they can even be constructed in a series of concentric layers around the accelerator beam-pipe. MWPCs are extremely fast in recording particle events. They easily handle as many as a million particles per second, and are now a standard method of detection in particle physics experiments.

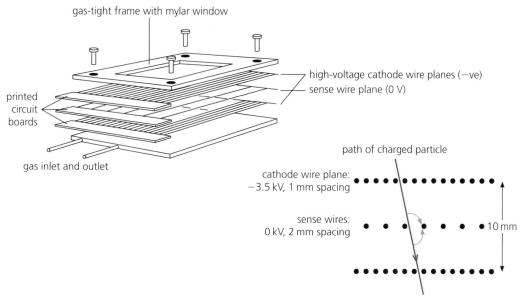

Figure 4.21 A multi-wire proportional counter or chamber (MWPC) consists of a sandwich of three layers of parallel wires contained within a gas-filled framework with thin plastic windows. A charged particle passing through the chamber ionises the gas along its path, releasing an avalanche of electrons, which move towards the nearest wire in the central plane. The position of the wire that collects the electrons locates a point along the particle's track

Drift chambers (Figure 4.22) are similar in construction to the MWPC but the wires in the central layer are further apart. These wires consist of alternate 'field' and 'sense' wires. The voltages across them produce an electric field in which the electrons 'drift' at a constant velocity towards the nearest sense wire, where they initiate a current pulse. The drift time is determined by an electronic timer triggered by a signal from a scintillation counter outside the chamber. The drift speed is known and the time taken to reach the sense wire can be measured, so the distance of the particle's path from the sense wire can be calculated. While not as fast as a MWPC, with drift chambers it has proved possible to measure the position of particle tracks to an accuracy of $50\,\mu m$ ($1\,\mu m = 10^{-6}\,m$).

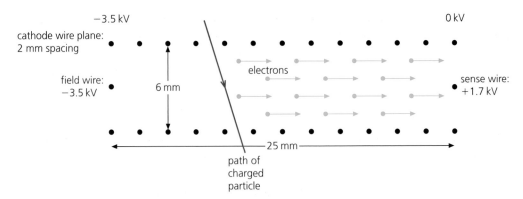

Figure 4.22 A drift chamber is similar in construction to a MWPC, but with the wires in the central plane spaced further apart. Electrons produced by the passage of a charged particle drift at a constant velocity towards the nearest sense wire. The drift time is electronically measured and is proportional to the distance between the track of the particle and the sense wire

Time projection chambers

This is a development of the drift chamber and was invented at the Lawrence Berkeley Laboratory in the USA. The basic arrangement is a cylinder of gas about 4 m long, divided into two halves by a high-voltage membrane held at a negative potential of about −150 kV. The ends of the cylinder are two plates of wires held at zero voltage. When a charged particle passes through the chamber, electrons caused by ionisation of the gas drift away from the membrane towards the ends of the cylinder. The time it takes for the electrons to arrive at the ends gives a measure of how far along the cylinder the ions were produced. An electron originating near the central membrane takes a longer time than one closer to the ends. As soon as the electrons hit the ends of the cylinder, pads on the end plates record the two-dimensional position of the electron and the time it takes to reach the end. These data are input to a computer, which then has enough information to reconstruct a three-dimensional image of a particle track.

Silicon detectors

The very high event rates encountered in modern particle physics experiments have led to the development of **silicon strip detectors**. These are strips of silicon (the same semiconductor material as in integrated circuits) with very narrow widths, typically a few micrometres across, arranged in parallel. When a charged particle passes through a strip, **electron–hole pairs** in the semiconductor material are formed, generating a current pulse that can be detected by electronic amplification. Silicon microstrip detectors are like solid-state versions of multi-wire proportional counters. Their principal advantages lie in the fact that they have a much greater density than a gas and require less ionisation energy (of the order of 1 eV to produce an electron–hole pair, compared with 30 eV to ionise a gas). They have a fast response and are easy to produce in very narrow strips <50 μm, which means they can give very precise position information. Their disadvantage is that the manufacturing limitations of silicon wafers mean that silicon detectors can only be made with areas of a few square centimetres compared with square metres for gaseous detectors. They are also susceptible to radiation damage and may need cooling to reduce electronic thermal noise.

Vertex detectors

This is a type of silicon detector containing an array of postage stamp sized **charge-coupled devices (CCDs)** similar to those which form the recording elements in video and digital cameras. Each of these chips might contain as many as 10^6 individual elements of silicon, called **pixels**, arranged in a series of layers. Particles travelling through the layers leave behind small deposits of electric charge in the elements they cross. The location of each deposit or 'hit' can be recorded electronically and a computer can 'connect the dots' to reconstruct the tracks of all the particles that have passed through the layers.

In particle events, charged particles are created in pairs of equal and opposite charge, each producing its own track in the detector. By drawing each path back to where it meets with one or more other paths, we can find the point where the charged particle is created, or the **vertex**, and in this way we can determine the origin of a single or group of particle decays (Figure 4.23). The pixels in a CCD are very small, permitting measurement of a charged particle's position within a few micrometres. A vertex detector gives the most accurate location of any outgoing charged particles as they pass through it.

These detectors have a very fast response and can be read directly by a computer, which can give information about a particle event almost immediately. The processing of particle events can thus be done 'on-line', allowing the computer controlling the experiment to decide whether or not an event is significant before making a permanent record of it.

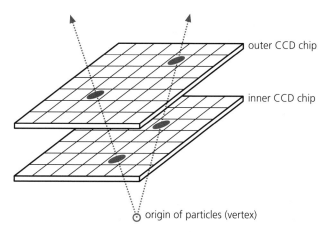

Figure 4.23 The principle of a vertex detector. Particles travelling through layers of CCD chips leave 'hits', which can be joined together to determine the origin of a particle decay, the 'vertex'

Calorimeters

A **calorimeter**, or 'calorie meter', is a device used in physics and chemistry experiments to measure the total heat of a reaction or process, which is of course an energy measurement since heat is a form of energy. (The calorie is a non-SI quantity of heat equal to about 4.2 J.) Particle physicists use the term 'calorimeter' for any device that is designed to measure the amount of energy carried by particles, not by an increment in temperature but by total absorption in a material. They are different from most other detectors, as the particle itself is changed by the detector, and they can also detect uncharged particles. Calorimeters are often of a sandwich construction made up of a number of layers or modules of absorbing material interspersed with counters. There are two types used in particle physics, as detailed below.

Electromagnetic calorimeters measure the energy carried by electrons, positrons and photons. When particles travel through the calorimeter modules, they interact with the energy absorption material, producing more particles of lower energy. These produce yet more lower-energy particles in a cascade of particles called a **shower**. The amount of absorption material is chosen so that all the particles in the shower are brought to rest. The amount of charge deposited in the calorimeter in a single shower and the depth to which the shower develops are proportional to the total energy of the particle that initiated the shower. In this way a calorimeter can measure energies for both charged and neutral particles. In electromagnetic calorimeters the electromagnetic shower consists of particles such as electrons, positrons and photons, and the absorption material can consist of thick sandwiches of lead, which act as an absorber, together with plastic scintillators. Particles entering an electromagnetic calorimeter will interact with the calorimeter modules, and the scintillators in the sandwich detect the energy of the shower at each stage of the cascade until all the energy is absorbed by the calorimeter.

Hadron calorimeters measure the energy carried by **hadrons**, which are strongly interacting particles such as protons that interact via the strong force. The construction of these calorimeters is similar to the electromagnetic type, but absorbers of lower Z number such as iron are used instead of lead. Showers of other hadrons are produced, which are fewer in number and travel shorter distances. As before, their energies are detected by scintillators and can be related to the total energy of the incoming particle.

The detectors used in modern particle physics use a combination of those described above to image particle events. Some of them are enormous in size. Box 4.2 describes the construction of the CDF detector at Fermilab.

Box 4.2 The CDF detector at Fermilab

The detectors used in particle physics can be enormous in size. At Fermilab, the CDF detector (which stands for Collider Detector at Fermilab) is three storeys high and weighs 5000 tonnes; it discovered the top quark in 1995. Figure 4.24 shows its construction.

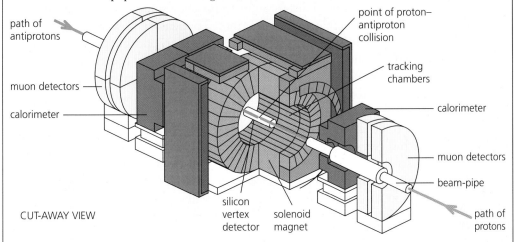

Figure 4.24 The CDF detector

Beams of protons and antiprotons travelling at nearly the speed of light hit each other at collision energies of 1.8 GeV in Fermilab's Tevatron synchrotron collider accelerator, producing showers of secondary particles such as quarks, electrons, muons, neutrinos and their antimatter counterparts. The detector's job is to observe as many collisions as possible, to identify and record the particles that come flying out, and to store the information for later study.

At the heart of the CDF is a silicon vertex detector that can locate the path of a particle to within 15×10^{-6} m, essential for detecting short-lived events. This is inside a gas-filled tracking chamber, which contains wires that pick up the ionisation due to charged particles. Surrounding the central part of the detector are calorimeters and a solenoid magnet that makes the charged particles move in curved paths. Each layer in the detector collects and transmits different kinds of information about the moving particles that interact with the material in that layer. This information is transmitted through circuitry for 100 000 electronic channels of information, where it is analysed by computers. During collider operations at the Tevatron, control room crews operate the detector in eight-hour shifts, 24 hours a day.

Quarks, which we look at in detail in Chapter 7, come in six different 'flavours' known as up, down, strange, charm, top and bottom that distinguish them from each other. One of these, the top quark, was discovered in the CDF from an event that involved the creation of a top–antitop pair of quarks. The decay of these quarks gives rise to jets in an event characterised by a highly energetic muon and electron.

Out of 10^{18} or so collisions created in the CDF, 12 events had been isolated that appeared to involve the creation of a top–antitop pair. In 1999, the Tevatron and CDF (together with another detector at Fermilab called DZero) were upgraded, allowing more top–antitop events to be produced, and the initial discovery was confirmed.

Luminosity and reaction rate

Luminosity

By using accelerators of higher energies, particle physicists are able to produce more particles of greater mass by collisions. But they also want to make sure that there are enough encounters between particles in order to produce sufficient numbers of particle reactions in a reasonable amount of time. To measure how frequently particles in a collision experiment collide with each other, particle physicists talk about the **luminosity** of the particle beam. Note that this is *not* the same concept as 'luminosity' measured in $W\,m^{-2}$ or photons $m^{-2}s^{-2}$ that is used in the intensity of a light source (although there are some similarities).

Particle accelerators commonly deliver beams in pulses or 'bunches'. Consider an accelerator that delivers a bunch of particles of type x each containing a number n_x at a rate of f bunches per second (Figure 4.25). If the cross-sectional area of the beam is A, then it will strike an area A of the target. The target contains particles of type y, of which there are n_y lying in the path of the beam. Each of the particles y has an area $\sigma\,m^2$, whereas in this simplified model the particles x have negligible area (the particles y are sufficiently far apart so that they do not overlap each other when viewed from the beam of particles x).

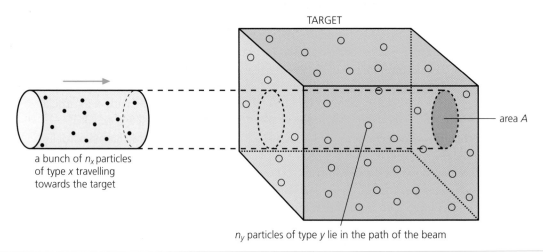

TARGET

area A

a bunch of n_x particles of type x travelling towards the target

n_y particles of type y lie in the path of the beam

Figure 4.25 The concept of 'luminosity' applied to fixed target experiments

Now the fraction of the target area occupied by the particles y is given by $n_y\sigma/A$, which is the same as the probability that a particle x undergoes a collision with a particle y. Remember, though, that there are fn_x particles x arriving every second, so the total number of collisions per second, or the **collision frequency** R, is

$$R = \frac{fn_x n_y \sigma}{A}\,Hz$$

The number of collisions per second therefore depends on the cross-sectional area of the target particle, the numbers of particles involved and the beam area. The quantity

$$L = \frac{fn_x n_y}{A}$$

is defined as the **luminosity** of the particle beam and has units of $m^{-2}s^{-1}$.

Reaction cross-sections and the barn

You might wonder how you can talk of the 'area' of a particle – after all, elementary particles such as electrons have no measurable diameter. At such small scales, particle physicists understand σ to be related to the effective *range* over which two particles can influence each other. But even this may not be a fixed quantity for two particles of the same type.

Now assuming that each collision produces a particle reaction, then the collision frequency R can be regarded as the **reaction rate**, which is the number of *reactions* per second, not just the number of collisions per second. The reaction rate can be easily measured using particle detectors, whereas the luminosity depends only on the properties of the beam and the target material. Particle physicists therefore make a working definition of σ called the **cross-section,**

$$\sigma = \frac{\text{reaction rate } (s^{-1})}{\text{luminosity } (m^{-2}s^{-1})}$$

So the cross-section of a particle reaction is the detected reaction rate divided by the luminosity of the accelerator used in the experiment. The important point here is that we now refer to the cross-section of a *reaction* rather than that of a particle, thus making it unnecessary to explain what we mean by the 'area' of a particle.

The cross-sections of particle reactions are very small – typically a few $10^{-28}\,m^2$. Particle physicists therefore use a more convenient unit for reaction cross-section called the **barn (b)**, such that

$$1\,b = 10^{-28}\,m^2$$

and the relationship between reaction rate, luminosity and cross-section can be expressed as

$$R = \sigma L$$

Fixed target and colliding beam experiments

The energetics of particle decays and reactions

When we consider the decays and reactions of elementary particles, the energy, linear momentum and total angular momentum must be conserved. By a **particle decay** we mean a single particle decaying into two or more product particles with its total energy shared among the decay products. A **particle reaction** is when one particle collides with another after being given kinetic energy via a particle accelerator. In this latter situation, the energy for the product particles comes not only from the rest mass of the initial particle, but also from the kinetic energy supplied by the accelerator. In both cases the relevant quantum numbers such as charge (together with some others that we will meet later) must be the same before and after the event. Subsequently, we will see how they help us to decide which decays are possible.

The energy of the accelerated particles as well as those of the reaction products is usually very large compared with the energies required for nuclear reactions, and this is why particle physics is sometimes called **high-energy physics**.

Particle decays

The energy available for a single particle decay, assuming that the decaying particle is initially at rest, is the difference between the rest energy of the initial particle m_ic^2 and the total rest energy m_fc^2 of the particles that are produced in the decay. This is called the **Q-value**:

$$Q = m_ic^2 - m_fc^2$$

A particle decay can only occur if Q is *positive*, that is, the rest energy of the incident particle must be greater than that of the product particles. The energy Q is shared among the decay products as *kinetic energy* in such a way as to conserve linear momentum.

WORKED EXAMPLE 4.4

A particle called the neutral lambda (Λ^0) decays into a proton (p) and a negative π meson or pion (π^-) via the decay

$$\Lambda^0 \rightarrow p + \pi^-$$

What is the combined energy of the proton and the negative π meson?
[Rest energies of Λ^0, p and π^- are 1116 MeV, 938 MeV and 140 MeV respectively. These are their energies expressed in MeV using the mass–energy relation $E = (\Delta m)c^2$, for example $m_pc^2 = 938$ MeV.]

The expression $Q = m_ic^2 - m_fc^2$ for this decay is

$$Q = m_{\Lambda^0}c^2 - (m_p + m_{\pi^-})c^2$$
$$= 1116\,\text{MeV} - (938 + 140)\,\text{MeV} = 38\,\text{MeV}$$
$$\therefore \quad KE(p) + KE(\pi^-) = 38\,\text{MeV}$$

which is the total kinetic energy shared among them.

Particle reactions

One objective in a particle physics experiment involving an accelerator is to create varieties of new particles from the *reaction* between an accelerated particle and a target:

accelerated incident particle + target → new product particles created

and so we need to know the minimum amount of energy or 'threshold energy' needed to 'manufacture' them. As the accelerated particles are moving at very high velocities, we need to use the equations obtained from the Special Theory of Relativity to analyse their dynamics.

Fixed target

Let's first consider the case of an accelerated particle hitting a target particle at rest (Figure 4.26).

Before collision

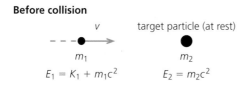

After collision

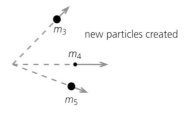

Figure 4.26 The reaction between two colliding particles, as seen from the laboratory reference frame

Suppose we have a reaction where an incident particle of mass m_1 collides with a target particle of mass m_2, creating a stream of new particles of masses $m_3, m_4, m_5, \ldots$

$$m_1 + m_2 \rightarrow m_3 + m_4 + m_5 + \cdots$$

Using the relativistic equations derived in Chapter 3, the incident particle will have a total energy E_1, a kinetic energy $K_1 = E_1 - m_1c^2$ and a momentum $p_1 = \sqrt{E_1^2 - m_1^2c^4}$. In this situation the target particle m_2 is in a frame of reference that is at rest with respect to the observer in a laboratory; this is called the **laboratory reference frame**. The Q-value for the reaction will be

$$Q = (m_1 + m_2)c^2 - (m_3 + m_4 + m_5 + \cdots)c^2$$

If Q is positive, then energy is available to the product particles as kinetic energy, so that m_3, m_4 and m_5 have more combined kinetic energy than the initial particles m_1 and m_2. If Q is negative, then some of the initial kinetic energy of m_1 is turned solely into rest energy.

WORKED EXAMPLE 4.5a

A particle accelerator accelerates a particle called a negative K meson or kaon (K^-) onto a fixed target of protons (p), producing the reaction

$$K^- + p \rightarrow \Lambda^0 + \pi^0$$

where Λ^0 and π^0 are reaction product particles called the neutral lambda and neutral π meson. What is the Q-value for this reaction?

[Rest energies of K^-, p, Λ^0 and π^0 are 494 MeV, 938 MeV, 1116 MeV and 135 MeV respectively.]

Using

$$Q = (m_{K^-} + m_p)c^2 - (m_{\Lambda^0} + m_{\pi^0})c^2$$

and substituting their rest energies in MeV/c^2 (the c^2 factor cancels), we have

$$Q = (494\,\text{MeV} + 938\,\text{MeV}) - (1116\,\text{MeV} + 135\,\text{MeV})$$
$$= 181\,\text{MeV}$$

This positive Q-value tells us that there is enough *rest energy* in the K^- and the proton to create the Λ^0 and the π^0. In fact, there is 181 MeV (plus the KE of the incident particle) left over for the kinetic energy of the Λ^0 and π^0.

WORKED EXAMPLE 4.5b

Using the same fixed target arrangement as in Worked Example 4.5a, find the Q-value of the reaction

$$\pi^- + p \rightarrow K^0 + \Lambda^0$$

[Rest energies of π^- and K^0 are 140 MeV and 498 MeV respectively.]

As before

$$Q = (m_{\pi^-} + m_p)c^2 - (m_{K^0} + m_{\Lambda^0})c^2$$
$$= (140\,\text{MeV} + 938\,\text{MeV}) - (498\,\text{MeV} + 1116\,\text{MeV})$$
$$= -536\,\text{MeV}$$

This time the Q-value is *negative*. This tells us that energy must be supplied from the accelerator in the form of kinetic energy for the incident particle for the K^0 and Λ^0 to be produced. For negative Q-values we require a **threshold energy** that the incident particle must have in order for the reaction to proceed. The threshold energy must be greater than the magnitude of the Q-value, since not only do we need to create the new particles but we must also give them sufficient kinetic energy to ensure that linear momentum is conserved.

It can be shown that the **reaction threshold energy** K_{th} is given by

$$K_{th} = (-Q) \times \frac{\text{total mass of all the particles involved in the reaction}}{2 \times \text{mass of target particle}}$$

WORKED EXAMPLE 4.6

What is the threshold kinetic energy needed to produce neutral π mesons (π^0) via the following particle reaction? [The rest energies of p and π^0 are 938 MeV and 135 MeV respectively.]

$$p + p \rightarrow p + p + \pi^0$$

The Q-value is

$$Q = (m_p + m_p)c^2 - (m_p + m_p + m_{\pi^0})c^2$$
$$= -m_{\pi^0}c^2 = -135 \, \text{MeV}$$

and so the threshold energy is

$$K_{th} = (-Q)\frac{4m_p + m_{\pi^0}}{2m_p}$$

$$= -(-135 \, \text{MeV}) \times \frac{4 \times 938 \, \text{MeV} + 135 \, \text{MeV}}{2 \times 938 \, \text{MeV}}$$

$$= 280 \, \text{MeV}$$

This means that protons with kinetic energies of at least 280 MeV are needed to produce neutral π mesons for study, which is well within the energies of modern particle accelerators.

So far we have been considering particle reactions where a stationary target is hit by an accelerated particle, but suppose we look at the reaction in a different reference frame. In Figure 4.27 we look at the reaction from a reference frame where two particles with equal and opposite momenta collide. In this frame, called the **centre-of-mass** or **centre-of-momentum reference frame**, the total momentum is *zero*. One of the problems with fixed target experiments is that the reaction process is not very 'efficient' energetically. Much of the incident energy is 'wasted' as kinetic energy in the reaction products. For particle physicists, such effects in fixed target experiments lead to much of the accelerator energy being transferred to kinetic energy of the reaction products rather than going into creating new particles that they wish to study.

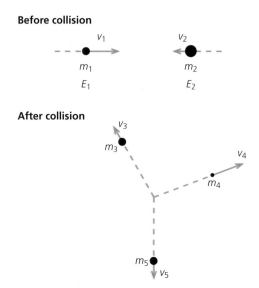

Figure 4.27 The same reaction as Figure 4.25, between two colliding particles, but this time as seen in the centre-of-mass reference frame. Viewed from this frame, the total momentum is zero both before and after the collision

Colliding beam

By colliding two beams of particles together in a colliding beam accelerator, the total energy of the two particles can be used in the interaction. This is equivalent to a centre-of-momentum frame where the final momentum is zero and no energy is wasted in kinetic energy of the products. The particle physicist can then simply observe the particle decays created at high energies. This is the reason for the development of colliding beam accelerators outlined earlier, and most modern accelerators are now of this type.

In a colliding beam experiment, particles are accelerated in bunches travelling in opposite directions and are then brought together at various points in the accelerator to collide. However, the reaction rate in a colliding beam experiment will be much lower than that in a fixed target one. For a particle reaction to occur, the particles must hit each other at a reasonably fast rate so that looked-for events can be readily detected. Fixed target experiments generally achieve a greater hit rate than colliding beam experiments because the *density* of a solid or liquid target is much greater than the density of the accelerated beams, although the energy available to create particles is larger in the latter case.

The concept of luminosity is also commonly applied to colliding beam experiments. Figure 4.28 represents the situation. Since accelerators normally fire particles at each other in short pulses, many bunches of particles will encounter each other at collision points as they travel round the ring. In this case you can think of one type of particle bunch as a 'target' and the other as the 'beam', but it doesn't really matter which. Each bunch of particles x encounters one bunch of particles y, so each particle x 'sees' a target of n_y particles y. As before, if the cross-sectional area of the colliding beams is A, then the fraction of the target area occupied by the particles y is $n_y\sigma/A$. Every time the beams collide, there will be $n_x n_y\sigma/A$ hits per pulse (since the chance of a particle x scoring a hit is also $n_y\sigma/A$). If the beams pulse with a frequency f Hz, then, similarly, the total number of hits per second R is also

$$R = \frac{fn_x n_y\sigma}{A}$$

and the luminosity L is

$$L = \frac{fn_x n_y}{A}$$

The point to note here is that collision energies aren't everything – the reaction rate is also an important factor in establishing whether enough reactions will occur in a reasonable amount of time to make an experiment practically feasible.

area A — n_x particles of type x

area A — n_y particles of type y

Figure 4.28 The concept of 'luminosity' applied to colliding beam experiments

WORKED EXAMPLE 4.7

The Super Proton Synchrotron (SPS) at CERN can be used to collide protons and antiprotons together to produce a particle called the W particle. The luminosity of the accelerator was $3 \times 10^{33}\,\mathrm{m}^{-2}\mathrm{s}^{-1}$ and the cross-section for the reaction was $0.53\,\mathrm{nb}$ ($1\,\mathrm{nb} = 10^{-9}\mathrm{b}$). Assuming that each collision produces a single W event, how many W events per day could particle physicists expect to observe?

The reaction rate

$$R = \sigma \times L = (0.53 \times 10^{-9} \times 10^{-28}\,\mathrm{b}) \times (3 \times 10^{33}\,\mathrm{m}^{-2}\mathrm{s}^{-1}) = 1.59 \times 10^{-4}$$

so there would be 1.59×10^{-4} W events per second. Since there are 86 400 seconds in one day, then the number of W events per day is $1.59 \times 10^{-4} \times 86\,400 = 13.7$.

Particle physicists can therefore expect to see the SPS producing about 14 W particles per day.

Their low cross-section accounts for the fact that few collisions produce a W particle. In fact, during its discovery, around 5 W particles were produced in 10^9 collisions.

Summary

◆ A **particle accelerator** is a machine that can accelerate charged particles to very high kinetic energies. **Electrostatic accelerators** use two points separated by a high *constant* potential difference to accelerate charged particles. Examples of electrostatic accelerators are the **Cockcroft–Walton** and **Van de Graaff** accelerators.

◆ In a **linear accelerator** (**linac**), particles are accelerated horizontally in an acceleration tube to high energies by means of a series of hollow cylindrical **electrodes** of increasing length placed in a line. Linear accelerators use the principle of **synchronous acceleration** to accelerate particles through a sinusoidally varying electric field at radio frequencies.

◆ A **cyclotron** uses the principle of synchronous acceleration but in addition uses a **magnetic field** to make the charged particles move in a **spiral** path. The **cyclotron frequency**

$$f_{cyc} = \frac{1}{T} = \frac{Bq}{2\pi m}$$

is the frequency needed to 'tune' the machine so that acceleration can take place. A variation of the cyclotron is the **synchrocyclotron**, where the magnetic field is steadily altered to take into account the relativistic mass increase of the accelerated particles as their speeds approach that of light. Synchrocyclotrons would not work if the Special Theory of Relativity was incorrect.

◆ **Synchrotrons** use an increasing magnetic field to keep not only the orbital frequency but also the orbital radius constant. **Quadrupole magnets** keep the particle beam tightly focused, while **bending magnets** ensure that they move in a curved path. Charged particles moving in curved paths in a magnetic field emit **synchrotron radiation**. Synchrotron radiation is equivalent to loss of kinetic energy of the particles circulating in the ring, and must be replaced by accelerating them around again. Most modern particle accelerators are of the synchrotron type.

◆ **Storage rings** are usually synchrotrons in which particles can be stored by circulating them in rings until they are needed in collisions. A typical storage ring circulates beams of electrons and positrons in opposite directions. A **collider** is an accelerator in which beams of particles can be made to collide. The particles are stored in storage rings and then brought together. Collider accelerators are generally of the synchrotron type, although a notable exception is the Stanford Linear Collider (SLC).

◆ Most particle detectors rely on **ionisation** processes for their operation, and may be divided into **counters**, which simply register the passage of a charged particle, and **track recording devices**, which provide images of the paths of particles. They may be **passive**, such as **nuclear emulsions, cloud chambers** and **bubble chambers**, or **active**, where electronic amplification is involved. Examples of active detectors are **scintillators** and **photomultiplier tubes, wire chambers, spark chambers, multi-wire proportional counters** (MWPCs), **drift chambers** and **silicon strip detectors**.

◆ Silicon strip detectors, MWPCs and drift chambers are superior in speed and position resolution. **Time projection chambers** are a type of drift chamber that can image a particle position in three dimensions. **Vertex detectors** are a type of silicon detector that can reconstruct the origin of a particle track to a very high degree of positional accuracy.

◆ **Calorimeters** are devices that can measure the total energy of particles by absorption in a material. **Electromagnetic calorimeters** measure the energies of particles such as electrons, positrons and photons, and **hadron calorimeters** measure the energy of strongly interacting particles such as protons.

◆ In particle physics experiments three important factors are the **luminosity** L of the particle beam, the **reaction rate** R and the **cross-section** σ. Reaction cross-section is measured in **barns** and is a measure of the ability of a particle beam to 'see' a target. The relationship between the reaction rate, luminosity and cross-section can be expressed as $R = \sigma L$.

◆ An important measure of whether there is sufficient energy for a **particle decay** or **particle reaction** to occur is the **Q-value**. If the Q-value for a reaction is positive, then particles can be created. If it is negative, then energy must be supplied from the accelerator to initiate the reaction. For negative Q-values, a certain **threshold energy** must be possessed by the incident particle for the reaction to proceed.

◆ Particle physics experiments may be '**fixed target**' or '**colliding beam**'. Colliding beam experiments, where the total momentum is zero, are more efficient in terms of energy going into making new particles, and no energy is lost as kinetic energy among the reaction products.

Questions

1 a Describe the main features of a *cyclotron*. What limits the energy of cyclotrons?
A cyclotron with dees 0.6 m in diameter is designed to accelerate protons. If the applied perpendicular magnetic flux density is 1.250 T, calculate:
i) the period,
ii) the frequency of the alternating voltage to be applied,
iii) the maximum velocity reached by the protons,
iv) their energy in MeV.

b Describe the main features of a *synchrotron*. What is *synchrotron radiation* and how does it limit the performance of synchrotrons? How do synchrotrons differ from linear accelerators?

In the Tevatron at Fermilab, protons of momentum 1 TeV/c (1 TeV $= 10^{12}$ eV) circulate in an orbit of radius 1 km. What is the strength of the magnetic field needed to hold the protons in this orbit? How many times per second do the protons go round the Tevatron? [Take the mass of a proton as $m_p = 1.67 \times 10^{-27}$ kg.]

2 a List the main components of a circular accelerator.

State why it is possible to achieve much higher energies using circular accelerators than using linear ones of a similar size. Explain why the particles in an accelerator must be contained within a vacuum tube.

What other two factors in the design of a circular accelerator limit the maximum energy available?

A charged particle moving in a plane perpendicular to a uniform magnetic field follows a circular path. Show that, if the speed of the particle is increased, the radius of the circular path increases but the period remains constant. *ULEAC, January 1996 (part)*

b In 1995 scientists at CERN noticed slight periodic variations in the energy of the particle beams. These variations occurred at the same frequency as the tides. It was decided that this was due to gravitational forces altering the shape of the Earth and hence the dimensions of the circular accelerator. Suggest why small changes in the circumference of the accelerator affect the energy of the particle beams.

Edexcel, January 2000 (part)

3 This question is about the acceleration of protons in the proton synchrotron at the European Centre for Nuclear Research (CERN) near Geneva.

You will find the data required below and in the question.

Universal gravitational constant	$G \approx 6.7 \times 10^{-11}$ N m^2 kg^{-2}
Electronic charge	$e \approx 1.6 \times 10^{-19}$ C
Mass of the electron	$m_e \approx 9.1 \times 10^{-31}$ kg
Mass of the proton	$m_p \approx 1.7 \times 10^{-27}$ kg
Speed of light in a vacuum	$c \approx 3.0 \times 10^8$ m s^{-1}
The Planck constant	$h \approx 6.6 \times 10^{-34}$ J s
The Boltzmann constant	$k \approx 1.4 \times 10^{-23}$ J K^{-1}
The Universal Gas constant	$R \approx 8.3$ J mol^{-1} K^{-1}
The Avogadro constant	$N_A \approx 6.0 \times 10^{23}$ mol^{-1}

Take $\dfrac{1}{4\pi\epsilon_0}$ to be 9×10^9 N m^2 C^{-2} and g to be 9.8 N kg^{-1}

Permeability of free space $\qquad \mu_0 = 4\pi \times 10^{-7}$ H m^{-1}

Particle accelerators are used to increase the energy of charged particles such as protons and electrons. The accelerated particles are made to collide with other particles in a 'target' in order to investigate the structure of matter. Sub-atomic particles created in these collisions can be observed and studied.

28 GeV proton synchrotron (PS)

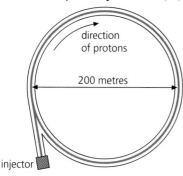

direction
of protons

200 metres

injector

Acceleration point

particle beam

vacuum
pipe

acceleration
cavity

alternating *E*-field
in this region
accelerates protons

diameter of ring	200 m
circumference of ring	628 m
number of accelerating points	14
average accelerating voltage	4 kV
At injection:	
energy of proton	50 MeV
At ejection:	
energy of proton	28 GeV
momentum of proton	1.6×10^{-17} kg m s^{-1}
speed of proton	almost 3×10^{8} m s^{-1}

Super Proton Synchrotron (SPS)	
diameter of ring	2.2 km
speed of proton	almost 3×10^8 m s^{-1}

One type of particle accelerator is the synchrotron. In this machine a magnetic field causes charged particles to travel in a circular path. The particles are accelerated by an electric field. As their momentum increases, the magnetic flux density is also increased to keep them travelling in a path of constant radius.

a Protons are injected into the 28 GeV proton synchrotron ring at CERN with an energy of 50 MeV (8×10^{-12} J).
Ignoring relativistic effects, show that:
i) the speed of a proton at injection is about 10^8 m s^{-1},
ii) a proton takes about 6 µs to travel round the ring at this speed,
iii) the momentum of a proton at injection is about 1.6×10^{-19} kg m s^{-1}.

b The accelerator ring is a vacuum pipe maintained at a very low pressure. It is 'filled' with protons by injecting a proton current of 100 mA for the 6 µs it takes for protons to make one revolution at the injection energy of 50 MeV.
i) Calculate the number of protons injected.
ii) Explain why the ring must be maintained at a very low pressure.

c Before the protons are accelerated, an electric field is used to group the protons in the ring into a number of bunches. The bunches of protons are then accelerated as they pass through each of 14 acceleration points spaced equally round the ring. An acceleration point is essentially a pair of electrodes between which an alternating voltage is applied.
i) Suggest why an alternating voltage is applied between the electrodes in order to accelerate the protons.

The proton bunches pass through the acceleration point when the voltage between its electrodes is about 4 kV.

ii) By how much does the energy of one proton increase in each revolution? (Give your answer in eV.)

The final energy of a proton is 28 GeV.

iii) Estimate the number of times a proton travels round the ring in acquiring this energy.

iv) Explain briefly why linear accelerators (as opposed to ring accelerators) are not used to accelerate protons to these energies.

d i) Show that the B-field required to maintain a proton of charge e in a circular path of radius r is proportional to the momentum p of the proton.

ii) Estimate the B-field required to maintain 50 MeV protons within the CERN proton synchrotron.

iii) Explain why the frequency of the accelerating voltage must be increased as the speed of the protons increases.

e When it reaches its maximum energy of 28 GeV, the momentum of a proton is $1.6 \times 10^{-17}\,\mathrm{kg\,m\,s^{-1}}$ and it is travelling at almost $3 \times 10^8\,\mathrm{m\,s^{-1}}$. The mass of any particle increases as its speed increases, although the effect only becomes important at speeds close to that of light.

Estimate:

i) by what factor the B-field must be increased during acceleration,

ii) by what factor the mass of the proton increases during acceleration.

Protons from this accelerator can be injected into the Super Proton Synchrotron (SPS). This machine is 2.2 km in diameter and the ultimate energy of protons from it is 400 GeV.

iii) Explain why the B-field is increased but the frequency of the accelerating voltage is kept almost constant, as protons are accelerated in the SPS.

charge on proton $1.60 \times 10^{-19}\,\mathrm{C}$
mass of proton $1.66 \times 10^{-27}\,\mathrm{kg}$
$1\,\mathrm{MeV} = 1 \times 10^6\,\mathrm{eV}$
$1\,\mathrm{GeV} = 1 \times 10^9\,\mathrm{eV}$

OCSEB, June 1990

4 a Outline the differences between fixed target and colliding beams experiments. Why is more energy available to produce new particles when particle beams are collided together?

b In a head-on collision between two protons of equal kinetic energy the following interaction was observed where π^+, π^-, K^+ and Λ are particles produced in the collision.

$$p + p \rightarrow p + 7\pi^+ + 7\pi^- + K^+ + \Lambda$$

Data: mass of $p = 938\,\mathrm{MeV}/c^2$
mass of π^+ or $\pi^- = 140\,\mathrm{MeV}/c^2$
mass of $K^+ = 494\,\mathrm{MeV}/c^2$
mass of $\Lambda = 1115\,\mathrm{MeV}/c^2$

Calculate the minimum kinetic energy in MeV of each proton for this interaction to occur.

Explain why this is the minimum possible value.

Why would this interaction not be observed if one of the protons were stationary and the other had twice your calculated minimum kinetic energy?

Part (b) only: ULEAC, June 1996 (part)

c The collision of a high energy proton with a stationary proton can produce the following interaction:

$$p + p \rightarrow p + p + p + \bar{p}$$

Mass of proton $= 1.0 \, \text{GeV}/c^2$

Kinetic energy of incident proton $= 6.0 \, \text{GeV}$

Calculate the initial total energy.

Calculate the momentum of the incident proton in this interaction given that $E^2 = m_0^2 c^4 + p^2 c^2$

Describe in detail how a beam of high energy protons is produced.

Edexcel, June 1999 (part)

d A particle called the *strange meson* K^0 can decay into two π-*mesons* (pions) π^+, π^-. If the K^0 has an initial kinetic energy of $1.00 \, \text{GeV}$, calculate the total energy of each pion produced.

Use your answer to calculate the momentum of each pion.

Useful data: $E^2 = m_0^2 c^4 + p^2 c^2$

Mass $K^0 = 0.498 \, \text{GeV}/c^2$

Mass $\pi^+ = \text{mass } \pi^- = 0.140 \, \text{GeV}/c^2$

Edexcel, June 2000 (part)

5 a When a charged particle passes through a bubble chamber containing liquid hydrogen, it produces hydrogen ions along its track, giving up its energy as it does so. If a proton of energy $1000 \, \text{MeV}$ passes through a bubble chamber $2 \, \text{m}$ long, estimate how many hydrogen ions are produced given that the energy loss of the proton is $0.4 \, \text{MeV} \, \text{cm}^{-1}$.

b In a plastic scintillator, a photon of light is produced for every $100 \, \text{eV}$ of energy lost by a charged particle passing through it. The scintillator has a light collection efficiency of 10% and is coupled to a photomultiplier tube of detection efficiency 25%. Protons of energy $1 \, \text{GeV}$ from the Tevatron pass through a plastic scintillator material $1 \, \text{cm}$ thick whose energy loss is $1 \, \text{MeV} \, \text{cm}^{-1}$.

How many photons are

i) produced inside the scintillator?

ii) detected by the photomultiplier tube?

Elementary particles

'If I could remember the names of all these particles, I'd be a botanist.'

Enrico Fermi (1901–1954)

'Who ordered that?'

Isidor Rabi (1898–1988), eminent American physicist who, in a fit of exasperation,
wondered how the muon fitted into nature

What are 'elementary particles'?

In Chapter 2, we discussed the concept of the atom, and showed that the atom itself is not a fundamental unit of matter but contains smaller particles such as the proton, neutron and electron. However, as we will see in Chapter 7, the proton and neutron are in fact composed of smaller entities called 'quarks', whereas the electron does not appear to have any deeper internal structure at all.

Particle physicists define particles of matter that do not seem to be composed of any smaller particles as **elementary**. The term **fundamental** is also used to describe them, with the words 'fundamental' and 'elementary' being used interchangeably in this context. So, for example, in the atom, the proton and neutron are not fundamental particles whereas the electron is. Nuclear physicists, on the other hand, who only describe matter at the nucleus level, would regard *their* 'fundamental particles' to be essentially protons and neutrons. However, from now on in this book, we will stick with the term 'elementary particles' to describe particles of matter that do not appear to have any structure other than their own, and do not appear to be composed of any smaller particles.

The particle 'zoo'

In Chapter 2 we looked at the discovery of the electron and the development of models of the atom. In Chapter 3 we saw that a new description of nature, the quantum theory, together with Special Relativity, was needed to account for new and unfamiliar experimental discoveries. In the early 1930s most physicists thought that the fundamental nature of matter was close to being understood. The atom was composed of just three particles, namely, proton, electron and neutron. The neutrino, although not observed, was postulated in the theory of beta decay (see below); and it looked like quantum mechanics could account for the structure of the nucleus. To physicists, then, it looked like a complete theory for the fundamental nature of matter was finally coming together – or was it?

From the mid-1930s onwards, physicists began to see that there existed many more particles than just the electron, proton and neutron, and that our understanding of the fundamental nature of matter was far from complete. The earliest discoveries in particle physics were made from studies of **cosmic rays**. These are highly energetic particles that originate from outer space and interact with atoms in the Earth's atmosphere. In recent years, using particle accelerators, particle physicists have found hundreds of different particles. Almost all of these particles are **unstable**. They decay spontaneously into other particles, which themselves may be unstable. Examples of stable particles are those of everyday matter such as the proton and electron, and also particles such as the photon and neutrino; but, as we shall see, even the proton might be unstable over very long time scales.

'Long-lived' has a very different meaning in particle physics compared with our everyday experience. We might think that someone who receives a telegram from the Queen on their 100th birthday is long-lived. However, in the particle world, some particles have lifetimes as short as 10^{-20} seconds, and if a particle lives for 10^{-10} seconds it is considered very old! If a particle lives long enough so that particle physicists can accelerate it or perform measurements on it, then it is regarded as stable. In this chapter, we will look at the many different particles that have been found, and see how it is possible to classify them in a logical way.

The quantum theory of fields

We saw earlier that objects can be regarded as waves or particles, and that this wave–particle duality can be applied in a way that depends on how the physical phenomena are observed. However, modern quantum theory introduces the idea that all particles can be thought of as **fields** spreading through space, which, as we saw in Chapter 1, is a concept that was anticipated by Michael Faraday in his investigations into electromagnetism.

If you are familiar with the concept of an electron as a wave extending through space (as we saw in Chapter 3), then it's not too hard to imagine an **electron field** doing the same thing; in a similar way, a photon is thought of as a materialisation of an electromagnetic field. Protons and neutrons are manifestations of **proton fields** and **neutron fields**. The individual wavefunctions of these particles localise the particle fields to a region of space depending on how they interact with their neighbours, and this is what we observe as a 'particle'. Conceptually, you need simply to think of a field as 'action at a distance' without worrying too much at this stage about what the field is 'made of'. We will see how this quantum field idea works and how it enables us to describe interactions between particles in Chapter 6. For the particle physicist, the discovery of new particles was an essential first step in constructing a **quantum field theory** that can describe how matter and energy interact with each other.

Nuclear beta decay

A useful place to start our investigation of elementary particles is with a form of radioactivity, **nuclear beta decay**. In beta decay, a neutron inside the nucleus decays into a proton, emitting a beta particle (electron) by the process

$$n \rightarrow p + e^-$$

The emitted electron is *not* one of the atom's orbital electrons. Also it is *not* an electron that may already exist inside the nucleus, as the Heisenberg Uncertainty Principle does not allow electrons with the energies observed in beta decay to exist there. The electron is *created* inside the nucleus from the rest-mass energy differences between the proton and the neutron. Physicists soon became aware that there was something rather odd about the nature of such beta decay. We shall now look at those peculiarities.

Spin

The first of these was to do with **spin**. This is a property of quantum particles that is related to the concept of rotation in classical mechanics. Objects that rotate have **angular momentum** (Box 5.1) and angular momentum is a conserved quantity. The concept of 'spin' was first introduced by two Dutch physicists Samuel A. Goudsmit (1902–1978) and George E. Uhlenbeck (1900–1988) in an attempt to explain some anomalies in atomic spectra. It was found that the Bohr model could not explain the 'splitting' of spectral lines which was observed in some atomic spectra. Instead of a single line, sometimes two lines were shown very close together as if a single line had been split down the middle, known as the **Zeeman effect**.

In order to explain this 'fine structure' in the spectral lines, Goudsmit and Uhlenbeck put forward the idea that the electron spins on its axis in its motion round the nucleus just as the Earth spins around its north–south axis as it orbits the Sun. The electron in its motion around the nucleus is a moving charged particle and therefore generates its own magnetic field, so the atom therefore behaves like a tiny magnet. The electron spinning on its axis is also an electric charge in motion, so sets up its own magnetic field, which physicists refer to as the **magnetic moment** of the electron. Depending on the direction in which the electron is spinning, the electron magnetic moment can either add to or subtract from the total magnetic field of the atom. The effect of this is to alter slightly the energy of the electron orbit for different spin directions, causing a splitting of the spectral lines associated with a particular Bohr orbit.

However, the observation that the lines split into just two components and not a continuous range suggests that some quantum principle is at work. Quantum theory tells us that the electron's spin is quantised and is measured in half-integer units of $\hbar$ ($\hbar = h/2\pi$, see Box 5.1), so that $\text{spin}_{\text{electron}} = \pm\hbar/2$. Depending on the direction in which it is spinning, an electron can have two values of spin:

$$\text{spin}_{\text{electron}} = +\hbar/2 \qquad \text{(spin up)}$$

or

$$\text{spin}_{\text{electron}} = -\hbar/2 \qquad \text{(spin down)}$$

It turns out that all particles have spin. Protons and neutrons have half-integer multiples of $\hbar$ just like the electron, whereas particles such as photons have integer or zero units of $\hbar$. Particle physicists tend to take the $\hbar$ as read, and refer to electrons as 'spin-$\frac{1}{2}$' particles and to photons as 'spin-0' particles. We will see later that spin plays an important role in classifying particles.

Box 5.1 Angular momentum and spin

The *linear momentum* of a mass m travelling with a uniform velocity v is given by $m \times v$ and is measured in kg m s^{-1}. The total linear momentum of a body is *conserved*, that is, its total momentum remains constant. If a point mass m is rotating about an axis at a distance r from the axis, we can define a vector quantity called the *angular momentum L*, which is given by $L = m \times v \times r$, the moment of momentum, where $m \times v$ is the tangential momentum at any point (Figure 5.1). For a rotating extended body, the angular momentum can be thought of as the sum of the elemental angular momenta of each part of it about the rotation axis.

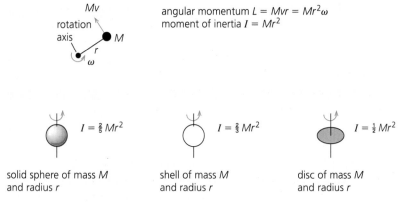

Figure 5.1 Moments of inertia of spinning nuclei can give us information about how the individual nucleons are distributed in an atomic nucleus

Using angular measure we can write L as $L = m \times (r \times \omega) \times r$ [since $v = r \times \omega] = \omega \times mr^2$, where the quantity mr^2 is known as the **moment of inertia** I and is a measure of the distribution of mass about the axis of rotation. Just as mass is a measure of a body's in-built opposition to linear motion, that is, its inertia, then I is the corresponding quantity for rotational motion. We can thus write angular momentum as

$$L = I\omega$$

where ω is the angular velocity measured in radians per second (rad s^{-1}). The expression for I depends on the shape of the body (Table 5.1) and on where the axis of rotation is taken, and is measured in kg m^2. Note that L has the units of $\text{kg m}^2 \text{s}^{-1} = \text{N m s}$.

Table 5.1 Values for I for some simple shapes

Shape	Moment of inertia I
Small mass M in a circular orbit of radius r	Mr^2
Solid sphere of mass M and radius r rotating about its diameter	$\frac{2}{5}Mr^2$
Thin spherical shell of mass M and radius r rotating about its diameter	$\frac{2}{3}Mr^2$
Uniform disc of mass M and radius r rotating about a perpendicular axis through its centre	$\frac{1}{2}Mr^2$

Like linear momentum, angular momentum is also conserved, and the **Law of Conservation of Angular Momentum** states that the total angular momentum of a rotating body remains the same. This law may be seen in action when watching a spinning ice-skater. As the skater pulls her arms towards her body, she reduces her moment of inertia and starts spinning much faster. This happens because her angular momentum L is constant, so, as the effective radius of her arms becomes less, a reduction of I from I_1 to I_2 causes ω to increase from ω_1 to ω_2 where

$$I_1\omega_1 = I_2\omega_2$$

so as she folds them in, her tangential velocity has to increase to keep L the same.

The *total* angular momentum of a body may be made up by

1 its spin angular momentum about an axis through its centre of mass, and
2 its orbital angular momentum about some external axis.

Sub-atomic particles such as the electron can be considered as 'spinning' and so have both spin angular momentum and orbital angular momentum (Figure 5.2). These ideas are important in atomic physics as well as particle physics. However, there are important differences between classical and quantum mechanics. Whereas in the classical mechanics of rotating bodies angular momentum can take a continuous range of values, quantum mechanics tell us that the angular momentum or **spin** of particles is *quantised*. Quantum mechanical spin is measured in units of $\hbar$, which is the Planck constant h divided by 2π:

$$\hbar = \frac{h}{2\pi} = 1.05 \times 10^{-34}\,\text{J s}$$

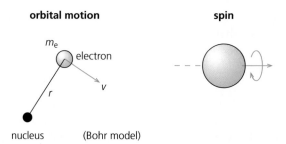

orbital motion   **spin**

m_e

electron

r v

nucleus   (Bohr model)

Figure 5.2 Quantum particles can be considered to be 'spinning'

In the Bohr model, an electron in an atom would have an orbital angular momentum (around the nucleus) given by $m_e rv$. The quantum description of the angular momentum of an electron combines both the orbital motion of the electron around the nucleus and the spin of the electron. The spin is quantised and is measured in units of $\hbar$

Quantum spin has another strange property, which is shown by the property of *fermions* or 'spin-$\frac{1}{2}$' particles. Normally if an object rotates through 360° it is back where it started. If, for example, you rotate through 360° then your view of the universe is the same as it was before. However, if a fermion rotates through 360°, it enters a different quantum state than the one it started out in. The fermion has to rotate through another 360° (making 720° in all) in order to get back to where it started, that is, its original quantum state. It's almost as if the fermion views the universe differently during the first rotation and then the same as before in the second. To the fermion there is a difference in what would be, to us, identical views of the universe that we would see if we rotated twice.

In addition, the *orientation* of a spinning quantum particle is also quantised, which is why we speak of 'spin-up' or 'spin-down' quantum states.

In beta decay, there is an apparent violation of the law of conservation of angular momentum. The total spin angular momentum of the initial particle before the decay occurs *must* equal the total angular momentum of all the particles produced afterwards. This is not satisfied by the nuclear reaction $n \rightarrow p + e^{-}$.

The beta particle energy spectrum

Another serious difficulty is that the beta particles emitted have been shown to have a continuous range of energies up to some maximum value E_{max}. Figure 5.3 shows a typical beta particle energy **spectrum**, and some end-point energies E_{max} for beta particles from various isotopes.

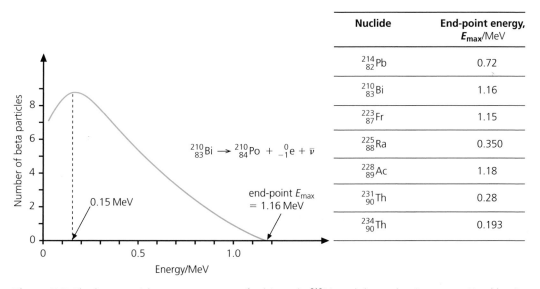

Nuclide	End-point energy, E_{max}/MeV
$^{214}_{82}$Pb	0.72
$^{210}_{83}$Bi	1.16
$^{223}_{87}$Fr	1.15
$^{225}_{88}$Ra	0.350
$^{228}_{89}$Ac	1.18
$^{231}_{90}$Th	0.28
$^{234}_{90}$Th	0.193

Figure 5.3 The beta particle energy spectrum for bismuth, $^{210}_{83}$Bi, and the end-point energy E_{max} (that is, the maximum kinetic energy) for beta particles emitted from various isotopes

In the neutron decay $n \rightarrow p + e^-$, the Q-value of the reaction works out as 0.782 MeV. Except for the very small recoil energy of the proton, *all* of the electrons should have this energy, but we find that they all have *less than* this value in a continuous fashion from 0 eV right up to this maximum energy. This is indeed very extraordinary. There are no alpha particles or gamma rays emitted along with the electron, and it is difficult to see how an electron could be emitted from the nucleus with an energy less than the maximum possible one. It appeared that beta decay violated the principle of conservation of energy!

An explanation was found by the Austrian physicist Wolfgang Pauli (see page 167). Pauli suggested that a *second* particle, with zero rest mass, was emitted during beta decay. The variability of the energy of the beta particles is accounted for by this extra particle ejected by the nucleus, which can have a range of energies corresponding to the beta particle energy spectrum. Pauli called this particle a **neutrino** (meaning 'little neutral one') and it is this particle that carries off the 'missing energy' from the beta emission spectrum. Sometimes the electron gets nearly all the energy, and sometimes the neutrino does. However, for all cases, the sum of the electron's energy and the neutrino's energy gives a constant value Q. As the conservation of electric charge is not violated during beta decay, this new particle must also have zero charge. In addition, in order to satisfy conservation of angular momentum, it must have spin $\frac{1}{2}$.

At the time when this explanation was put forward by Pauli in 1931, there was no experimental evidence for the existence of this extra particle in beta decay. Pauli's idea was an act of faith based entirely on his belief in the integrity of the conservation laws.

The symbol for the neutrino is the Greek letter nu (ν), and so the beta decay process is now written as

$$n \rightarrow p + e^- + \bar{\nu}$$

Particle physicists now know that the neutral particle emitted along with the electron in beta decay is the neutrino's *antiparticle* counterpart, the **antineutrino**, which is why the symbol $\bar{\nu}$ is used instead of ν (antiparticles and their discovery are discussed later in this chapter).

Since it appears to have no internal structure, the neutrino is an elementary particle. The prediction of the existence of the neutrino by reference to conservation laws is a good example of how particle physics works. The neutrino was first observed in 1956 in experiments carried out alongside a nuclear reactor at Savannah River in the USA. This was done by noting that, in a nuclear reactor, the products of nuclear fission are radioactive and decay by beta emission. Since the activity inside a reactor is very high, it is reasonable to expect a large flux of neutrinos. Neutrinos interact very weakly with matter and only through the weak interaction (see Chapter 6), so by placing a scintillator near the reactor it was possible to detect a neutrino indirectly by making it initiate a reaction in the scintillator fluid and detecting the reaction products from the flashes of light produced.

Cosmic rays

Figure 5.4 A gold leaf electroscope (c. 1900). When charged by ionisation the gold leaves move apart

Cosmic rays were first discovered by the Austrian physicist Victor Franz Hess (1883–1964) in 1912. At this time gold leaf electroscopes (Figure 5.4) were used to study the intensity of ionising radiations by observing the deflection of the gold leaves. It was found that gold leaf electroscopes always exhibited a *natural leak* or residual ionisation, which was initially thought to come from radiation emitted by radioactive elements in the Earth's crust. To test this assumption, Hess made ascents in balloons carrying gold leaf electroscopes to high altitudes to see if this effect decreased. To his surprise he found that it *increased* with altitude, causing him to conclude that the natural leak was due to radiation from space. Modern work using rockets and satellites shows that this cosmic radiation increases with height up to about 22 km and then drops slowly to heights of 40 km or so, after which it remains approximately constant.

Primary cosmic rays are highly energetic atomic nuclei that are thought to originate from stars and supernovas in intergalactic space. The nuclei consist mostly of protons, with about 10% being alpha particles and less than 1% being nuclei of atoms heavier than helium. Since nearly all these particles undergo nuclear reactions with atoms in the upper atmosphere, what we observe at ground level are **secondary cosmic rays**. These are the reaction products and consist of a 'soft' component made up of electrons and gamma rays and a 'hard' component comprising muons, which can penetrate large thicknesses of lead and rock. Cosmic rays sometimes appear in the form of **showers** (Figure 5.5) in which significant increases in counts are recorded over large areas of the Earth's surface.

One way of seeing cosmic ray tracks is by exposing them to nuclear emulsions. Figure 5.6 shows one such track. An incoming cosmic ray collides with a nucleus in the emulsion and produces a 'star' of different particles. Cosmic rays have extremely high energies. Some events occurring in air showers indicate that the energy of some cosmic rays is in excess of 10^{15} eV (1000 TeV). This is very large compared with the highest energies of 1 TeV currently achievable in today's particle accelerators.

Figure 5.5 A schematic cosmic ray shower. Some particles such as muons and neutrinos can penetrate deep into the Earth and be detected 600 m underground by a neutrino detector, such as the IMB. The IMB was designed to look for proton decay (see Chapter 8), but is now used to study neutrinos. It is named after the three US research institutions that operate it (Irvine, Michigan, Brookhaven)

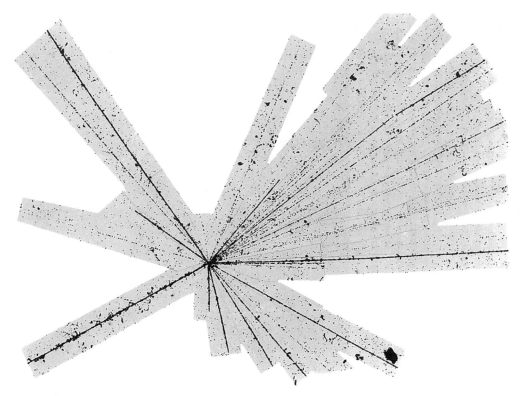

Figure 5.6 A cosmic ray consisting of a magnesium nucleus collides with a bromine nucleus in photographic emulsion, producing a 'star' of various particles. The track of the incoming magnesium nucleus is the thick line top left and is about 0.18 mm long

Antimatter and the discovery of the positron

We have seen that, in 1897, J. J. Thomson discovered the particles now called electrons (Chapter 2) and, in 1924, de Broglie showed that electrons could also

Figure 5.7 Paul Adrien Maurice Dirac (1902–1984)

be considered as waves (Chapter 3). A British physicist, Paul Adrien Maurice Dirac (1902–1984), sought to combine the Special Theory of Relativity with quantum theory to give a complete description of the electron (Figure 5.7). Previously, these two theories were considered quite distinct and separate.

In particular, using Einstein's proposition that matter and energy were equivalent ($E = mc^2$), Dirac showed that quantum behaviour involves the creation and destruction of particles, with matter being converted into energy and vice versa. Dirac's theoretical investigations contained some startling ideas. They predicted the existence of **antimatter**. This is matter that is identical to ordinary matter except that it has the opposite set of quantum properties (such as electric

153

charge) to the matter we know in the everyday world. An electron and an anti-electron, for example, have the same mass but opposite electric charge. When they interact, they produce total **annihilation**, in which both the particles disappear and are converted into pure energy as a pair of energetic photons.

The existence of antimatter was dramatically confirmed in 1932 by the American physicist Carl David Anderson (1905–1991). Anderson was observing the tracks of cosmic ray particles in a cloud chamber which was operated between the poles of a large electromagnet. He noticed that some tracks were curved in the magnetic field in the way that would be expected from the track of an electron, while other tracks produced the same degree of ionisation but curved in the *opposite* direction. Since the only sub-atomic particles known at the time were the proton and the electron, Anderson concluded that the only way to account for this oppositely curved track was the presence of a particle with the same mass as an electron but with *opposite charge*.

Figure 5.8 This picture from Carl Anderson's cloud chamber was the first evidence for the positron – a particle with the same mass as the electron but with positive charge

Figure 5.8 shows the photograph that Anderson obtained from his cloud chamber. Going across the diameter of the picture is a lead plate 6 mm thick. The particle is moving from the bottom of the picture towards the top. Notice that the trajectory is less curved below than it is above the plate. This is because the particle loses energy as it passes through the lead and slows down. The magnetic field B acting perpendicular to its path produces a curvature r, which can then be measured. Using the relationship

$$\frac{mv^2}{r} = Bqv \quad \text{or} \quad \frac{mv}{q} = Br$$

values of mv/q can be determined, from which the charge-to-mass ratio q/m can be shown to have the same value as that of the electron. This evidence provided convincing proof of a positive electron or **positron**, and Anderson was awarded the Nobel Prize for Physics in 1936.

Physicists gradually came to see that *every* particle should have its corresponding **antiparticle**. As the electron has the anti-electron, so the proton has the antiproton. In general, particle physicists represent antiparticles by putting a bar over the symbol for the particle. Thus if the proton has the symbol p, the antiproton has the symbol $\bar{p}$ (pronounced 'p bar'), the antineutron $\bar{n}$, the antineutrino $\bar{v}$, and so on. In Chapter 2 we saw that electrons are beta particles. When writing down nuclear reactions or radioactive decays, an electron beta particle is commonly given the symbol $_{-1}^{0}\beta$ and a positron beta particle $_{+1}^{0}\beta$. In particle physics, the symbol e^- is used for the electron and e^+ for the positron, and likewise for muons (μ^-, μ^+).

Particle physicists now know that there are antiparticles for all the matter particles, and these can be produced routinely in accelerator experiments. It also seems that the laws of physics work nearly in the same way for antimatter as they do for matter, but this needs to be confirmed (see Box 5.2).

Box 5.2 First atoms of antihydrogen produced at CERN

Paul Dirac predicted the existence of antimatter, which was confirmed with the discovery of the **positron** by Carl Anderson in 1932, but what about anti-atoms? Ordinary atoms consist of electrons in orbit around an atomic nucleus. The simplest atom is that of hydrogen, which consists of a single proton and a single electron. An **antihydrogen** atom would consist of an antiproton with a positron in orbit around it, but these do not exist naturally (Figure 5.9).

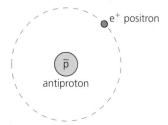

Figure 5.9 An antihydrogen atom

In 1995, at CERN, an accelerator called LEAR (Low Energy Antiproton Ring) succeeded in producing the first atoms of antihydrogen. Figure 5.10 overleaf shows how this was done.

LEAR is a synchrotron that can store and circulate antiprotons to energies of about 0.1 GeV. The antiprotons pass through a jet of xenon gas each time they go round – about three million times each second. Every so often an antiproton interacts with the electromagnetic field of a xenon nucleus and emits a photon. This photon then decays into an electron–positron pair, where, very rarely, the positron's velocity is sufficiently close to the velocity of the antiproton to form an atom of antihydrogen. Each anti-atom lives for about 40 billionths of a second before annihilating with ordinary matter, sending out a characteristic signal that antihydrogen has been created.

Hydrogen is abundant in the universe and much of what we have learnt about atoms has developed from studying this simplest atom. If the behaviour of an antihydrogen atom was in any way different from that of 'normal' hydrogen, then particle physicists would have to rethink their ideas about the symmetry between matter and antimatter. In order to do this, physicists at CERN will need to 'trap' antihydrogen and hold it for prolonged periods of time so that its properties can be measured and compared with those of ordinary hydrogen. This may be possible using special electrical or magnetic 'bottles'. Experiments are also being planned at CERN in an attempt to verify the symmetry between hydrogen and antihydrogen.

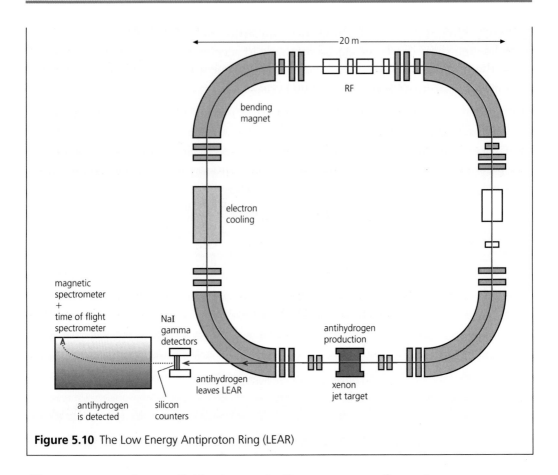

Figure 5.10 The Low Energy Antiproton Ring (LEAR)

Creation and annihilation of electrons and positrons

One feature of Anderson's discoveries was the observation that electron and positron tracks were often observed in **pairs**. Another question raised was what happened if an electron and positron combined? This can be explained if we consider that electron–positron pairs can be created from gamma rays of energy above a certain value. This **pair production** is perhaps the most striking example of mass–energy equivalence.

We can write pair production symbolically as:

$$\text{photon} + \text{photon} \rightarrow \text{particle} + \text{antiparticle}$$

The reverse process is also possible. A particle and antiparticle can collide and annihilate each other, producing two high-energy gamma ray photons:

$$\text{particle} + \text{antiparticle} \rightarrow \text{photon} + \text{photon}$$

For pair production to occur, two conditions must be satisfied:

1 Pair production must obey the laws of conservation of energy and momentum.
2 The energy of the photons must be greater than the mass–energy of the created particles.

WORKED EXAMPLE 5.1

The annihilation of an electron with an anti-electron may be represented by

$$e^- + e^+ \rightarrow 2\gamma$$

(that is, two gamma rays are produced). What is the energy of each gamma ray, and in what directions do they move?

[Mass of e^- = mass of e^+ = 9.1×10^{-31} kg or $0.511\,\text{MeV}/c^2$]

The sum of the rest energies of the e^- and e^+ (which have the same mass) is

$$m_{e^-}c^2 + m_{e^+}c^2 = 9.1 \times 10^{-31} \times (3 \times 10^8)^2 + 9.1 \times 10^{-31} \times (3 \times 10^8)^2$$
$$= 1.64 \times 10^{-13}\,\text{J} \quad \text{or} \quad 1.022\,\text{MeV}$$

Since energy is conserved, each gamma ray has an energy of

$$(1.022\,\text{MeV})/2 = 0.511\,\text{MeV}$$

The photons move in opposite directions in order to conserve momentum.

This example also illustrates the usefulness of expressing particle masses in units of MeV/c^2 (or GeV/c^2). Substituting for m_{e^-} and m_{e^+} the total rest energies, we get

$$\left(0.511\,\frac{\text{MeV}}{c^2}\right)c^2 + \left(0.511\,\frac{\text{MeV}}{c^2}\right)c^2 = 1.022\,\text{MeV}$$

The factor c^2 cancels out and does not need to be calculated explicitly if we use these units.

Note that the *creation* of an electron–positron pair from the energy of a gamma ray photon

$$\gamma \rightarrow e^- + e^+$$

is not always possible. Although, on energetic grounds, a gamma ray of $0.511\,\text{MeV}$ would be sufficient to provide the rest-mass energy of one electron (or positron), pair production requires at least *twice* this amount. In addition, the production of a single electron (or positron) would violate the law of conservation of charge.

It is obvious to us that we seem to live in a universe that is dominated by matter. While we can make antimatter in particle accelerators and observe it in cosmic rays, we do not observe the universe annihilating itself. It is possible that there might exist whole galaxies composed of antimatter, but fortunately they don't seem to come anywhere near us, and astronomers do not see galaxies that are colliding with each other destroying themselves in a flash of radiation.

This lack of symmetry that favours matter over antimatter in the observable universe is disturbing to particle physicists, who generally expect things to be equal and opposite! This question is addressed again in Chapter 8, where we consider what particle physics can tell us about the early universe.

Quantum electrodynamics (QED) and virtual photons

Dirac was interested in applying quantum mechanics to the classical dynamics of Maxwell's electromagnetic fields. By doing this, Dirac was able to show how photons and electrons interact with each other. This union of relativity, electromagnetism and quantum mechanics is called **quantum electrodynamics (QED)**.

QED introduced an even more surprising concept. The electromagnetic force is itself made up of particles. If, for example, two electrons interact via Coulomb's Law, then according to QED the electrons exchange a photon that 'carries' the electromagnetic force from one electron to the other. A photon is emitted by one electron and absorbed by the other. These photons are examples of **virtual particles** and are not like ordinary observable photons since they are not limited by energy or momentum conservation. To see why, recall the Heisenberg Uncertainty Principle that we discussed in Chapter 3. The principle can be written in terms of energy and time:

$$\Delta E \times \Delta t \geqslant \hbar$$

This mathematical statement is saying that the uncertainty in the measurement of the energy of a particle multiplied by the uncertainty of the time in which it can be measured is equal to the Planck constant divided by 2π. So over a very short time period, the uncertainty in the energy ΔE can be very large.

In particle physics, quantum particles can be thought of as being surrounded by clouds of virtual particles, which are emitted and usually reabsorbed by the parent particles (Figure 5.11a). These virtual particles have lifetimes that depend on the Uncertainty Principle. The more massive and energetic a particle, then the briefer its lifetime and the smaller its range. In fact, the lifetime of a virtual particle is related inversely to its mass. A key idea in understanding interactions between quantum particles is that particles are said to *interact* when one of them comes close enough for one or more virtual particles in its cloud to be absorbed by the other (Figure 5.11b).

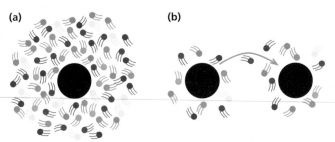

(a)　　　　(b)

Figure 5.11(a) Any quantum particle can be thought of as being surrounded by a cloud of virtual particles, which are emitted and reabsorbed by the parent particle
(b) Two quantum particles interact when they come close enough together for one or more of the virtual particles in one cloud to be absorbed by the other cloud

In QED the electromagnetic interaction between two charged particles (such as two electrons) is understood to be due to the exchange of **virtual photons**. A virtual photon is emitted by one electron and absorbed by the other. In this situation the virtual photon exists only as an unobservable intermediate state, and for this reason particle physicists call them 'virtual' because they are hardly there at all.

However, at this point you may think that, because everyday photons have zero rest mass, a virtual photon should therefore have an infinite lifetime, and thus an infinite range. So how can a virtual photon have a finite lifetime and finite range?

The answer is that the Uncertainty Principle permits the virtual photon to have an 'imaginary' mass of just the right value so that it exists long enough to mediate the electromagnetic interaction. This means that an electron can emit an energetic photon over very small time scales provided that energy and momentum are conserved in the interaction overall. So, although we do not see them, to work out how two electrons interact with each other we must add up the number of possible ways for the virtual photon to travel from one electron to the other, combine them, and find the highest probability that gives its path in space–time.

This idea that force between particles can be mediated by a carrier particle is now a very important concept in particle physics, but in the early days QED had serious theoretical difficulties. Heisenberg and Pauli noticed that some calculations that involved the interaction of the electron with itself yielded nonsensical results. This concept, called the **self-energy** of a particle, arises due to (in the case of an electron) the fact that the electron has an electric charge and therefore has an electric field associated with it. The strength of the electric field is inversely proportional to the square of the distance from the source, which is the electron, and because the electron is at zero distance from itself, then the strength of the interaction should be infinite. This difficulty gave rise to infinite values of energy and mass in the quantum calculations, making them meaningless. It can be shown that the same kind of problems arise with other interactions such as gravity and the strong force (see Chapter 6). As a result, the full development of QED was held back until a way was found to deal with these infinities some 20 years later by using a mathematical technique called **renormalisation**. We will return to QED again in Chapter 6 when we look at how particles interact with each other, and see how this idea of force-carrying particles can be extended to other interactions.

Mesons

π mesons (or pions)

In 1935, the Japanese particle physicist Hideki Yukawa (1907–1981) (Figure 5.12) was interested in the nature of the strong nuclear force that binds protons and neutrons together in the nucleus. Yukawa was familiar with Dirac's theory of QED and wanted to know exactly what 'glued' the nucleons together. With Dirac's success in explaining the electromagnetic force as an exchange of virtual photons, Yukawa wondered whether a similar mechanism might be at work inside the nucleus that could explain how the strong nuclear force worked. Using quantum field theory, he considered that neutrons and protons could be regarded as fields in which virtual particles were exchanged between them. It was this flow of particles that constituted the strong force.

The question was: What kind of particle was it? Since the strong nuclear force operates over a very short range (about the diameter of a proton), the lifetime of these particles had to be extremely short or otherwise they would extend beyond the nuclear radius, and the strong force would operate outside the boundaries of the nucleus. Yukawa calculated that the mass of these particles must be about 200 times that of the electron and one-tenth of that of the proton. In other words they were between the electron and proton in mass. These particles were called **mesons** from the Greek word for 'middle'. Yukawa postulated that the strong nuclear force was carried by a virtual particle now called the **π meson** or **pion**. The π meson came in three varieties in order to accommodate the possible interactions between nucleons during a nuclear reaction. Yukawa showed that a proton could change into a neutron by emitting a π meson with a positive charge or, equivalently, by absorbing a negatively charged π meson. Alternatively, the interaction between nucleons could leave them unchanged, which could only be explained by the existence of an uncharged π meson. The three charge states of the π meson were therefore positive (π^+), negative (π^-) and neutral (π^0).

False hopes that the π meson had been found experimentally were raised by Anderson (see page 162). However, Yukawa's meson *was* finally found in 1947 by Cecil Frank Powell (1903–1969) and Giuseppe P. S. Occ hialini (1907–1993) from the University of

Figure 5.12 Hideki Yukawa (1907–1981)

Bristol, who exposed nuclear emulsions to cosmic rays at high altitudes (Figure 5.13). The charged π meson decayed into a muon,

$$\pi^+ \to \mu^+ + \bar{\nu}$$
$$\pi^- \to \mu^- + \nu$$

and the muon in turn decayed into an electron by the reactions shown on page 162.

The π^+ and π^- were the first to be discovered, with a mass of $0.150\,u$ ($140\,MeV/c^2$). The π^0 was discovered later in accelerator experiments. It has a smaller mass of $0.145\,u$ ($135\,MeV/c^2$) and decays into two gamma photons.

$$\pi^0 \to 2\gamma$$

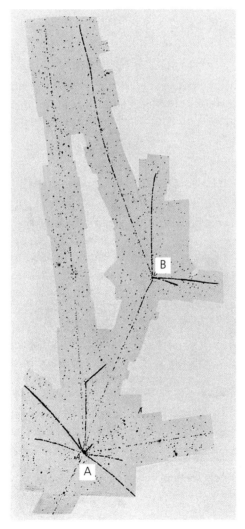

Figure 5.13 The creation of a π meson (at point A) and its subsequent decay (at point B) are recorded on photographic emulsion in a cosmic ray event. The distance between A and B is about 0.11 mm

Muons

Anderson continued to look for new particles in his cloud chamber. In 1937, five years after he discovered the positron, Anderson found another particle created in cosmic rays. This particle had both positive and negative charge states and a mass 200 times that of the electron. Anderson thought at first that he had discovered Yukawa's π meson. However, this particle did not seem to interact with atomic nuclei as Yukawa's π meson was expected to do. Also, the short lifetime of the π meson meant that it was not expected to live long enough to be detected at ground level. In addition, Anderson could find no trace of any neutral π meson. Particle physicists eventually came to the conclusion that this particle was not the π meson, and called it the **mu meson** (symbol μ) or **muon**. The muon has a lifetime of 2×10^{-6} s and decays into an electron or a positron via the reactions:

$$\mu^+ \rightarrow e^+ + \nu + \bar{\nu} \quad \text{(positive muon)}$$

or

$$\mu^- \rightarrow e^- + \nu + \bar{\nu} \quad \text{(negative muon)}$$

The muon is not found in everyday matter, and particle physicists are still not exactly sure how it fits into the scheme of things. It is not a force-carrying particle and is **no longer regarded as a meson**. Muons can provide a test for Special Relativity (see Box 3.1, page 61).

The discovery of more particles

After the Second World War many more elementary particles were discovered, initially in cosmic ray interactions and then with accelerators, which, as well as increasing in energy, were also being complemented by new and more sensitive detection methods. These particles both confirmed existing theories of particle physics, and inspired new ideas about the workings of matter.

In 1947, particles about 1000 times more massive than the electron were discovered in cloud chamber photographs of cosmic rays. These particles were initially called **V particles** because of their characteristic V-shaped tracks (Figure 5.14), and it was found that V particles came in two types. Some of these particles had decay products that always included a proton, and were called **hyperons** or 'heavy particles'. The others always had decay products that consisted of mesons, and were called **K mesons** or **kaons**.

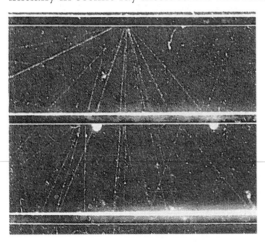

Figure 5.14 'V' particles. A cosmic ray (top) interacts in a 0.6 cm lead plate inside a cloud chamber. A number of particles are produced, including a neutral one that reveals itself by decaying into two charged particles in a 'V' formation

K mesons (or kaons)

The K meson was the first of many particles discovered in cosmic ray tracks that were later termed **strange**. Since it is cosmic rays that are colliding with atomic nuclei, K mesons are produced via the strong nuclear force, but the strange thing about them was that they seemed to live too long. Together with another particle, the sigma (Σ) (see page 165), they always seemed to be produced in pairs. It seemed impossible to produce only one of them at a time. For example, if a beam of energetic π mesons interacts with protons in a bubble chamber, say, then the reaction

$$\pi^+ + p \rightarrow K^+ + \Sigma^+$$

is frequently seen (where K^+ and Σ^+ are positively charged K meson and sigma particles), but the reaction

$$\pi^+ + p \rightarrow \pi^+ + \Sigma^+$$

which in the early days of particle physics did not violate any known conservation law, was never seen.

There was something else that was strange as well. Both these particles were fairly massive but decayed over comparatively long lifetimes. Particles produced via the strong interaction must have short lifetimes ($\leq 10^{-20}$ s), or else their influence could be felt beyond the nucleus; yet K mesons have average lifetimes of 10^{-10} s, which is exceptionally long for the decay mechanism to be attributed to it.

There are two charged K mesons, K^+ and K^-, and also two neutral ones, K^0, which have different lifetimes. K mesons decay in a variety of ways. There is a 63% chance that a charged K meson will decay into a muon and a neutrino, and a 21% chance of decay into a charged π meson together with a neutral π meson. Neutral K mesons decay most frequently into a positive π meson and negative π meson. Figure 5.15 shows a typical K meson decay.

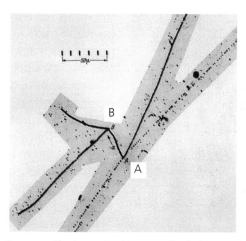

Figure 5.15 The decay of a K meson. The incoming K meson (top) decays at A into three π mesons. One of these moves slowly, leaving a thick track, and decays at B. The other two π mesons go off in opposite directions at high speed, leaving thinner tracks. The distance between A and B is 25×10^{-6} m (25 μm)

The lambda

In 1951, another particle was observed in cloud chamber experiments. There were V tracks that showed a neutral K meson and another particle some 2250 times more massive than the electron or 20% *more* massive than the proton. This particle was heavy and, because the track produced by its decay products resembled the Greek letter Λ, particle physicists called it the **lambda** (Λ). It exists only as a neutral particle Λ^0, often called lambda-zero, and is a hyperon. Like the K meson, the lambda lived for 10^{-10} s, which again is unexpectedly long for a particle produced via the strong interaction. The decay process for the Λ^0 is

$$\Lambda^0 \rightarrow p + \pi^-$$

Particle physicists now had two particles, one heavier than the other, whose lifetimes seemed to contradict what was understood about the nature of the strong nuclear force. This puzzle was resolved by the American physicist Abraham Pais (1918–2000) and the Japanese physicist Kazuhito Nishijima (1926–), who put forward the idea that the K meson and the lambda are always produced by the strong force in **pairs**. When a pair of strange particles separate from one another, then the strong force can no longer interact with them, their lifetimes are no longer restricted by the confines of the strong force and they consequently live for much longer. This idea was confirmed in 1954 in accelerator experiments that could produce beams of particles with sufficient energy to create the K meson and lambda.

Strangeness

In 1954, another American physicist, Murray Gell-Mann (1929–) (Figure 5.16), proposed a new property of quantum particles called **strangeness**, with its own quantum number S. In a similar way to electric charge, strangeness is conserved in particle decays, when the strong nuclear force is at work. The strangeness quantum number S is either zero or a positive or negative integer. π mesons and protons have zero strangeness. The K^+ and K^0 have S = +1, and the K^- and Λ^0 have S = −1. The assignment of strangeness is entirely arbitrary, in the same way that the electron is assigned to have a negative charge. We could just as easily have assigned it a positive charge, with the positron negative – it's just the way things were decided at the time! There is an important difference between charge and strangeness though. Charge is always conserved in any interaction, but strangeness is only

Figure 5.16 Murray Gell-Mann (1929–)

conserved in the strong and electromagnetic interactions. We will have more to say about the conservation of strangeness when we talk about conservation laws and particle physics in Chapter 6.

The xi and the sigma particles

Two more hyperons were found in cloud chamber tracks soon after the discovery of the Λ. Both of these were heavy and eventually decayed into a proton. The first of these was a negatively charged particle called the **xi-minus** (Ξ^-), represented by the Greek capital letter xi, Ξ. This particle was called a 'cascade particle' as it was observed to decay in a series of steps into a proton. The Ξ particle comes in two charge states, negative (Ξ^-) and neutral (Ξ^0). One decay mode of the Ξ decays to a proton via a Λ particle by the process

$$\Xi^0 \to \Lambda^0 + \pi^0$$
$$\hookrightarrow \pi^- + p$$

Soon after the discovery of the xi in 1952, another strange particle was found in cloud chamber tracks. This particle had a mass 1.27 times that of the proton and was positively charged. In 1953, both negatively charged and neutral versions were found in accelerator experiments. These were the **sigma** particles (Σ) mentioned earlier. They are represented by the Greek capital letter sigma as Σ^+, Σ^- and Σ^0, and all have strangeness quantum numbers of -1.

The omega-minus

By the beginning of the 1960s particle physicists were attempting to group the particles known at the time according to their properties in a way not unlike that of Mendeleyev, who grouped the chemical elements on the Periodic Table. This method, called the 'Eightfold Way', naturally associated their properties in groups of eight. This led to the prediction that there were further particles yet to be discovered. One of these was the **omega-minus** (Ω^-), a *baryon* (see below) of mass 1672 MeV/c^2 (which is more than 1.5 times the mass of the proton), strangeness $S = -3$, a lifetime of nearly 10^{-10} s, the same charge as an electron and a spin of $\frac{1}{2}$, making it a *fermion* (see page 169).

As we will see in Chapter 7, the discovery of the Ω^- was an important step in developing the quark model. The fact that particle physicists could predict its existence showed that their theories of matter and energy were following the right lines, especially that of *symmetry*, which we discuss in Chapter 6. The hunt for the Ω^- is described in Box 5.3.

Box 5.3 The hunt for the omega-minus

In 1962, Murray Gell-Mann, using the theory and symmetry of the Eightfold Way pattern, predicted the existence of a spin-$\frac{3}{2}$ particle with a strangeness of -3 and a mass of $1675\,\mathrm{MeV}/c^2$, which was dubbed the omega-minus (Ω^-). Particle physicists saw this as a challenge to the validity of the Eightfold Way and set about trying to find it. In 1964, a team at the Brookhaven National Laboratory, on Long Island in New York, using a beam of K mesons, found it in a hydrogen bubble chamber. Figure 5.17 is the first picture to show the production and decay of the Ω^-.

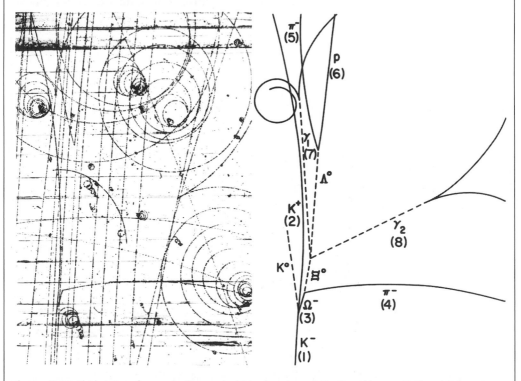

Figure 5.17 This picture from a hydrogen bubble chamber at the Brookhaven National Laboratory was the first evidence for the omega-minus (Ω^-) particle

A K$^-$ particle collides with a proton in the bubble chamber, producing an Ω^- and two other K mesons, a K$^+$ and an unseen K^0, which is represented by the dotted line. The Ω^- travels a short distance and then decays into a π meson (π^-), which bends sharply to the right, and a xi-zero or neutral xi (Ξ^0), which decays into three more neutral particles, a neutral lambda (Λ^0) and two gamma ray photons (γ), which are also marked by dotted lines. Finally, the gamma ray photons decay into electron–positron pairs (e$^+$ and e$^-$) and the lambda decays into a proton (p) and a negative π meson (π^-).

The measured mass of Ω^- agreed exactly with its expected value, and in this way revealed the power of symmetry as a mathematical tool able to predict the existence of particles.

Classifying particles: leptons, baryons and mesons

The π meson proved to be only the first of a great many new particles discovered after the Second World War, and it soon became necessary to put them into some kind of order. We have now seen that our picture of matter being made up of just a few particles is far from being correct. A great many different particles have been found in accelerator experiments as well as in cosmic rays, and we will look at some more of these later. Is there any way we can sort them out? The answer is, yes we can. Particle physicists divide all particles into three categories based on the way they interact with the basic forces. However, a good place to start is by using the concept of *spin* described earlier to divide quantum particles into two classes, together with another important principle in quantum theory called the *Exclusion Principle*.

The Pauli Exclusion Principle

One of the things that may have occurred to you about the Bohr model is that there appears to be nothing to stop all the electrons occupying the same orbit. However, this cannot be true, for if all electrons were to occupy a single orbit, then electronic transitions would not occur and we would not observe spectral lines characteristic of each element. There must be some principle at work that 'spreads' electrons out across the various orbits so that electronic transitions can happen. This puzzle was solved in 1925 by the Austrian physicist Wolfgang Pauli (1900–1958) (Figure 5.18), who put forward his Exclusion Principle. The **Pauli Exclusion Principle** states that:

No two electrons can be in the same quantum state in an atom.

By a **quantum state** we mean quantum numbers that can be assigned to an electron, such as its principal quantum number n, its charge and its spin. It is the Pauli Exclusion Principle that determines the electron shell structure of

Figure 5.18 Wolfgang Pauli (1900–1958)

atoms, so making chemistry possible. For example, only two electrons are allowed in the ground state as this is the only number that gives each electron a unique quantum state. More electrons are allowed in higher energy levels because there is a greater range of unique quantum states to choose from. Table 5.2 illustrates the quantum states of the atoms up to potassium.

Table 5.2 Electronic structure of atoms up to potassium

Element	Z	Quantum numbers of last-added electron			
		n	l	m	m_s
H	1	1	0	0	$+\frac{1}{2}$
He	2	1	0	0	$-\frac{1}{2}$
Li	3	2	0	0	$+\frac{1}{2}$
Be	4	2	0	0	$-\frac{1}{2}$
B	5	2	1	-1	$+\frac{1}{2}$
C	6	2	1	0	$+\frac{1}{2}$
N	7	2	1	$+1$	$+\frac{1}{2}$
O	8	2	1	$+1$	$-\frac{1}{2}$
F	9	2	1	0	$-\frac{1}{2}$
Ne	10	2	1	-1	$-\frac{1}{2}$
Na	11	3	0	0	$+\frac{1}{2}$
Mg	12	3	0	0	$-\frac{1}{2}$
Al	13	3	1	-1	$+\frac{1}{2}$
Si	14	3	1	0	$+\frac{1}{2}$
P	15	3	1	$+1$	$+\frac{1}{2}$
S	16	3	1	$+1$	$-\frac{1}{2}$
Cl	17	3	1	0	$-\frac{1}{2}$
A	18	3	1	-1	$-\frac{1}{2}$
K	19	4	0	0	$+\frac{1}{2}$

Four quantum numbers are found to be sufficient to designate an electron's quantum state inside an atom:

1 The **principal quantum number** n denotes the specific energy level of the electron.
2 The **orbital quantum number** l can take any integral number from 0 to $n-1$; it is related to the angular momentum of an electron in its motion around the nucleus.
3 The **magnetic quantum number** m is responsible for the splitting of electron energy levels due to an external magnetic field, as seen in the Zeeman effect (see page 146).
4 The **spin quantum number** m_s for the electron has values of $\pm\frac{1}{2}$.

In summary, possible values of the quantum numbers are as follows:

$$n = 1, 2, 3, 4, \ldots$$
$$l = 0, 1, 2, 3, \ldots, (n-1)$$
$$m = -l, -(l-1), \ldots, (l-1), l$$
$$m_s = -\tfrac{1}{2}, +\tfrac{1}{2}$$

While Pauli was concerned with the quantum states of electrons, quantum theory says that the Pauli Exclusion Principle can be applied to other quantum systems, which are defined by their component particles and corresponding quantum numbers, and we will see how this can help us to classify quantum particles.

Fermions and bosons

The dynamics of large numbers of particles are analysed using statistical methods. All particles that have half-integer spin quantum numbers, that is $\pm \frac{1}{2}\hbar$, are called **fermions** as they obey statistical laws called **Fermi–Dirac statistics** first set out by the two physicists Enrico Fermi (1901–1954) (Figure 5.19) and Paul Dirac (whom we met earlier). *All fermions obey the Pauli Exclusion Principle.*

Particles that have integer or zero spin quantum numbers, that is, 0 or $\pm \hbar$, are called **bosons** in honour of the Indian physicist Satyendra Nath Bose (1894–1974), who together with Albert Einstein discovered the statistical laws called **Bose–Einstein statistics** that describe the dynamics of these particles. Bosons *do not* obey the Pauli Exclusion Principle and there is no limit to the number of bosons that can have the same quantum numbers.

We can now make a general statement called the **Exclusion Principle** as applied to all quantum particles:

Figure 5.19 Enrico Fermi (1901–1954), who was awarded the Nobel Prize for Physics in 1938

No two fermions can occupy the same quantum state, that is, they cannot have the same set of quantum numbers as each other.

Fermions are the particles that make up what we consider to be the real world, for example, nucleons and electrons. Fermions are also conserved. In other words, in an interaction involving fermions, the total number of each kind of fermion remains the same provided we count an antiparticle as a 'minus one' particle. Bosons are not conserved. Photons, for example, are created and disappear during the emission and absorption of light by atoms.

WORKED EXAMPLE 5.2

The antiproton ($\overline{p}$) was discovered in 1955 in Berkeley, California, using a proton synchrotron called the Bevatron that could accelerate protons up to an energy of 6.4 GeV. Antiprotons were produced by the reaction

$$p + p \rightarrow p + p + p + \overline{p}$$

Show that the number of fermions in this reaction is conserved.

The proton is a fermion and so is the antiproton. On the left-hand side of the reaction, the number of fermions is 2. On the right-hand side, there are three protons plus an antiproton. However, the antiproton counts as a 'minus one' particle, so that the total number of fermions on the right-hand side is $3 + (-1) = 2$, which is the same number as on the left-hand side. Therefore the number of fermions is conserved in this reaction.

Leptons

We said earlier that particle physicists divide all particles into three categories based on the way they interact with the basic forces. The first of these are the **leptons**. The word 'lepton' means 'light particle' but refers to all particles that do not experience the strong nuclear force. There are six leptons, together with their associated antiparticles. All leptons are fermions. These are listed in Table 5.3.

Table 5.3 The six leptons

Particle	Symbol	Mass/ MeV/c^2	Charge/e	Spin/$\hbar$
Electron	e	0.511	−1	$\frac{1}{2}$
Muon	μ	105.658	−1	$\frac{1}{2}$
Tau	τ	1784	−1	$\frac{1}{2}$
Electron neutrino	ν_e	0*	0	$\frac{1}{2}$
Muon neutrino	ν_μ	0*	0	$\frac{1}{2}$
Tau neutrino	ν_τ	0*	0	$\frac{1}{2}$

Quantum numbers spans Charge/e and Spin/$\hbar$.

* There are now compelling reasons to think that neutrinos have non-zero mass – see Chapter 8.

We have already mentioned neutrinos in connection with beta decay. Neutrinos are ghostly particles which only interact via the weak force and emerge from nuclei when they break up. They are produced in large numbers as the Sun converts hydrogen into helium by nuclear fusion, and can pass through the Earth without difficulty since they interact so tenuously with matter. Particle physicists recognise three types of neutrino: the **electron neutrino** ν_e, the **muon neutrino** ν_μ and the **tau** (rhymes with 'wow') **neutrino** ν_τ. The electron neutrino is the one involved in beta decay and, as we mentioned earlier, was discovered in 1956 by experiment using nuclear reactors. The other two are emitted in the decays of the muon and the tau. The muon neutrino, for example, appears in the muon decay

$$\mu^- \rightarrow e^- + \nu_\mu + \bar{\nu}_e$$

The first direct evidence for the tau neutrino was announced on 21 July 2000 at Fermilab. In 1997, using the Tevatron accelerator, particle physicists produced an intense neutrino beam, which they expected to contain tau neutrinos. A detector called DONUT gathered data from six million potential interactions, from which 1000 were candidates for a tau neutrino event. Of these, just four provided direct evidence for the existence of the tau neutrino, and this discovery is described in Box 5.4.

The **tau** (τ) lepton or **tauon** was discovered in 1975 by the American physicist Martin L. Perl (1948–) and his associates who were colliding electrons and anti-electrons together using the SPEAR electron–positron collider at SLAC.

There are also six lepton antimatter particles, the positron, antimuon, antitau, and the three antineutrinos. All of these have exactly the same mass and spin as their matter counterparts but opposite electric charge.

There is now evidence to suggest that the neutrino may not be massless. If neutrinos do have mass, then this has important cosmological implications, as we will see in Chapter 8.

Baryons

Particles with half-integral units of spin (that is, $\frac{1}{2}\hbar$, $\frac{3}{2}\hbar$, $\frac{5}{2}\hbar$, ...) that interact with the strong force are called **baryons**. All baryons are fermions. Well known baryons include the proton and neutron, and baryons are the most massive of the elementary particles. Some baryons are listed in Table 5.4.

Table 5.4 The eight spin-$\frac{1}{2}$ baryons

Particle	Symbol	Mass/ MeV/c^2	Quantum numbers		
			Charge/e	Spin/$\hbar$	Strangeness
Proton	p	938.3	+1	$\frac{1}{2}$	0
Neutron	n	939.6	0	$\frac{1}{2}$	0
Lambda-zero	Λ^0	1115.6	0	$\frac{1}{2}$	−1
Sigma-plus	Σ^+	1189.4	+1	$\frac{1}{2}$	−1
Sigma-zero	Σ^0	1192.5	0	$\frac{1}{2}$	−1
Sigma-minus	Σ^-	1189.4	−1	$\frac{1}{2}$	−1
Xi-zero	Ξ^0	1314.9	0	$\frac{1}{2}$	−2
Xi-minus	Ξ^-	1321.3	−1	$\frac{1}{2}$	−2

As for leptons, every baryon has an antibaryon. The most important baryons are the proton and the neutron, since they make up most of the mass of ordinary atoms. It is for this reason that particle physicists refer to the everyday matter in our universe as **baryonic matter**.

Box 5.4 The discoveries of the tau lepton and tau neutrino

The **tau** (rhymes with 'wow') lepton is an electrically charged particle that is much heavier than the electron or muon. At SLAC in 1974, the American physicist Martin Perl and his colleagues were using the SPEAR electron–positron collider to collide electrons and positrons with a total energy of 8 GeV. Attached to the accelerator was a detector called the MARK 1, which consisted of an array of spark chambers wrapped in concentric cylinders around the accelerator beam-pipe. The entire assembly was encased in an electromagnet to bend the paths of the charged particles created in the annihilation of the electrons and positrons. Other types of detector both inside and outside the electromagnet helped reveal the identity of the particles so that electrons, muons and π mesons could be separately identified.

Perl and his colleagues announced the evidence for the production of events containing both electrons and muons. This was possible if a new heavy lepton was able to decay, producing either electrons or muons along with associated neutrinos. German particle physicists at DESY using an electron–positron collider called DORIS were able to corroborate the results from the MARK 1 at SPEAR.

When these decays were first observed in the MARK 1, they appeared rather odd because the electron and positron annihilate, but a muon and an electron emerge *from* the annihilation! Perl was able to show that this indicated the existence of a new heavy lepton of mass $1.784\,\text{GeV}/c^2$ and a lifetime of $3 \times 10^{-13}\,\text{s}$. Like all particles, the tau has its corresponding antiparticle, the antitau. In common with all leptons, the tau does not experience the strong force, but because of its large mass, it can sometimes decay into hadrons.

An electron and positron annihilate to produce a tau (τ^-) and antitau (τ^+)

$$e^+ + e^- \rightarrow \tau^+ + \tau^-$$

The taus are not directly seen in the detector, because after travelling only a very small fraction of a millimetre, a tau decays to produce other particles. About 17% of the total number of tau leptons decay by the process

$$\tau^- \rightarrow e^- + \nu_\tau + \bar{\nu}_e$$

where the ν_τ is the **tau neutrino**. The neutrinos pass through the detector without interacting and we only see the electron. A further 17% of tau leptons decay to produce a muon, which can be detected, and neutrinos (which are not seen) by the decay

$$\tau^+ \rightarrow \mu^+ + \nu_\mu + \bar{\nu}_\tau$$

The remaining 66% of tau leptons decay to produce neutrinos and hadrons.

The existence of the tau lepton was confirmed in experiments using the TASSO drift detector chamber at DESY, and its discovery was important in confirming the concept of *generations* between leptons and quarks, as we will see in Chapter 7.

Particle physicists needed about three years of painstaking work analysing the tracks of the tau lepton and its decay in the emulsion layers in a detector at Fermilab called DONUT (Direct Observation of the NU Tau) in order to reveal the tau neutrino 25 years after its existence was inferred by Perl and his colleagues at SLAC in 1975. The observation of the tau neutrino was announced at Fermilab in July 2000, by experimenters using DONUT (Figure 5.20). At the DONUT target station are iron plates sandwiched with layers of special three-dimensional photographic emulsion which recorded the particle interactions. About 1 out of 10^{12} tau neutrinos interacted with an iron nucleus and produced a tau lepton, which left a 1 mm tell-tale track in the emulsion.

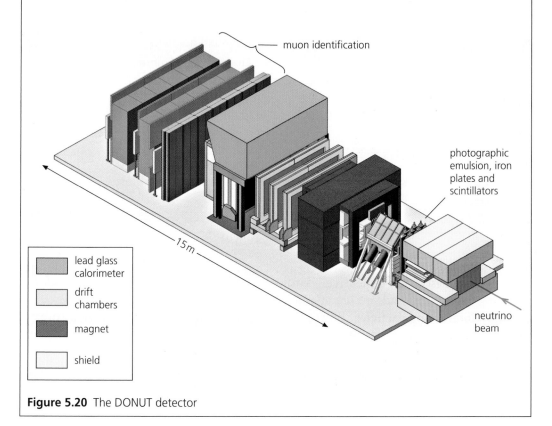

Figure 5.20 The DONUT detector

Mesons

We have already met two mesons, the π meson (pion) and K meson (kaon), and the mesons are another group of numerous elementary particles all with integer spin. Remember that mesons are bosons all having spins of 0, 1, 2, etc. Table 5.5 shows some selected mesons all with spin 0.

Table 5.5 Some mesons with spin zero

Particle	Symbol	Mass/ MeV/c²	Quantum numbers		
			Charge/ε	Spin/ℏ	Strangeness
Neutral π meson	π^0	135.0	0	0	0
Positive π meson	π^+	139.6	+1	0	0
Negative π meson	π^-	139.6	−1	0	0
Positive K meson	K^+	493.7	+1	0	+1
Negative K meson	K^-	493.7	−1	0	−1
Neutral K meson	K^0	497.7	0	0	+1
Anti-neutral K meson	$\overline{K}^0$	497.7	0	0	−1
Eta	η	547.5	0	0	0
Eta prime	η'	957.8	0	0	0

Baryons and mesons are collectively known as **hadrons** from the Greek word *hadros* meaning 'bulky'. All hadrons are affected by the strong force and are said to be **strongly interacting particles**. Baryons and mesons are distinguished by their spin and also by other properties. Baryons have half-integer spin and are *fermions*, whereas mesons have integer spin and are *bosons*.

Deciphering particle tracks

In Chapter 4 we discussed the different kinds of detectors used to detect particle events. The paths of particles seen in cloud and bubble chambers and computer reconstructions from multi-wire proportional chambers often appear as a maze of whirls, spirals and forked tracks. It takes considerable skill to interpret these tracks and understand what they mean. Figure 5.21 shows the result of a collision between a proton and an antiproton in a bubble chamber at CERN. Several particles are created in this collision.

An antiproton p̄ enters the chamber from above, entering a perpendicular magnetic field. Since the antiproton has a negative charge, its path is slightly curved as shown. At point A the antiproton hits a proton p at rest in the bubble chamber liquid. The proton and antiproton annihilate each other, and two K mesons (K^0, K^-) and two π mesons (π^+, π^0) are created. By coincidence these particles produce further events in the chamber. Since the K^0 is electrically neutral, it leaves no visible track but its path can be reconstructed because after a short time it decayed spontaneously at B into two π mesons (π^+, π^-). One of these (the π^+) then decayed at C into an antimuon (μ^+), which in turn decayed at D into a positron (e^+) and a neutrino (not visible).

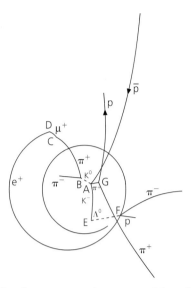

Figure 5.21 Drawing of the tracks of a proton–antiproton particle collision

The K^- that was originally created in the $p + \bar{p}$ collision, at A, hits another proton at rest at E. This collision produces a lambda particle (Λ^0) and a π meson (π^0), neither of which are visible but whose trajectories can be inferred. At F the Λ^0 decayed into a proton and a π meson (π^-). Finally, one of the π mesons (π^+) created at A had an elastic collision with an adjacent proton at rest at G, and they move apart from each other in opposite directions following curved paths because of their positive charge.

We can summarise the sequence of these events as follows:

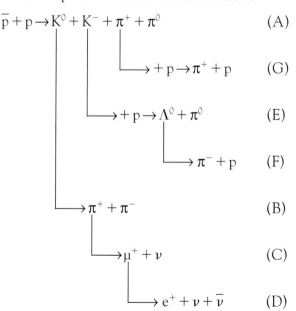

$$\bar{p} + p \rightarrow K^0 + K^- + \pi^+ + \pi^0 \qquad \text{(A)}$$

$$\rightarrow + p \rightarrow \pi^+ + p \qquad \text{(G)}$$

$$\rightarrow + p \rightarrow \Lambda^0 + \pi^0 \qquad \text{(E)}$$

$$\rightarrow \pi^- + p \qquad \text{(F)}$$

$$\rightarrow \pi^+ + \pi^- \qquad \text{(B)}$$

$$\rightarrow \mu^+ + \nu \qquad \text{(C)}$$

$$\rightarrow e^+ + \nu + \bar{\nu} \qquad \text{(D)}$$

From these examples you can see that it takes great experience on the part of the particle physicist to interpret what has happened during a particle event. When photographing particle tracks was the common way to record events in bubble chambers and other passive detectors, people called 'scanners' were instrumental in discovering many new particles; they were not particle physicists themselves, but simply individuals trained to look for certain patterns of tracks in the photographs. These days the reconstructions of particle events in electronic detectors are analysed by computers programmed to look for certain particle decay patterns.

Resonances

If a particle lives long enough so that particle physicists can accelerate it or perform measurements on it, then it is regarded as stable. If a particle lasts for only 10^{-23} s, however, it cannot be directly observed. To see why, consider this. The maximum speed at which a particle can travel is the speed of light, so in a time interval of 10^{-23} s, it can travel at most a distance of 10^{-23} s $\times 3 \times 10^8$ m s$^{-1} \approx 10^{-15}$ m. This is of the order of the size of an atomic nucleus, and such particles can only be observed indirectly, through the way they decay and interact with other particles.

The existence of such a short-lived particle is sometimes referred to as a **resonance**. A typical example of a resonance is the neutral rho (ρ^0) meson. This is formed in the reaction

$$\pi^- + p \rightarrow \rho^0 + n$$

The ρ^0 has an extremely short lifetime and rapidly decays to two π mesons, so particle physicists actually observe

$$\pi^- + p \rightarrow \pi^+ + \pi^- + n$$

If this reaction proceeded without the step involving the rho meson, the energy would be smoothly distributed among the pions and neutrons. The momenta and energy of the two pions created can be mathematically combined and plotted against the number of reaction events, and we would expect a smooth energy spectrum not unlike the beta particle spectrum that we saw in Figure 5.3. However, the existence of the rho meson makes its presence known by a small 'peak' protruding above an otherwise smooth energy spectrum. This peak is analogous in appearance to the resonance peak in the amplitude of forced mechanical oscillations. The ρ^0 is called a **resonant particle**.

Mesons contain *quarks* (see Chapter 7), and in resonant particles such as the ρ^0 meson the constituent quarks are excited to higher energy levels. When they return to their initial states, particles such as π and K mesons, which we can observe directly, are emitted and carry away the excess energy.

In this chapter we have introduced some elementary particles and shown how they may be classified in terms of their interactive and quantum properties. To understand how particles interact with each other we must search for some underlying patterns to which they all conform. In Chapter 6 we will see that *conservation laws* and the concept of *symmetry* play key roles in bringing order to the particle world.

Summary

◆ **Elementary particles** are strictly **quarks** and **leptons**, which are 'point-like' and appear to have no further internal structure. However, this term is loosely applied to **baryons** and **mesons**. The term **fundamental particles** is synonymous with elementary particles. The earliest discoveries in particle physics were made from studies of very high-energy **cosmic rays**. Many hundreds of different particles have been found, almost all of which live for only short periods of time.

◆ All particles can be thought of as **fields** extending through space. The wavefunctions of particles localise these fields to a particular region of space, which is what we observe as a particle. The description of matter in terms of particle fields is called **quantum field theory**.

◆ **Spin** is a property of quantum particles, which is analogous to **angular momentum** in classical physics. It is quantised and measured in units of $\hbar$ where $\hbar = h/2\pi$. **Nuclear beta decay** provided the first evidence of the existence of the **neutrino**, which interacts with matter only by the weak interaction.

◆ **Antimatter** is matter that has the opposite set of quantum properties (such as electric charge) to ordinary matter. The existence of antimatter was confirmed in 1932 with the discovery of the **positron**. When matter and antimatter meet, they are converted into pure energy by **annihilation**. Every particle that exists has a corresponding **antiparticle**. A mystery in particle physics is why the universe in which we live is dominated by matter and not antimatter. In 1995 the first atoms of antihydrogen were produced at CERN.

◆ **Quantum electrodynamics** (QED) is a theory that can explain the electromagnetic force by the exchange of **virtual photons** which 'carry' the force from one electron to another. An early difficulty with QED was calculations involving the **self-energy** of a particle, which were later resolved by a technique called **renormalisation**.

◆ **Mesons** are particles that were initially thought to have masses between those of the electron and proton (although some are more massive than the proton). The application of QED to the strong nuclear force led to the discovery of the π **meson** or **pion** as an exchange particle between nucleons. Another particle called the **muon** was also discovered which is 200 times more massive than the electron and was initially thought to be the π meson. It is not a force-carrying particle and is now *not* regarded as a meson.

◆ A new quantum property of particles, that of **strangeness**, was invented to explain the unusually long lifetimes of particles produced via the strong interaction. Particles are assigned a **strangeness quantum number** S, which is either zero or a positive or negative integer. Strangeness quantum numbers are *conserved* in all interactions involving the strong force.

◆ The development of particle accelerators and detectors after the Second World War led to many more particles being discovered. Cloud chambers played an important role in the discovery of **V particles** in cosmic rays, so-called because of the characteristic V-shaped tracks that were seen. The decay products of some of these particles sometimes included protons, and these particles were called **hyperons**. Others always had decay products made up of mesons and were called **K mesons** or **kaons**. Other particles found were the **lambda, xi** and **sigma** particles. The **omega-minus** particle was predicted to exist by grouping together known particles with similar properties.

◆ The **Pauli Exclusion Principle** states that no two electrons can exist in the same quantum state in an atom. The **Exclusion Principle** can be applied to *any* quantum system that is defined by its component particles and corresponding numbers. By using the Exclusion Principle and spin, we can describe the dynamics of large numbers of particles statistically using **Fermi–Dirac statistics** for quantum particles that have half-integer spin quantum numbers and obey the Pauli Exclusion Principle. Such particles are called **fermions**. Particles that obey **Bose–Einstein statistics** and have zero or integer spin quantum numbers are called **bosons** and do not obey the Pauli Exclusion Principle. The Exclusion Principle is responsible for the electron shell structure of atoms.

◆ Particles can be classified into **leptons, baryons** and **mesons**. **Leptons** are particles that *do not* experience the strong nuclear force. They are six leptons together with six corresponding **antileptons** and all are spin-$\frac{1}{2}$ particles. **Baryons** are particles that interact with the strong force and are the most numerous group of particles. Baryons have half-integral units of spin and include the proton and neutron. Leptons and baryons are fermions.

◆ **Mesons** are another numerous group of particles and all have integer values of spin. All mesons are bosons. Baryons and mesons both feel the strong force and are collectively known as **hadrons**. All hadrons are affected by the strong nuclear force and are said to be **strongly interacting particles**.

◆ Skill is required to interpret the tracks from particle events. In the early days of particle physics this was done by specially trained people looking for certain kinds of tracks. Modern methods involve sophisticated computers that are programmed to detect certain particle decay patterns from electronic detectors. The existence in a decay chain of a very short-lived particle that cannot be observed directly is called a **resonance**. A **resonant particle** is a particle that only exists as a resonance, inferred by the appearance of a peak in an otherwise smooth particle decay energy spectrum.

Questions

1 **a** Explain what are meant by an elementary particle. What are the differences between leptons, baryons and mesons?

   b $^{23}_{10}$Ne decays to $^{23}_{11}$Na by emitting a negative beta particle (electron) and an *antineutrino* $\bar{\nu}$, which is a particle of zero (or negligible) mass and zero charge.

     i) Write down the nuclear equation for this decay.

     ii) What is the maximum kinetic energy of the emitted electron?

     [Mass of $^{23}_{10}$Ne $= 22.994\,465\,$u, mass of $^{23}_{11}$Na $= 22.989\,768\,$u]

2 **a** What is *spin*? Explain how spin can be used to classify particles.

   b Calculate the angular momentum of:

     i) the Earth, radius 6400 km, mass 6×10^{24} kg [for sphere $I = \frac{2}{5}Mr^2$]

     ii) a uniform disc of moment of inertia 10 kg m^2 rotating about its centre with an angular velocity of 20 rad s^{-1}

     iii) a proton where $L_{proton} = \frac{1}{2}\hbar$, where $\hbar = h/(2\pi)$.

   c How does angular momentum at the atomic and sub-nuclear level differ from angular momentum in everyday large objects?

3 **a** Write an equation for the decay of a neutron.

    Sketch a graph showing how the number of beta minus particles emitted per second from a radionuclide will vary with the kinetic energy of the beta minus particles. Describe briefly without experimental details the principles behind the detection of neutrinos.

Edexcel, June 2000 (part)

   b What is the nature of a beta plus (β^+) particle?

    Name another particle emitted during β^+ decay.

   c Some nuclides change by absorbing an orbital electron. Suggest how this process of *electron capture* alters the composition of the nuclide.

4 **a** What are cosmic rays? How do the energies of cosmic rays compare with the maximum energies achievable by modern particle accelerators?

   b Cosmic rays bombard the Earth constantly from all directions. If all cosmic ray protons reach the surface of the Earth, and the average rate is 1500 per square metre of the Earth's surface, what would be the corresponding current intercepted by the total surface area of the planet?

    [Radius of Earth $= 6400$ km]

5 **a** What are antiparticles?

    How much energy would be released if the Earth were annihilated by a collision with an anti-Earth?

    [Mass of Earth $= 6 \times 10^{24}$ kg]

   b At CERN (see Box 5.2) antiprotons can be produced at a maximum rate of 6×10^{10} per hour. How long would it take to make enough protons for 1 kg of antihydrogen gas? If the maximum number of antiprotons that are allowed to accumulate in the accelerator ring is 1.2×10^{12}, what mass of antiprotons is this?

    [Mass of antiproton $= 1.67 \times 10^{-27}$ kg]

6 a State the Pauli Exclusion Principle. How is this principle related to the quantum state of a particle?

b According to the Pauli Exclusion Principle, the maximum number of electrons that an atom can have in a particular electron shell is given by $2n^2$, where n is the *principal quantum number* = 1, 2, 3, etc., with $n = 1$ corresponding to the ground state. What are the maximum numbers of electrons that can be accommodated in each of the first five shells of an atom?

7 Explain the difference between *fermions* and *bosons*.
Distinguish between *baryons*, *mesons*, *hadrons* and *leptons*. Name the six leptons associated with ordinary matter. Are leptons thought to be truly elementary particles? Which of these four categories of particles are fermions and which are bosons?

8 a What is meant by *strangeness*? In which interactions is strangeness always conserved?

b By examining strangeness S, determine which of the following decays or reactions proceed via the strong interaction:

i) $K^0 \rightarrow \pi^+ + \pi^-$

ii) $\Lambda^0 + p \rightarrow \Sigma^+ + n$

iii) $\Lambda^0 \rightarrow p + \pi^-$

iv) $K^- + p \rightarrow \Lambda^0 + \pi^0$

[$S = 0$ for p, n, π^0, π^+, π^-; $S = +1$ for K^0; $S = -1$ for K^-, Λ^0, Σ^+]

Fundamental interactions, symmetries and conservation laws

'Nature seems to take advantage of the simple mathematical representations of the symmetry laws. When one pauses to consider the elegance and the beautiful perfection of the mathematical reasoning involved and contrast it with the complex and far reaching physical consequences, a deep sense of respect for the power of the symmetry laws never fails to develop.'

Chen Ning Yang (1922–), Chinese-born American theoretical particle physicist, quoted in his Nobel Prize address

What is 'force'?

You may have held the poles of two bar magnets together and, depending on their polarity, felt either the attractive or repulsive force between them. But exactly what is this force? What is 'going on' in the space between the magnets? Precisely how does matter interact with itself? To answer these questions we need to take a closer look at what we mean by 'force' and in particular what we mean by an 'interaction'.

An **interaction** is the exchange of energy and momentum between two bodies. If you like playing rugby or netball, then you already have an idea of what this means. Force is the rate of change of momentum with time. When two rugby players pass a ball, they are interacting with each other. There is a change in momentum as the ball leaves the first player, and a second change in momentum as the other player catches it (Figure 6.1). When this happens, both players experience a force, which is 'carried' by the rugby ball. Even though the two rugby players are not physically touching, they are nonetheless feeling an 'action at a distance'. We will see that this concept is very similar to the way particles experience forces.

Figure 6.1 The concept of 'force'. Particle physicists regard the steady attractive or repulsive force between quantum particles to be mediated by other short-lived particles travelling between them. We can see how this concept – the exchange of a particle – works by considering the simple approximate analogies shown here
(a) Repulsion: When two rugby players pass the ball to each other, a net repulsive force occurs between them due to the recoil that they experience when passing or catching the ball
(b) Attraction: We can explain attraction taking place by thinking about the recoils experienced by two Australians throwing and catching a boomerang

In Chapter 5 we caught a glimpse of what 'force' might be when we introduced quantum electrodynamics (QED) and the quantum theory of fields. In this chapter we are going to develop these ideas further and see how the interactions between particles can be represented diagrammatically. We will also look at the importance of *symmetry* in particle physics, and how conservation laws can govern particle interactions.

The forces of nature

Particle physicists know of four basic forces in nature, which govern the interaction between matter and energy:

1 **The gravitational interaction.** Gravity attracts all matter, but is too weak to have significant effects at sub-atomic scales.
2 **The electromagnetic interaction.** This plays an important role in the forces between charged particles and is responsible for the phenomenon of electromagnetism.
3 **The strong interaction.** This force binds atomic nuclei together and stops them flying apart due to the mutual repulsion of their protons, and is the dominant interaction in reactions and decays of most of the fundamental particles. However, leptons do not experience the strong force.
4 **The weak interaction.** The weak interaction is a force that is involved in nuclear beta decay and other radioactive processes. All particles experience the weak interaction.

Gravity and the electromagnetic interaction have an unlimited range, which is mainly why they are so familiar in everyday life. The strong interaction has a range no larger than that of an atomic nucleus, and the weak interaction much less than this, so they are not observed directly.

The relative strength of a force decides the time scale over which it can act. If two particles are brought together so that any of these four forces can act in an interaction, then, because of its lesser strength, the weak force requires a longer time to cause a particle decay or reaction than does the strong force. The **characteristic time** is a measure of how long the force needs to act to produce an effect. The lifetime of a particle decay process is often a clue to the type of interaction going on, with the strong force acting for much shorter characteristic times ($\sim 10^{-23}$ s) than the others.

Fundamental particles can interact with each other through any of these four forces. All of them can interact through the gravitational and weak forces, but not all of them (the leptons do not) interact with the strong force. When two particles interact through the strong force, then we can often neglect the effects of the weak and electromagnetic forces in decay processes since, because of their relative strengths, these effects are much smaller. However, it should be noted that this is not always true, and in some nuclear processes in stars that involve the fusion of protons, the weak force has an important role to play. The gravitational force is far too small to be of any consequence in particle reactions and decays. Table 6.1 shows the properties of the basic interactions.

We saw in Chapter 5 that particles can be classified in terms of the forces that act on them. All hadrons (baryons and mesons) interact with the strong force. Particle physicists like to say that a particle 'feels' a force, and so all hadrons are said to feel the strong force, and all particles feel the gravitational and weak forces.

Table 6.1 Properties of the basic interactions

Interaction	Strength (relative to strong force)	Range/m	Characteristic time
Gravitational	10^{-38}	∞	years
Electromagnetic	10^{-2}	∞	10^{-14}–10^{-20} s
Strong	1	10^{-15}	$<10^{-22}$ s
Weak	10^{-6}	10^{-18}	10^{-8}–10^{-13} s

The weak interaction and beta decay

We have already discussed the electromagnetic and strong interactions. The electromagnetic force is the force between two charges and is described by Coulomb's Law. The strong nuclear force binds nucleons together in the nucleus, and we have already discussed some particles involved with the strong force in Chapters 2 and 5. But what is the weak interaction?

In Chapter 5 we discussed nuclear beta decay. We saw that beta-minus decay involves the changing of a neutron into a proton and the emission of an electron and an antineutrino

$$n \rightarrow p + e^- + \bar{\nu}$$

and beta-plus decay can be written as

$$p \rightarrow n + e^+ + \nu$$

Note that, while a free neutron can decay into a proton, a proton can decay into a neutron only within a nucleus. Free protons appear to be stable and not to decay into other particles. This is an interesting observation that points to some difference in the internal structures of a neutron and a proton. Using QED, the Italian physicist Enrico Fermi (see page 169) proposed that beta decay could be explained by introducing a new force, the **weak interaction**, which enabled a neutron to transform itself into a proton (or vice versa) in beta decay. The weak interaction does not participate in the forces that bind together the nucleons in the nucleus, and it is much weaker than the strong or electromagnetic interactions – hence its name. In Chapter 7 we will see how the weak and electromagnetic interactions are in fact closely connected with each other and can be regarded as a single force, and that beta decay can be explained in terms of the *quark model*.

QED revisited

In Chapter 5, we saw how Dirac combined quantum mechanics, electromagnetism and relativity to produce the theory of quantum electrodynamics (QED), which explained how the electromagnetic force could be regarded as the exchange of 'virtual photons' between charged particles. However, the theory of QED had difficulties. When particle physicists tried to develop QED, they hit a snag in the mathematics. According to QED, an electron moving through space is continuously emitting and absorbing virtual photons. It is interacting with its *own* electromagnetic field, the source of which is its own electric charge. This **self-energy**, as it is called, gives rise to infinitely large quantities in the wavefunctions that describe the electron's behaviour, rendering the calculations meaningless. The development of quantum field theory was held back by this mathematical problem for some 20 years until, in the 1940s, the two American particle physicists Richard P. Feynman (1918–1988) and Julian Schwinger (1918–1984) and also the Japanese physicist Sin-Itiro Tomonaga (1906–1979) found a way to solve it. Their concept was called **renormalisation**.

A mathematical description of renormalisation is well beyond the scope of this book, but the idea involves a mathematical 'sleight of hand' by which the infinities in the wavefunction can be dealt with by supplying an infinity on one side of an equation that can cancel out an infinity on the other. The science writer John Gribbin has summed up this business of subtracting infinities in renormalisation as follows:

> 'You might think that infinity minus infinity if it means anything at all ought to be zero. But infinity is a funny thing. Imagine making infinity by adding up all of the integer numbers $(1 + 2 + 3 + 4 \ldots)$. You could also make infinity by doubling every integer and then adding up all the resulting numbers. You might think that the second infinity is twice as big as the first one. But it isn't, because it includes only all the even numbers $(2 + 4 + 6 + 8 \ldots)$, while the first infinity includes both odd numbers and even numbers! If you subtract the second infinity from the first one, in this case you will be left with infinity once again – the sum of all the odd numbers $(1 + 3 + 5 + 7 \ldots)$. What is needed in QED is to be able to subtract one infinity made up of (infinity), from another infinity, made up of (infinity plus a little bit), to leave behind (a little bit).'

John Gribbin, *Q is for Quantum*, Weidenfeld & Nicolson, 1998

Renormalisation performs the same trick. In this case the 'little bit' left over is what makes the wavefunction for the electron meaningful and allows particle physicists to describe its behaviour. Renormalisation cannot be made to work with just any set of equations and infinities. Only certain kinds will work or, when you subtract them, you will be left with either zero or infinity. The important point to understand about renormalisation is that it was a key mathematical breakthrough that allowed QED and the quantum theory of fields to be fully developed as a theory for describing the way particles interact with each other.

Feynman diagrams

The quantum theory of fields tells us that the Coulomb force between charged particles comes about due to an exchange of photons. If, for example, we have two electrons separated by a distance, then each electron emits a succession of photons, which propagate to and are then captured by the other electron. The photons in effect carry the 'Coulomb force' from one electron to another, and as we saw in Chapter 5 they are called *virtual* because they cannot be observed directly. Virtual photons are assigned an 'imaginary mass' so that they can exist for the short time Δt permitted by the Uncertainty Principle, and so that the excess of energy needed to create them is less than the uncertainty in the energy

$$\Delta E \approx \frac{\hbar}{\Delta t}$$

This action of emission, propagation and absorption of photons can be represented on a **Feynman diagram**, named after Richard Phillips Feynman who first invented them (Figure 6.2).

In a Feynman diagram, vertices are used to represent the mathematical expressions describing particle interactions. They are **space–time diagrams** in which charged particles are generally represented by straight lines and photons by wavy ones.

Unlike conventional distance–time graphs, a space–time diagram has time along the y-axis and distance along the x-axis. As well as being a pictorial representation, Feynman diagrams also represent a shorthand notation for the mathematical expressions that govern the interaction in the same way as a graph is a pictorial description of a mathematical function. In Chapter 3, we discussed the fact that, as Feynman showed, in order to calculate the quantum probability that a particle will go from A to B, you have to take into account *every* possible path, *not* just the most direct route as would be described by classical mechanics. Some of these paths would be more probable than others, but they cannot be ignored, and their probability amplitudes must be added up to give a probability which is close to the path that would be taken as calculated by Newtonian mechanics (Figure 6.3).

Figure 6.2 Richard Phillips Feynman (1918–1988)

Figure 6.3 The idea of path integrals
(a) Classical Newtonian mechanics says that a particle will follow a single path from A to B
(b) Quantum mechanics says that a quantum particle will take *every* possible path from A to B, each with a certain probability. The Newtonian path emerges as the result of adding up all the probabilities

This method of adding quantum probabilities is called the **path integral**, and the Feynman diagram represents the average over all the ways particles can interact with each other by the exchange of bosons. While Feynman diagrams can be used as a mathematical tool for quantum field calculations, in this book we will only be using them as a simple pictorial model to illustrate various particle interactions. Some examples of Feynman diagrams are shown in Figure 6.4.

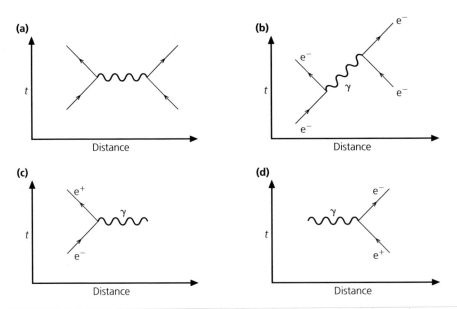

Figure 6.4 Some examples of Feynman diagrams involving the electromagnetic interaction
(a) Feynman diagram which represents two charged particles (straight lines) that interact with one another by exchanging a photon (wavy line) that 'carries' the force from one to the other
(b) Feynman diagram representing the exchange of a virtual photon (γ) between two electrons (e⁻). The electron on the left emits the photon, and the electron on the right absorbs it
(c) Feynman diagram representing the annihilation of an electron (e⁻) and a positron (e⁺)
(d) Feynman diagram representing the creation of an electron–positron pair

Note that in **(c)** and **(d)** the positrons are shown as going backwards in time. The reason for this is that, if you take a particle (for example, an electron) and replace it with its antiparticle (that is, a positron) and then look in a mirror and reverse time, the two will appear identical. So one way of thinking about antiparticles is that they behave as if time was running backwards!

QED is found to be one of the most successful and accurate theories in all of physics. Some of the predictions of QED have been confirmed experimentally to one part in a billion. It is because of the successes of QED that the concept of 'exchange particles', or **field quanta** as they are called, as a mechanism for what 'force is', has been extended to the weak and strong interactions. However, before we look at how this was done, we need to look at another important concept that is central to understanding particle interactions, that of *symmetry*.

Symmetry

The physical world in which we live is undeniably a complex one. Yet, for such a diverse universe, the laws of physics are remarkably simple. Gravity, for example, which operates right up to the largest scales of the universe, is represented by a simple inverse square law. Electromagnetic phenomena are described by a series of concise mathematical formulas as expressed by Maxwell's equations. Why should this be so? It turns out that nature displays a marvellous symmetry on a mathematical level, which greatly simplifies our description of the physical world.

A **symmetrical structure** is simply a structure that exhibits a repetition of form. The symmetry may be geometrical with respect to a line or a point, mathematical (in an algebraic sense) or even musical like a fugue. A snowflake has symmetry, whereas a cloud does not. The snowflake has geometrical symmetry relative to **lines of symmetry** drawn through its centre (Figure 6.5). Symmetry has to do with how objects remain the same if we attempt to change them in some way.

You may have already met geometrical symmetries such as that of a square (Figure 6.6) or a triangle (see Figure 6.10, page 191). In mathematics you may have performed simple operations on them, such as rotations, reflections and translations. Such operations are called **transformations**.

Figure 6.5 Lines of symmetry of a snowflake

Not all transformations preserve symmetry. A reflection transformation through the diagonals of a rectangle does not produce an identical mirror image, although reflections through lines connecting the mid-points of its parallel sides do preserve symmetry (Figure 6.7).

In particle physics we also find symmetries, but they are non-geometric. An example is the following **charge symmetry**. Suppose we have two charges, one positive and the other negative (Figure 6.8a). We then measure the Coulomb force between these charges to have some magnitude F_q. If the position of the charges is reversed (Figure 6.8b), the Coulomb force between the charges remains unchanged.

187

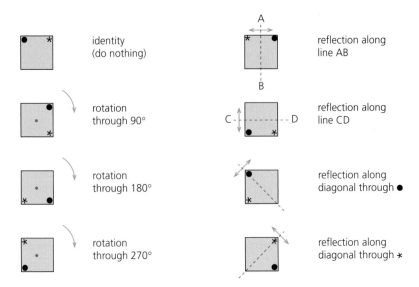

Figure 6.6 A square has eight symmetries, each characterised by a simple operation called a 'transformation'

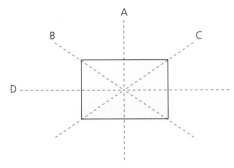

Figure 6.7 Not all transformations preserve symmetry. For this rectangle, lines A and D preserve symmetry, whereas lines B and C do not

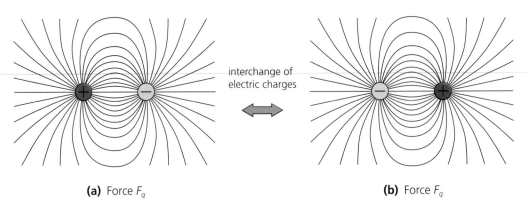

(a) Force F_q interchange of electric charges   **(b)** Force F_q

Figure 6.8 Charge symmetry. Charges will exert a force F_q on each other as given by Coulomb's Law. If the positions of the charges are interchanged, the force between the charges remains the same

Earlier we discussed the concept of particles being represented as quantum fields, and, in 1954, Chen Ning Yang (1922–) and Robert Mills (1924–), two American theoretical particle physicists, made an important breakthrough. They found that geometrical symmetries could be used to describe the relationship between forces and particles in terms of quantum field theory. This led to a major understanding in the way nature assembles matter at the most fundamental level, including the prediction of new and exotic particles that could be looked for experimentally using particle accelerators. Particle physicists began to see that among the chaos of the universe is a deep underlying symmetry, which is the key to why the laws of physics can be so simple and yet the world so complicated.

Gauge symmetries

In mathematical terms, symmetry is the maintenance of certain measurable quantities when we apply some kind of transformation. Electromagnetism can be described in terms of symmetry. The velocity of light stays the same regardless of changes in motion. Another example is that of gravity. Consider the gravitational potential energy (GPE) that a ball of mass m has by virtue of its position on a staircase. Suppose it moves down one step. It loses an amount of GPE equal to $mg\Delta h$, where Δh is the difference in height between the two steps. Now it doesn't matter *where* you measure the GPE from, the difference in GPE between the two steps is always the same (Figure 6.9). In other words the standard of measurement we apply, or the *gauge*, does not affect the rules governing the behaviour of electromagnetic or gravitational fields. Particle physicists call this principle **gauge invariance**, and electromagnetism and gravity are examples of **gauge theories** that display **gauge symmetry**.

The four fundamental interactions that we listed earlier are all gauge theories that display gauge symmetry. Why are gauge theories important in particle physics? Yang and Mills found that, by using a branch of mathematics called **group theory** (Box 6.1), many of these gauge symmetries can be manipulated together with various transformations, to reveal clues about the nature of the interactions between particles, and to enable the particle physicist to bring a level of order to the fundamental interactions of matter.

As we will see as we read on, symmetries give rise to *conservation laws*. Physicists are always interested in conservation laws, since they represent an invariance inherent in complex and changing phenomena. Invariant quantities provide a kind of reference point, in the same way, for example, that a fixed amount of money in an economy can be spent or exchanged in a multitude of ways, but which must all add up to the total amount of money circulating. Theoretically, the total amount of money in the economy remains the same – it is *invariant*. In this way, governments can know how much they have to spend, and how different parts of the economy are performing. Similarly, conservation laws in particle physics provide a way of establishing invariance and predicting how a particular system of particles will behave.

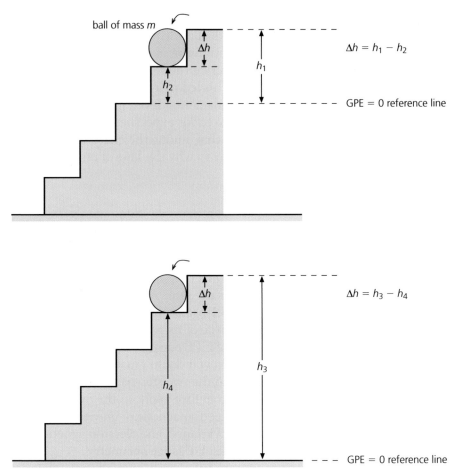

Figure 6.9 Gravity as a gauge theory. The position from which we measure the gravitational potential energy has no effect on the change $mg\Delta h$ when the ball moves down a step

Broken symmetries

Suppose you waved at yourself in a mirror 99 times and then, on the 100th occasion, your other hand waved back! As well as being extremely shocked, a possible explanation that might occur to you is that, under some conditions, the laws of reflection may not always be true. In other words, the reflection symmetry is broken. This might also lead you to think that some hitherto hidden electromagnetic interaction, that you didn't know about before, was at work.

As we will see later, particle physicists have observed a number of analogous situations in their experiments. Theoretical particle physicists are also able to investigate analytically what happens when a particular symmetry gets broken. They also get very excited when it is announced that a physical symmetry has been found to be violated – it tells them that there must be a new particle or a new property of nature waiting to be discovered!

Box 6.1 Group theory

Symmetry is described by a branch of mathematics called **group theory**. To see how group theory works, let's consider the six symmetries of an equilateral triangle whose vertices are labelled 1, 2 and 3 (Figure 6.10).

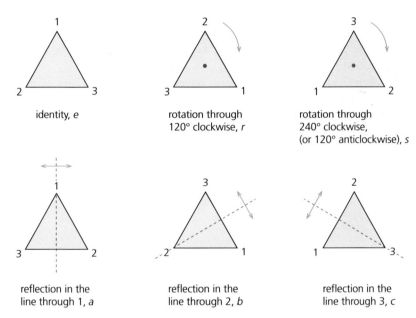

Figure 6.10 The six symmetries of the equilateral triangle

These are:

- *e*, the identity (this simply means 'do nothing')
- *r*, rotation clockwise through 120°
- *s*, rotation anticlockwise through 120° (or 240° clockwise)
- *a*, reflection in the line through 1
- *b*, reflection in the line through 2
- *c*, reflection in the line through 3

We can represent these symmetries as transformations in the form of a **symmetry group** governed by a **group operation** × as shown in Table 6.2.

The group has six members, but note that the group operation × does not mean 'multiply' here. To understand how it works, consider the operation *e* × *e*. This means perform the transformation 'do nothing' followed by the transformation 'do nothing'. The effect on the triangle of doing nothing twice is to do nothing to the triangle! So we write the effect of *e* × *e* as equal to *e*. This result is inserted in Table 6.2 at the intersection of row *e* and column *e*.

Now consider the operation $b \times s$. This means 'reflect in the line through 2' followed by 'rotate clockwise through 120°'. The effect of this is equivalent to another member of the group, c, which is the operation 'reflect in the line through 3'. This is found at the intersection of row b and column s in Table 6.2. You should be able to convince yourself that other combinations of group operations yield the results shown in Table 6.2. (The best way to do this is to cut out a paper equilateral triangle, label and mark it with lines of symmetry and move it around.) You should note that in group theory the operation $\times$ is *not* commutative. For example, for the operations we have just looked at for the symmetry group in Table 6.2, $b \times s = c$ is *not equal* to $s \times b = a$.

Notice that each combination yields another member of the group. There are no combinations that give results that are outside the group. A group with this property is said to be **closed**. If a group that represents an object has N rows and N columns, then N is called the **dimension** of the group. The dimension of the group in Table 6.2 is 6.

Table 6.2 The symmetry group for the equilateral triangle

		Second transformation				
$\times$	e	r	s	a	b	c
e	e	r	s	a	b	c
r	r	s	e	b	c	a
s	s	e	r	c	a	b
a	a	c	b	e	s	r
b	b	a	c	r	e	s
c	c	b	a	s	r	e

(First transformation labels the rows; columns are the Second transformation.)

What we have given here is an example of a *geometrical* symmetry group. In particle physics, we also encounter symmetry groups. The symmetries are to do with the quantum properties of a system of particles, such as their charge and spin, and are more abstract and so harder to visualise. However, the same mathematical principles apply. Groups that appear in particle physics are closed, with 'room' for only certain kinds of particles with specific properties. These groups represent symmetries inherent in the laws of physics, which help us to understand how forces arise and can also be used to infer the existence of new particles. Particle physicists specify these groups as a 'group name' followed by the dimension in brackets. An example is the group that represents the electromagnetic interaction, which is called U(1), where 'U' stands for 'Unitary group' and the dimension of the group is 1.

The concept of symmetry and its manipulation using group theory has proved to be an outstanding mathematical tool that particle physicists use to understand the nature of matter. The group representing the electromagnetic interaction, U(1), and the group representing the weak interaction, called SU(2) ('Special Unitary', dimension 2), when combined together form a composite symmetry group SU(2) $\times$ U(1), which predicted the existence of particles that carry the weak force (see Chapter 7).

Conservation laws in particle physics

Noether's Theorem

One of the most important consequences of symmetries is that they give rise to **conserved quantities**. This principle was first put forward by the German mathematician and theoretical physicist Emmy Noether (Box 6.2). She discovered that the symmetries in the mathematical equations that describe a physical system automatically produce conserved quantities. The symmetry is preserved if a physical law remains invariant under transformations of translation through space, translation through time and rotation about an axis.

For example, under translation of space, the law of gravity is the same in London as it is in Edinburgh. Under time, the law is still the same at midday as it is at midnight. It is also unaltered by the fact that London and Edinburgh rotate around the Earth's axis of rotation.

You are probably already familiar with conserved quantities like the laws of conservation of energy and of momentum. The law of conservation of energy arises from the fact that energy equations are invariant with translations in time – the total energy expenditure in a system of particles remains the same regardless of what time it is. Conservation of momentum is due to the equations being invariant to translations in space. If two *identical* games of snooker are played that are separate in space but in which all the motions of the balls are the same, then you would expect the same player to be 'snookered' whether the game was played in London or Edinburgh. These conservation laws of physics are closely connected with the properties of space and time, and in general physicists believe them to be inviolable.

However, in particle physics, other conservation laws come into play. For example, charge is conserved in particle reactions. In nuclear equations, we balance the A and Z numbers, which tell us the elements produced in the nuclear reaction, and this might lead us to believe that nuclear processes conserve *both* the proton and neutron numbers. However, this is *not* true in beta decay, that is, when a neutron changes into a proton

$$n \rightarrow p + e^- + \bar{\nu}_e$$

This process *does not* conserve neutron number or proton number, but it does conserve the total neutron *plus* proton number; that is, the nucleon number A is the same on both sides of the reaction. This might lead us to be cautious about making any generalisations involving particle decays. However, conservation laws can provide us with a way of understanding why some particle decays occur and others do not.

From considerations of symmetry and using Noether's Theorem, particle physicists have established a number of other conservation laws that enable them to predict what particles will do when they interact with each other and what we can expect to see.

Box 6.2 Emmy (Amalie) Noether

Emmy Noether (1882–1935) (Figure 6.11) is generally regarded as one of the greatest female mathematicians of all time. Her work on group theory and the mathematical description of symmetry forms the basis of much of modern theoretical particle physics.

Noether was the daughter of the mathematician Max Noether (1844–1921), who was Professor of Mathematics at the University of Erlangen in Germany. She studied at Erlangen and at the University of Göttingen, and received her PhD in 1907. This was despite much opposition to allowing women to register as research students at the time. Her key work was carried out in Göttingen from 1927 onwards. In 1933, when the Nazis came to power in Germany, she fled to the USA, where she was offered a position at the Institute for Advanced Study at Princeton.

In an important result called **Noether's Theorem** she discovered that symmetries imply conservation laws. For example, the conservation of energy comes about due to the equations describing energy being symmetrically invariant with translations in time. Similarly, it can be shown that conservation of momentum results from invariance in translations in space. The converse (that a conserved quantity implies a symmetry) is not always true. Noether's Theorem is a very important and fundamental result for particle physics, and whenever a conserved quantity in particle physics is found, particle physicists look very hard for a corresponding symmetry. In this way, Emmy Noether's work has helped to bring direction and order to particles and their interactions, long after her death in 1935.

Figure 6.11 Emmy Noether (1882–1935), mathematician

Conservation of lepton number *L*

When we discussed beta decay in Chapter 5, you will have noticed that, in β^- decay, an antineutrino, and never a neutrino, is emitted with the electron, whereas, in β^+ decay, it is always a neutrino emitted with the positron. We assign each of these decays involving leptons a **lepton number** *L*. Leptons such as the electron, muon, tau and their neutrinos have lepton number +1. Their antiparticles have lepton number −1, whereas mesons and baryons are assigned a lepton number of 0. Leptons and their lepton numbers are summarised in Table 6.3.

Table 6.3 Families of leptons and lepton number

Family	Particle	Symbol	Lepton number	Antiparticle	Symbol	Lepton number
ELECTRON	Electron	e^-	+1	Positron	e^+	−1
	Electron neutrino	ν_e	+1	Electron antineutrino	$\bar{\nu}_e$	−1
MUON	Muon	μ^-	+1	Antimuon	μ^+	−1
	Muon neutrino	ν_μ	+1	Muon antineutrino	$\bar{\nu}_\mu$	−1
TAU	Tau	τ^-	+1	Antitau	τ^+	−1
	Tau neutrino	ν_τ	+1	Tau antineutrino	$\bar{\nu}_\tau$	−1

There are in fact three separate conservation laws for lepton number, which correspond to the three varieties of lepton, e, μ and τ. To see how this works, consider β^- decay:

$$n \rightarrow p + e^- + \bar{\nu}_e$$
$$L = 0 \rightarrow 0 + 1 + (-1) = 0$$

The neutron and proton are baryons, so they have $L = 0$, and the electron e^- and electron antineutrino $\bar{\nu}_e$ have $L = +1$ and −1 respectively. You should see that the total lepton number (zero) in β^- decay remains the same for both sides of the decay. Similarly, for β^+ decay:

$$p \rightarrow n + e^+ + \nu_e$$
$$L = 0 \rightarrow 0 + (-1) + 1 = 0$$

and, once again, total lepton number is conserved.

When we are observing leptons, we must make a note of each type of lepton (e, μ and τ) separately. There is a distinction between electron- and muon-type leptons, which is shown by the following processes. Experiments with particle accelerators have been carried out where beams of muon antineutrinos $\bar{\nu}_\mu$ are incident on a proton target and the following reaction takes place:

$$\bar{\nu}_\mu + p \rightarrow n + \mu^+$$

Now if there was no difference between electron- and muon-type leptons (that is, μ^+ and e^+), then the following reaction should also be possible:

$$\bar{\nu}_\mu + p \rightarrow n + e^+$$

However, this reaction is *never* observed. This indicates that there must be fundamental differences between the types of leptons, and we need to take this into account when we calculate the total lepton number of a particle reaction. We give the symbols L_e, L_μ and L_τ to these separate lepton numbers for the different families of leptons. Worked Example 6.1 illustrates how we do this.

WORKED EXAMPLE 6.1

Show that lepton numbers for electron-type leptons, muon-type leptons and tau-type leptons must each remain constant in the following reactions:

a $\bar{\nu}_e + p \rightarrow e^+ + n$
b $\nu_\mu + n \rightarrow \mu^- + p$
c $\mu^- \rightarrow e^- + \bar{\nu}_e + \nu_\mu$

a All leptons in this decay are from the electron family.

$$\bar{\nu}_e + \quad p \rightarrow e^+ + \quad n$$
$$L_e = -1 + 0 \rightarrow -1 + 0 = -1 \qquad \text{lepton number is conserved}$$

b All leptons in this decay are from the muon family.

$$\nu_\mu + \quad n \rightarrow \mu^- + \quad p$$
$$L_\mu = 1 + \quad 0 \rightarrow 1 + \quad 0 = +1 \qquad \text{lepton number is conserved}$$

c This is a muon decay that involves leptons from *both* the electron and muon families. We have to consider lepton numbers for both electron and muon neutrinos and show that the lepton numbers L_e and L_μ are conserved for both kinds of leptons separately.

$$\mu^- \rightarrow \quad e^- + \quad \bar{\nu}_e + \quad \nu_\mu$$
$$L_e = 0 \rightarrow \quad 1 + \quad (-1) + 0 = 0 \qquad \text{lepton number is conserved for electron family}$$

$$L_\mu = +1 \rightarrow \quad 0 + \quad 0 + \quad 1 = +1 \qquad \text{lepton number is conserved for muon family}$$

It is important to understand that, when we have a decay involving leptons of different families, then we apply lepton numbers to that family only. Leptons in other families are assigned the value 0. Hence, for the decay in c where we consider L_e, the μ^- and ν_μ are *not* members of the electron family of leptons, and are therefore assigned the value 0. Similarly, for conservation of L_μ, the e^- and $\bar{\nu}_e$ are not members of the muon family of leptons, so are given lepton number 0.

In the decay in part c of Worked Example 6.1, both L_e and L_μ are conserved, so muon decay by this process is possible. This leads us to a general statement about a conservation law involving leptons. We state the law of **conservation of lepton number** as follows.

In processes involving particle reactions, the lepton numbers for electron-type leptons, muon-type leptons and tau-type leptons must remain the same.

By imposing the constraint that lepton number must be conserved in particle reactions, particle physicists can explain why neutrinos and sometimes antineutrinos (or sometimes both in the case of muon decay) must be observed.

Conservation of baryon number *B*

We can state another conservation law, this time involving baryons. We assign a baryon number (B) of $B = +1$ to baryons, $B = -1$ to antibaryons, with all non-baryons (mesons and leptons) having $B = 0$. We state the law of **conservation of baryon number** as:

In any particle reaction, the total baryon number remains the same.

The conservation of nucleon number A is a special case of conservation of baryon number in which all the baryons happen to be either protons or neutrons. While nuclear physicists use the symbol A to represent nucleons, particle physicists always use the symbol B to represent *all* baryons, including nucleons. Conservation of baryon number is illustrated in Worked Example 6.2.

WORKED EXAMPLE 6.2

a The antiproton $\bar{p}$ was discovered by means of the following particle reaction:

$$p + p \rightarrow p + p + p + \bar{p}$$

Confirm that baryon number is conserved.

b Is the following reaction possible?

$$p + p \rightarrow p + p + \bar{n}$$

a
$$p + p \rightarrow p + p + p + \bar{p}$$
$$B = 1 + 1 \rightarrow 1 + 1 + 1 + (-1)$$

The total baryon number is $+2$ on both sides, therefore B is conserved, and the reaction is possible. (This is the same as Worked Example 5.2 in Chapter 5, where we saw that fermions are conserved in reactions.)

b
$$p + p \rightarrow p + p + \bar{n}$$
$$B = 1 + 1 \rightarrow 1 + 1 + (-1)$$

The right-hand side of the equation is $+1$ but the left-hand side is $+2$. The total baryon number is *not* conserved, since B is not the same, and this reaction is therefore forbidden.

No violation of the law of conservation of baryon number has ever been observed, although in Chapter 8 we will see that *Grand Unified Theories* suggest that the proton might decay in a manner that could violate this law.

Conservation of strangeness

In Chapter 5 we introduced a new quantum number called *strangeness*. Recall that strangeness (S) was introduced to explain why some hadrons like the K mesons (K) and the lambda (Λ^0) had unexpectedly long lifetimes when interacting with the strong force. All particles are either strange or non-strange. In K meson decay, the K^0 and the K^+ are assigned strangeness $S = +1$, and K^- has $S = -1$, whereas π mesons and leptons are non-strange particles and have $S = 0$. In addition, particle physicists find that the K meson and the sigma (Σ) are always produced in pairs. For example, if a beam of π mesons is incident on the protons in a bubble chamber, then the reaction

$$\pi^+ + p \rightarrow K^+ + \Sigma^+$$

is often observed. However, the reaction

$$\pi^+ + p \rightarrow \pi^+ + \Sigma^+$$

is never observed, even though it conserves charge and baryon number (the Σ^+ and p are baryons with $B = +1$; the π^+ is a meson with $B = 0$).

Strangeness can be used to explain the properties of certain particle decays. Particle physicists have found that strangeness is always conserved in interactions involving the strong force or electromagnetic force, but in the weak interaction the strangeness is either conserved or changes by $+1$. We can state this as the law of **conservation of strangeness**:

For particle interactions governed by the strong or electromagnetic forces, the total strangeness must remain the same.
For interactions governed by the weak force, the strangeness either remains the same or changes by one unit.

Strangeness conservation can explain why strange particles are always produced in pairs. π mesons are produced in nuclear collisions and, since there is no 'meson conservation law', in theory any number could be produced. But the K mesons and other strange particles are always produced in pairs. For example, for any general K meson, we could have

$$p + p \rightarrow p + p + K + K$$

Since the proton is a non-strange particle with $S = 0$, this reaction could be explained if the first K meson in this reaction is a K^+ or K^0 with $S = +1$ and the other is a K^- or $\overline{K}^0$ with $S = -1$. Similarly, the reaction

$$\pi^+ + p \rightarrow K^+ + \Sigma^+$$

can be explained if the Σ^+ has $S = -1$. If we do this, we see that the total strangeness number (in this case 0) is the same on both sides and is therefore conserved. This phenomenon, where strange particles are produced in pairs, is called **associated production**.

Conservation of strangeness also tells us what forces particles interact with and which reactions *cannot* happen. For example, baryons come in strange and non-strange kinds. The Λ^0 is a strange particle that is observed to decay by the process

$$\Lambda^0 \rightarrow p + \pi^-$$

with a lifetime of about 10^{-10} s. But, as it is a baryon, we would expect this strongly interacting particle to decay to other strongly interacting particles with characteristic lifetimes of 10^{-23} s. However, if we assign $S = -1$ to the Λ^0, strangeness is not conserved (the p and π^- are non-strange) and so the decay cannot go by the strong interaction. It must therefore be due to the weak interaction, which gives its relatively long lifetime of 10^{-10} s.

Conservation of strangeness number tells us why the decay

$$\Sigma^0 \rightarrow \Lambda^0 + \gamma$$
$$S = -1 \rightarrow -1 + 0 = -1$$

can happen by the strong interaction, but the decay

$$\Lambda^0 \rightarrow n + \gamma$$
$$S = -1 \rightarrow 0 + 0 \neq -1$$

cannot.

The weak interaction can change the strangeness by *at most one unit*. Consider, for example, the decay of the Ξ^0 which has $S = -2$:

$$\Xi^0 \rightarrow n + \pi^0$$
$$S = -2 \rightarrow 0 + 0$$

Here the change in strangeness number is greater than one unit, so this decay cannot happen at all.

Do not be mystified by the use of the term 'strangeness' to describe particles. There is nothing inherently strange about them in the intuitive sense. They were only called 'strange' because their behaviour was initially unexpected, and strangeness is simply a quantum property of particles just like charge or spin. Particle physicists do not know what lepton number, baryon number and strangeness really mean. They are merely useful concepts that emerge by considering mathematical symmetries and their associated conservation properties, which enable us to describe the ways in which particles can decay into other particles. The name 'strangeness' stuck for historical reasons, however. We will see later that particle physicists have a fondness for labelling the properties of particles with whimsical names!

Symmetry transformations in physical processes

In our discussions concerning particles we have used terms like 'reactions', 'decays' and 'processes' to describe how matter changes from one form to another. A particle (or nuclear) reaction is a change in the identity of a particle (or nucleus) leading to the production of other particles or nuclei. Reactions and decays are physical processes in which something changes as opposed to remaining the same. A particle reaction is an example of a specific process – one that is governed by a **physical law**. But how we observe it depends on the way we *look* at it.

Consider viewing a clock with moving hands. This is a physical process, and you can compare the direction of movement of the hands with the way your fingers curl round your thumb in your left hand. Now look at the image of the clock in a mirror (Figure 6.12). You now see a different process. The hands of the clock move in the opposite direction – in the way your fingers curl round your thumb in your *right* hand. In the real world clock hands move clockwise, but in the 'mirror world' they move *anticlockwise*. Now it is important to understand that a clock with hands that move anticlockwise is a perfectly possible process. Any competent clockmaker could make you a clock whose hands move backwards (whether it would be of any practical use is another question), but such a clock could be constructed without violating any known law of physics. This leads us to ask the following question:

Is there any physical process whose mirror image is impossible?

If the answer to this question is 'yes', then it means that nature makes a physical distinction between left and right.

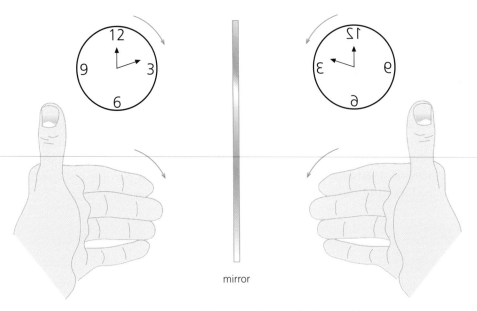

Figure 6.12 A clock whose hands move 'anticlockwise' is a perfectly possible process

Parity

One symmetry that is encountered in nuclear and particle physics is called **parity** and is denoted by the symbol **P**. Parity is akin to looking at a physical process in a mirror. A measurement of a particle reaction is usually observed from a coordinate reference frame, which may be chosen in an arbitrary way. We refer to its **space inversion** when the particle system is viewed from a 'mirror' perspective. For example, if we started looking at a system using a right-handed coordinate set, then under **P** we now view it from a left-handed set (Figure 6.13). If the particle system behaves in the same way under a space inversion, then it is invariant under **P**, and we say that the parity is *preserved*. In terms of the quantum mechanical wavefunction ψ, **P** applied to a one-dimensional wavefunction (considering just the x coordinate) is written as

$$\mathbf{P}(\psi(x)) = \psi(-x)$$

This simply means that replacing a physical variable x by $-x$ does not change the value of the function, just like for example the function $y = x^2 = 4$ when $x = +2$ or $x = -2$.

Under **P**, the wavefunction may either remain unchanged, so that $\mathbf{P}(\psi) = \psi$, in which case we say that it is in an *even* parity state; or it may change sign, so that $\mathbf{P}(\psi) = -\psi$, in which case it is in an *odd* parity state.

To see how parity is useful, think about how light is emitted from an atom. If we assume that the quantum state of an electron in an atom has a wavefunction that has a definite parity, which can be either odd or even, and that the electromagnetic force governing the emission of light respects parity, then within the atom *only* transitions that preserve the same definite parity of the quantum system will be allowed. In this way, parity conservation helps us to decide which electron transitions are possible and which are excluded.

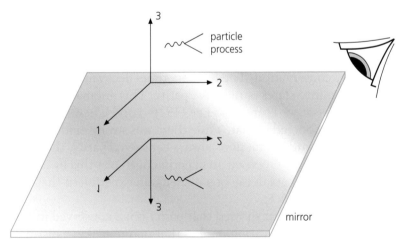

Figure 6.13 A space inversion is when a particle system or process is viewed in a mirror. If the particle process behaves in the same way, then we say that its parity is preserved

In particle decays, each particle is assigned an odd or even parity (-1 or $+1$). The total parity of the decay is found by *multiplying* the individual parities of the particles together to give either a positive ($+1$) or a negative (-1) value of one. Note that, unlike lepton number or strangeness, which are *additive*, parity is a *multiplicative* quantum number, and we must multiply the various parities together to find the net parity.

(In fact, there are two kinds of parities involved in particles, one pertaining to its wavefunction, which we have just seen, and a parity that is intrinsic to the particle itself. The net parity of a system of particles is the product of all these parities, but the same principle applies.)

The main point to understand about parity is this:

Parity is the conserved quantum number that expresses the conservation of symmetry between left and right in a process when the physical law involved is invariant under space inversion.

Processes in which parity is *not* conserved would look different in a mirror image world.

WORKED EXAMPLE 6.3

Confirm that the η (eta) particle, which has a lifetime of 8.0×10^{-19} s, can decay via the strong interaction through the process

$$\eta \rightarrow \pi^+ + \pi^- + \pi^0$$

The lifetime of the η makes this a strong interaction process. The η has odd parity (-1). The π mesons also have odd parity, so

$$\eta \quad \rightarrow \quad \pi^+ \quad + \quad \pi^- \quad + \quad \pi^0$$
$$P = (-1) \rightarrow (-1) \times (-1) \times (-1) = (-1)$$

The parity is the same on both sides, so this decay is allowed by conservation of parity.

However, some puzzling features of the decay of the K^+ meson hinted that all was not well with parity conservation. The K^+ sometimes decayed into states of positive parity and sometimes into states of negative parity. This anomaly was investigated by the two American physicists Tsung Dao Lee (1926–) and Chen Ning Yang (whom we met earlier). They concluded that, while there was strong evidence that the electromagnetic and strong interactions always conserved parity, the weak interaction might not. They proposed an experiment to test this by checking for parity conservation in the beta decay of a nucleus, an experiment that was carried out by C. S. Wu *et al.* (Box 6.3).

The result of this experiment showed that parity is *not* conserved in the weak interaction, and its non-conservation implies that nature makes a distinction between left and right. The mirror-image process of the K^+ decay is *not* the same as its non-mirror counterpart!

Box 6.3 Nature distinguishes between left and right

In 1956, particle physicists knew of two recently discovered particles, the 'theta' (θ^+) and the 'tau' (τ^+), which decayed by the following processes:

$$\theta^+ \rightarrow \pi^+ + \pi^0$$
$$\tau^+ \rightarrow \pi^+ + \pi^+ + \pi^-$$

In the first case two π mesons are produced, and in the second we observe three π mesons. Detailed studies of these decays showed that the lifetimes, spins and masses of the θ and τ were the same, and their decays were slow, with characteristic times of the order of the weak interaction. Now if this is the case, then particle physicists might conclude that the θ and the τ are the same particle decaying in different ways. But there's a problem. If the laws of physics, including the weak interaction, are invariant under space inversion, then parity should be conserved. The π meson has odd parity, so that for the θ with two π mesons the total parity is $(-1) \times (-1) = (+1)$, which is even; and for the τ the total parity is $(-1) \times (-1) \times (-1) = (-1)$, which is odd. If this is the same particle, then the parities are different. Lee and Yang suggested that the reason for this is that parity might *not* be conserved under the weak interaction.

To test for parity conservation in the weak interaction, an experiment was performed by the Chinese-American physicist Chien Shiung Wu (1912–1997) and her colleagues using nuclei of cobalt-60, $^{60}_{27}\text{Co}$. A cobalt-60 nucleus decays by β^- emission via the process

$$^{60}_{27}\text{Co} \rightarrow {}^{60}_{28}\text{Ni} + {}^{0}_{-1}\beta + \bar{\nu}$$

and behaves like a tiny magnet since it has both charge and spin. The idea was to line up cobalt-60 nuclei using an external magnetic field and see if more electrons are produced in one direction than in the opposite direction. The cobalt-60 had to be cooled to temperatures of 0.01 K to prevent random thermal motions of the nuclei destroying their alignment with the external magnetic field. An idealisation of Wu's experiment is shown in Figure 6.14.

We can represent the external magnetic field by means of a circular current loop (remember that an electric current produces a magnetic field). Figure 6.14a shows the experiment and Figure 6.14b its mirror image. When the current is flowing in the directions shown (both in the 'real world' and 'mirror world'), more electrons are always detected at A than at B. In the mirror, the current appears to be reversed but more electrons still appear at A than at B. If this experiment was invariant under space inversion, then we would expect more electrons to appear at B in the mirror world. This asymmetric result is the significance of Wu's experiment. Since space inversion is not preserved, the parity of the system is not preserved. The cobalt-60 was found to emit electrons preferentially in a direction opposite to that of the nuclear spin (Box 5.1). Since the weak interaction is responsible for beta decay, it seems that it can tell 'left' from 'right' and we are led to conclude from Wu's experiment that *parity is not conserved under the weak interaction*. Note what we mean by 'left' and 'right'. Particles don't have left and right hands, and neither does the weak interaction! Left and right are merely conventions used to distinguish a process from its mirror image, and they just happen to be convenient and familiar ones for us to use.

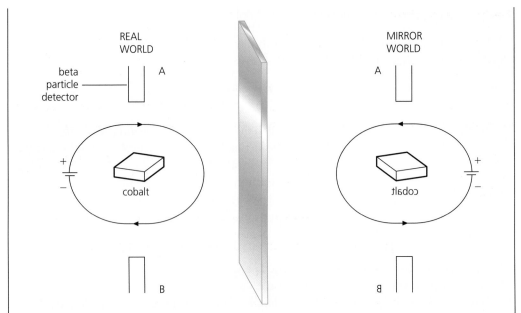

Figure 6.14 An idealisation of the experiment by C. S. Wu and her collaborators. In the mirror, the current appears to be reversed, but more electrons are still detected at A than at B

The electron produced by a decaying muon also violates parity. Muons have spin, and a real decaying muon releases an electron to the left of the direction of spin – so a true mirror image would emit electrons to the *right* of the direction of spin. Instead, we find that it too emits electrons to the *left* (Figure 6.15).

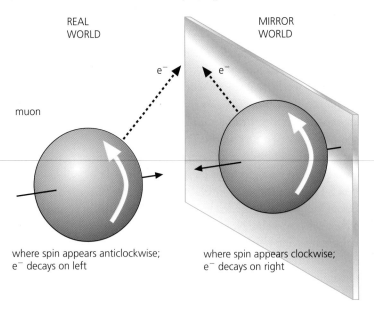

Figure 6.15 Parity violation in muon decay

Charge conjugation

In Chapter 5 we saw that Dirac's prediction of antimatter and the subsequent discovery of antiparticles show that there is a physical difference between matter and antimatter inasmuch as they have opposite electric charges. But can we *distinguish* matter from antimatter? What we want to know is, when particle physicists study a process that involves particles of matter, would the *same* process be possible with particles of antimatter? If the answer is 'no', then we have found a physical distinction between matter and antimatter.

We don't have large lumps of antimatter available on which to experiment, but particle physicists have a ready collection of antiparticles to use in an 'anti-process'. For example, the process of a neutron decaying into a proton

$$n \rightarrow p + e^- + \bar{\nu}$$

would have the anti-process of an antineutron decaying into an antiproton

$$\bar{n} \rightarrow \bar{p} + e^+ + \nu$$

Notice how the charges are opposite. This operation of replacing a process with its antiparticle counterparts is called **charge conjugation** and is given the symbol **C**. In a charge conjugation version of a process involving a system of particles, the mass, lifetime, spin, and position and velocity coordinates *do not* change. The charge conjugate version is simply obtained by replacing all the particles by their corresponding antiparticles.

Particle physicists have found that the electromagnetic and strong interactions *do not* distinguish matter from antimatter, and there is no evidence that the gravitational interaction does so either. However, when the weak interaction was found to distinguish left from right by violating **P**, they wondered whether the weak interaction might be able to tell the difference between matter and antimatter as well. The conversion of the neutron into a proton by the emission of a beta particle is a weak decay process, and so this provided a method by which to see if the weak interaction does in fact distinguish matter from antimatter. Amazingly, it turned out that the spin of the neutrino in weak decays showed that there was a loss of invariance under charge conjugation!

To see why, quantum theory tells us that, in every weak decay in which a neutrino is emitted, the spin of the neutrino is related by the left-hand rule illustrated in Figure 6.16a. In other words, the neutrino from the decay of an antineutron is spinning in a sense that appears *clockwise* as it travels towards you, but the antineutrino from the decay of a neutron is spinning in a sense that is *anticlockwise* as it travels towards you (Figure 6.16b). Particle physicists call this 'handedness' of a quantum particle, which is related to its spin, **helicity**.

Consider the helicity for neutron decay and its charge conjugate counterpart:

$$n \rightarrow p + e^- + \bar{\nu} \quad \text{(antineutrino, right-handed helicity)}$$
$$\bar{n} \rightarrow \bar{p} + e^+ + \nu \quad \text{(neutrino, left-handed helicity)}$$

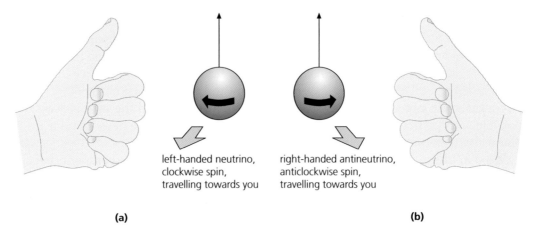

left-handed neutrino,
clockwise spin,
travelling towards you

right-handed antineutrino,
anticlockwise spin,
travelling towards you

(a)

(b)

Figure 6.16 Rules for spinning neutrinos

In another weak decay in which neutrinos are produced, we have:

$$\pi^- \rightarrow \mu^- + \bar{\nu} \qquad \text{(antineutrino, right-handed helicity)}$$
$$\pi^+ \rightarrow \mu^+ + \nu \qquad \text{(neutrino, left-handed helicity)}$$

While other particles can exist in either state, particle physicists always see right-handed spinning antineutrinos and left-handed spinning neutrinos.

We can now make a physical distinction between matter and antimatter. If you begin with a beam of neutrons and observe them changing into protons, you will always find right-handed antineutrinos. But the charge conjugate version of this process will produce *left-handed* neutrinos. Under **C** spin should not change but in this process it does. So we have a physical distinction between left and right. Left-handed neutrinos exist but right-handed neutrinos do not. The symmetry between processes starting as charge conjugate versions is violated. So we are led to the conclusion that *weak processes do distinguish matter from antimatter.*

CP invariance

Particle physicists have found that symmetry can be restored in weak interaction processes where *both* **C** and **P** transformations are applied together. Such a process is said to be **CP invariant**. In the late 1950s and early 1960s it was hoped that the non-invariances in **C** and **P** that had been found in weak processes would cancel each other out so that the combination of **CP** invariance would always be preserved. However, they found to their surprise that even this symmetry was violated!

The process involving the decay of the neutral K meson (K^0) provides a very sensitive test of **CP** invariance. Recall (Box 6.3) that Lee and Yang explained the θ–τ puzzle by theorising that the K^+ meson cannot decay both into two π mesons (θ decay) and into three π mesons (τ decay) unless the law of conservation of parity is violated. The K^0 or *neutral* K meson as a single particle was *not* allowed under the

CP theory to decay by two-π-meson and three-π-meson processes. This was confirmed when further experiments revealed that there were *two* versions of the neutral K meson, which decayed by similar two-meson and three-meson processes:

- the short-lived K_S^0 meson, which decayed via

$$K_S^0 \to \pi^+ + \pi^- \qquad (\theta\text{-type decay})$$

- the long-lived K_L^0 meson, which decayed via

$$K_L^0 \to \pi^+ + \pi^- + \pi^0 \qquad (\tau\text{-type decay})$$

The long-lived neutral K meson K_L^0 has a half-life about 600 times greater than the K_S^0 meson. However, in 1964, a team at the Brookhaven National Accelerator Laboratory in the USA found a small number of K_L^0 meson events in which the long-lived neutral K meson decayed into *two* π mesons by θ decay. The number of K_L^0 θ decays was small, about 0.3% of all K_L^0 decays, but has been confirmed in many experiments involving neutral K mesons. This was a totally unexpected result, and could be explained if the weak interaction *does not* conserve **CP** invariance, although the violation happens rarely.

The K_L^0 meson is also known to decay by the processes

$$K_L^0 \to \pi^+ + e^- + \bar{\nu} \qquad (\text{spin of electron is left-handed})$$

and

$$K_L^0 \to \pi^- + e^+ + \nu \qquad (\text{spin of positron is right-handed})$$

in which electrons and positrons are produced.

An important point to note here is that, in all these processes we have been talking about involving **C** and **P** transformations and their results under different interactions, we are taking the *average* of a large number of results. This is because decay processes are statistical. For example, in radioactive decay, *on average* nuclei will decay with the expected half-life characteristic of a particular nucleus, although some will decay with a shorter life and some with a longer one.

Particle physicists at Brookhaven studied 10 million K_L^0 decays and found that on average there were always slightly more positrons produced than electrons. The electrons come out spinning towards you in a left-hand sense, whereas the positrons spin towards you in a right-hand sense. This asymmetry means that you could tell anyone in the universe what you mean by 'left' and what you mean by 'electron' since the electrons from K_L^0 decay spin in a left-hand sense and are produced slightly less often than positrons.

The failure of **CP** invariance in the weak interaction means that there is a small tendency for the decay of the K_L^0 to increase the number of positrons in the universe, since there are, on average, very slightly more positrons than electrons produced in these decays. The gravitational, electromagnetic and strong nuclear interactions are **CP** invariant since they are invariant separately under the **C** and **P** transformations.

Time reversal symmetry

Despite its exotic sounding title, this simply means that the behaviour of a quantum system is the same if we reverse the direction of motion of the particles in the system, rather like running a film backwards. Time reversal is denoted by T and transforms the time coordinate of the wavefunction to its negative value. If we denote the one-dimensional *time-dependent* wavefunction as $\psi(x, t)$, then

$$T(\psi(x, t)) = \psi(x, -t)$$

which has the effect of reversing the direction in time of all the processes in the system. Under T, particle reactions are equally capable of proceeding forward in time as well as backward. A general theorem of quantum field theory (the proof of which is beyond the scope of this book), called the **CPT** Theorem, asserts that any violation of **CP** invariance must be compensated by a violation of time reversal invariance. The **CPT Theorem** states that:

Every known law of physics is invariant under the combination of the symmetry transformations of charge conjugation, space inversion (parity) and time reversal.

It is because of this invariance across all the fundamental interactions that particle physicists say that **CPT** invariance is an **exact symmetry**. The violation of **CP** invariance by weak processes thus has important consequences for T, as it implies that some microscopic processes are not reversible.

In large-scale systems we are used to seeing asymmetries in time in terms of thermodynamic processes – for example, a hot cup of coffee will eventually get cold as time flows into the future. But an asymmetry in time in submicroscopic processes was totally unexpected by particle physicists. Its existence is only inferred by the **CPT** Theorem; its occurrence is rare, and only in conjunction with the decay of the long-lived neutral K meson. This is the only evidence we have, to date, that nature distinguishes the past from the future on the submicroscopic level, and why it should do so is completely unknown.

We may also add that every law of physics that is **CP** invariant is also T invariant. An intuitive definition of the **CPT** Theorem might be expressed as a film of a process run backwards as seen through a mirror, representing a theoretically possible process in which all the particles have been replaced by their antiparticles. Using the **CPT** Theorem, particle physicists have inferred a time *asymmetry* in the laws governing K_L^0 decay, which means that nature makes a rare distinction between past and future. So far, no other effects of this kind have been detected with any of the other interactions, but this is not ruled out.

So what can these symmetry violations tell us about nature? Firstly, they tell us that the physical laws of nature are not exactly the same for particles and antiparticles. For every particle there is a corresponding antiparticle, and we can in principle make antimatter. This has been done recently where particle physicists at CERN have made atoms of antihydrogen (see Box 5.2, page 155). As far as we know, atoms of antimatter appear to work in the same way as those of ordinary matter, but now that we have seen that **CP** invariance fails with weak interactions we have a way of distinguishing between them.

The second thing is that symmetry invariance and violations can enable us to predict how particles will behave, and more importantly, as we will see in the next chapter, symmetries (and their lack) can be used to infer the existence of other particles and point to a way of unifying the forces of nature.

An arrow of time

If the laws of physics were symmetrical under time reversal, then we might expect the world (which is governed by these laws) to be approximately symmetrical under time reversal as well. But it is obvious that life isn't like that! A falling plate that breaks never repairs itself. If two liquids are mixed together, they will never return to their unmixed state by themselves. Stars emit light rather than absorb it. You and I grow older, rather than younger. All these are examples of processes that are not reversible in time, and are all large scale.

However, we have seen that the decay of the K_L^0 meson provides a microscopic distinction between past and future by relying on the **CPT** Theorem, because the symmetry associated with the **CP** transformation was lost, which implies that **T** is not reversible. Remember that **C** invariance fails because more positrons than electrons are produced in decays, and **P** invariance fails because the positrons are right-handed and the electrons are left-handed. Together, **CP** invariance fails because there are more right-handed positrons than left-handed electrons. The consequence is that **T** invariance must also fail, in order to preserve the overall **CPT** exact symmetry as required by the **CPT** Theorem.

Particle physicists wish to know if there exists some physical law that explains why time should flow in a particular direction. This is what we regard intuitively as an 'arrow of time' – in other words, the direction in which time flows from the past to the future that we see in complex processes involving large systems of atoms and molecules.

Thermodynamics, which is the study of processes that involve heat changes, can provide us with one way of defining a direction in time. Thermodynamics tells us how a system of isolated particles such as air molecules change in time when they are given an initial amount of energy. The particles always tend to disorder – from being more organised to becoming less organised. This is just like, for example, steam emerging from a boiling kettle, which doesn't stay as steam, but diffuses across the room, cooling as it does so. This tendency to disorder is quantified as an increase in **entropy**. Entropy is a measure of the degree of disorder. In energy terms, a change in entropy ΔS may be defined as the ratio of the amount of heat received by a system (ΔQ) to the absolute temperature (T) at which the heat is absorbed, that is,

$$\Delta S = \frac{\Delta Q}{T}$$

The total entropy of any isolated system (by 'isolated' we mean that there is no external input of energy) can never decrease in any change. It must either remain constant or increase.

In the universe as a whole, the radiated energy given out by stars goes from a concentrated and organised state to one that is spread out and becomes less organised (although still very energetic) as time passes. Overall, the entropy of the universe is *increasing*, tending towards a maximum corresponding to complete disorder of the particles in it. Implicit in this prediction is the assumption that the universe as a whole is an isolated system and there is no extra-universe input of energy. In this way the direction of time is compared with that of increasing disorderedness. At some distant time in the future, all the particles in the universe will be in a disordered unconcentrated state of minimum energy.

In the quantum world, things are not so clear-cut. While the decay of the K_L^0 meson is not time invariant as a process, it hardly affects how we perceive our direction of time. However, if time was reversed, then the decay would proceed differently. In fact, we will see in Chapter 8 that processes which are asymmetric such as the decay of the K_L^0 could explain why we live in a universe of matter rather than antimatter. In fact, quantum processes can define an arrow of time in which some quantum states become more complex than others as time progresses. Feynman diagrams, which describe particle interactions, work just as well backwards in time as they do forwards in time. So it seems that the arrow of time that we know of is a large-scale (macroscopic) phenomenon, but at the quantum level, things may be quite different.

Finally, we need to ask, why is nature nearly symmetrical? To quote Feynman:

'So our problem is to explain where symmetry comes from. Why is nature so nearly symmetrical? No one has any idea why. The only thing we might suggest is something like this: There is a gate in Japan, a gate in Neiko, which is sometimes called by the Japanese the most beautiful gate in all Japan; it was built in a time when there was great influence from Chinese art. The gate is very elaborate, with lots of gables and beautiful carving and lots of columns and dragon heads and princes carved into the pillars, and so on. But when one looks closely he sees that in the elaborate and complex design along one of the pillars, one of the small design elements is carved upside down; otherwise the thing is completely symmetrical. If one asks why this is, the story is that it was carved upside down so that the gods will not be jealous of the perfection of man. So they purposely put an error in there, so that the gods would not be jealous and get angry with human beings.

We might like to turn the idea around and think that the true explanation of the near symmetry of nature is this: that God made the laws nearly symmetrical so that we should not be jealous of His perfection!'

Feynman Lectures by the Division of Physics, Mathematics and Astronomy at the
California Institute of Technology

Summary

◆ An **interaction** is the exchange of energy and momentum between two bodies. The four fundamental interactions in nature are the **gravitational, electromagnetic, strong** and **weak**. Each interaction has a **characteristic time** over which it can act. The characteristic time is a measure of how long it takes for an interaction to have an effect on a particle, and is related to its strength.

◆ An interaction can be thought of as the exchange of **field quanta** between particles. The theory of exchange forces is provided by QED, which has been developed using a key mathematical technique called **renormalisation**. **Feynman diagrams** are space–time diagrams that show a pictorial representation of the way particles interact with each other. They can also be used as a notation for the mathematical equations describing the interaction. A **path integral** is a way of adding up all the quantum probabilities in position that a particle will have in travelling from one point to another, and the Feynman diagram represents the most probable path the particle will take.

◆ **Symmetry** is a repetition of form and may be geometrical or mathematical. A symmetry may be represented by a **group** on which mathematical operations called **transformations** may be applied. A **gauge symmetry** is a symmetry in which the standard of measurement used does not affect the rules governing particle interactions. **Gauge theories** are theories of matter and forces that display gauge symmetry. The four fundamental interactions are all gauge theories that display gauge symmetry.

◆ **Noether's Theorem** tells us that symmetries give rise to **conserved quantities**. In particle physics, three important conserved quantities are the **conservation of lepton number, conservation of baryon number** and **conservation of strangeness**. Conservation of strangeness can explain why strange particles are produced in pairs by **associated production**.

◆ A **physical process** is a process governed by a **physical law**. Three symmetries that apply to physical processes are **parity (P), charge conjugation (C)** and **time reversal (T)**. Parity and charge conjugation are not preserved by the weak interaction. Symmetry can be restored in the weak interaction where both charge conjugation and parity are applied together, a symmetry called **CP invariance**. However, it has been shown that the weak interaction violates even this symmetry by very rare decays of the long-lived neutral K meson K_L^0. The **CPT Theorem** states that every law of physics is invariant under the symmetry transformations of charge conjugation, parity (space inversion) and time reversal, and is known as an **exact symmetry**. The violation of **CP** invariance by weak processes means that time reversal must be violated in some processes involving the weak interaction, to ensure that the exact symmetry of **CPT** is always maintained.

◆ An **arrow of time** may be defined with respect to **entropy** in **thermodynamic** processes insofar as entropy either remains constant (reversible processes) or increases (irreversible processes). The entropy of the universe is increasing with time. On a quantum level, the decay of the K_L^0 meson is not time invariant as a process, and would proceed differently if time was reversed, allowing us to define a direction in time using the weak interaction.

Questions

1 Find all the lines of symmetry in the following shapes:
a a square, **b** a rhombus, **c** the letter 'A', **d** a circle.

2 Particle physicists in 1996 succeeded in making atoms of antihydrogen (see Box 5.2, page 155). These are just like atoms of normal hydrogen except that they have an antiproton for a nucleus about which orbits a single positron instead of an electron. Would you expect the spectrum of antihydrogen to be the same as that of normal hydrogen or would it be different?

3 a Explain what is meant by the *invariance* of a physical law.
   The force between two charges may be expressed as *Coulomb's Law*:

$$F = \frac{1}{4\pi\epsilon_0} \frac{Q_1 Q_2}{r^2}$$

   Verify that Coulomb's Law is invariant under
   i) space inversion $(r \rightarrow -r)$
   ii) charge conjugation $(q \rightarrow -q)$.
b The expression for relativistic momentum is written as

$$p = \frac{m_0 v}{\sqrt{1 - \left(\dfrac{v}{c}\right)^2}}$$

   Is this expression invariant under space inversion?

4 In these questions, refer to the values of B, L, Q and S for the various particles given in the text and on pages 284–5.
   a Which of the following reactions violates conservation of lepton number? [Don't forget that L is conserved *separately* for each of the three types of leptons (e, μ, τ).]
   i) $\mu^- \rightarrow e^- + \bar{\nu}_e + \nu_\mu$ iii) $\mu^- \rightarrow e^- + \gamma$
   ii) $\bar{\nu}_\mu + p \rightarrow \mu^+ + n$
   b Which of the following reactions is forbidden because it would involve the non-conservation of charge?
   i) $p + p \rightarrow p + p$ iii) $p + p \rightarrow p + p + \pi^+ + \pi^- + \pi^+$
   ii) $p + p \rightarrow p + p + \pi^+ + \pi^-$

c Which of the following reactions is forbidden because it would involve the non-conservation of baryon number?

i) $p + p \rightarrow p + p + n$

ii) $p + p \rightarrow p + p + p + \bar{p}$

iii) $\bar{\Lambda}^0 \rightarrow p + \pi^-$

iv) $p \rightarrow \pi^0 + \pi^+$

d By checking to see whether Q, B and S are each conserved, decide which one of the strong interactions i) to iii) can occur.

i) $\pi^- + p \rightarrow n + p$

ii) $\Sigma^- + p \rightarrow K^0 + n$

iii) $K^- + p \rightarrow \bar{K}^0 + n$

e Each of the following reactions violates one (or more) conservation laws. Name the conservation law(s) violated in each case.

i) $\nu_e + p \rightarrow n + e^+$

ii) $p + p \rightarrow p + n + K^+$

iii) $p + p \rightarrow p + p + \Lambda^0 + K^0$

iv) $\pi^- + n \rightarrow K^- + \Lambda^0$

v) $K^- + p \rightarrow n + \Lambda^0$

5 a The two decays shown below may both conserve energy and momentum.

$$\mu^+ \rightarrow e^+ + \nu_e$$
$$\mu^+ \rightarrow e^+ + \nu_e + \bar{\nu}_\mu$$

The first cannot occur but the second can.

Which conservation law forbids the first decay?

Show how this conservation law allows the second decay.   *Edexcel, June 1999 (part)*

b A muon can decay by means of the weak interaction shown below.

$$\mu^- \rightarrow e^- + X + \nu_\mu$$

Use the laws of conservation of charge and lepton number to deduce the identity of X.

Give one reason why this decay must take place by means of the weak interaction.

What would be a typical decay time for this interaction?   *Edexcel, January 2001 (part)*

6 The following strong interaction was first observed in 1964 at Brookhaven National Laboratory in the USA:

$$K^- + p \rightarrow K^0 + K^+ + X$$

where X is a hadron, which until then had not been observed. By using conservation of Q, B and S, deduce:

a the charge of X in coulombs,   **b** whether X is a meson or baryon,

c the strangeness of X.

7 a What is a Feynman diagram?

b i) An electron and positron can interact by the mechanism represented by this Feynman diagram. What interaction is represented by the Feynman diagram?

ii) Draw a Feynman diagram that represents the exchange of a π^+ between a neutron and a proton.

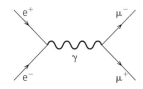

Part (b) only: ULEAC, June 1996 (part)

[Note that you will find Feynman diagrams in exam papers that do not follow the convention of antiparticles going backwards in time.]

The Standard Model of particle physics

7

'*Three quarks for Muster Mark*
Sure he hasn't got much of a bark
And sure any he has it's all beside the mark.'

James Joyce (1882–1941), *Finnegans Wake*

Putting it all together

The unification or 'joining together' of the four basic interactions has long been a central objective of particle physics. While we have separate theories for the electromagnetic, weak, strong and gravitational forces, the 'Holy Grail', as it were, is a comprehensive theory that would incorporate all these interactions and reveal some deep connection between them all while at the same time accounting for their observed differences.

The first such unification occurred with electricity and magnetism. We saw in Chapter 1 that Faraday, who had no mathematical training, carried out investigations with magnets and coils of wire that showed a link between magnetism and electricity. By thinking about Faraday's experiments, Maxwell showed mathematically that a changing magnetic field produces an electric current, and that a magnetic field is produced by a steady electric current. **Maxwell's equations**, as they are called, are too complex to summarise here, but there are four of them, which unmistakably show that electric and magnetic fields are so intimately associated that one acts as a source for the other. Maxwell published his findings in a paper in 1854 entitled 'A Dynamical Theory of the Electromagnetic Field', in which his four equations clearly showed that electricity and magnetism were just different aspects of the same basic force. Physicists immediately saw that it now made sense to talk of a single electromagnetic force rather than separate electric and magnetic ones. Maxwell's theory is an entirely classical one. But, with the advent of the quantum theory, has it been possible to make any further progress in unifying the fundamental interactions?

The answer to this question is, 'Yes, it has – up to a point'. Particle physicists have been able to construct a theory of matter called the **Standard Model**, which has been very successful in uniting electromagnetic, weak and strong forces. But it is still incomplete, particularly in its ability to unify gravity. The ultimate aim is to construct a **Theory of Everything** (TOE) – a theory that can explain all known physical phenomena, including gravity, in quantum terms.

The Electroweak Theory – broken symmetries

After Maxwell, the next step towards unification was taken in 1967 by Steven Weinberg (1933–), Sheldon Lee Glashow (1932–) and Abdus Salam (1926–1996).

In Chapter 6 we saw that the decay of the long-lived neutral K meson K_L^0 revealed a *broken symmetry* in the weak interaction. To see why broken symmetries are significant, think of the following analogy. Snowflakes are formed as ice crystals in clouds under a certain physical process. When we look at them under a microscope, we always observe them to have geometrical symmetry (Figure 6.5, page 187). However, suppose we woke up one snowy day and found, when we looked at the snowflakes, that they were all random shapes and sizes. The symmetry we expected to see had been broken. Some possible explanations that might occur to us could include the idea that a different physical law was at work in the cloud, which occasionally produced odd-looking snowflakes under certain conditions, or that random shaped snowflakes were really the norm, but, owing to local weather conditions, we nearly always observe symmetric flakes. Could we suppose that there might be a single cause for both symmetric and asymmetric snowflakes?

The broken symmetry in the weak interaction led Weinberg and Glashow to wonder if the weak interaction might be somehow linked to the electromagnetic one. In other words, do we see the weak interaction at work because some form of symmetry between the electromagnetic and weak interactions has been lost? If so, then it would be reasonable to assume that the weak force was mediated by carrier particles in a similar fashion to the electromagnetic force, as described by QED. By using an analogy with the carriers of the electromagnetic force (photons), Weinberg and Glashow postulated the existence of carrier particles that transmit the weak force between particles.

The mathematical details of the Electroweak Theory are beyond the scope of this book. In summary, the electromagnetic and weak forces can be explained by gauge theories, and their particles as representing a group whose members behave in a definite way governed by mathematical transformations of the group. Weinberg, Glashow and Salam's work involved combining together the group representing the electromagnetic interaction, called U(1), and the group representing the weak, called SU(2) ('Special Unitary', dimension 2), to form a composite symmetry group SU(2) × U(1), which predicted the existence of three bosons that carry the weak force. Glashow at first had difficulties in applying a gauge theory to the weak interaction. His equations gave infinite results and had to be renormalised. In addition, the carrier bosons, unlike the photons, carried *mass*, and were much more massive than the proton.

These problems were resolved by the contributions of Salam and Weinberg, who developed Glashow's original idea, and also Peter Higgs (1929–), who postulated the existence of a hitherto undiscovered boson that enables particles to acquire mass. Another important contribution came from the Dutch physicist Gerard 't Hooft (1946–), who found a way to renormalise the theory and make it work. The particles that were predicted by the new combined group theory were called the W and the Z. There are three of them, and their masses and charges are listed in Table 7.1.

Table 7.1 The W and Z particles

Particle	Charge/e	Mass/GeV/c^2
W$^+$	+1	80.4
W$^-$	−1	80.4
Z^0	0	91.2

Compared with the proton which has a mass of $0.938\,\text{GeV}/c^2$, these particles are nearly 100 times more massive! This theory of the weak force showed many conceptual similarities with that of QED, and was so successful that it revealed that the weak and electromagnetic forces are in fact different manifestations of a single **electroweak** force. Figure 7.1 shows the Feynman diagrams for the electroweak force.

The number of fundamental forces had now reduced from four to three, and in 1979 Weinberg, Glashow and Salam received the Nobel Prize for Physics for their work, Salam being the first Pakistani and Muslim to win the award. However, could these particles be observed experimentally? In the early 1980s accelerator energies had reached the level where these particles could be created, and they were found in 1983 with masses very close to those predicted by the Electroweak Theory (Box 7.1).

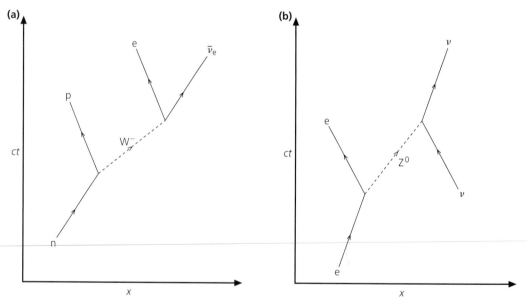

Figure 7.1 Feynman diagrams for the electroweak force
(a) The emission of a virtual W$^-$ particle by a neutron. The neutron transforms into a proton and the W$^-$ subsequently creates an electron and an antineutrino. This process is the beta decay of a neutron, that is,

$$n \rightarrow p + e + \bar{\nu}_e$$

(b) A Z^0 particle being exchanged between an electron and a neutrino, causing them to be scattered off each other

Box 7.1 The discovery of the W and the Z particles

Theoretical work by Weinberg, Glashow and Salam showed that the electromagnetic and weak interactions could be unified into a single electroweak force. They predicted that three new massive particles, the W^+, the W^- and the Z, should exist which carry the weak force responsible for radioactive decay.

In the late 1970s CERN's large proton–antiproton collider, the Super Proton Synchrotron (SPS), came into operation. The SPS was capable of colliding beams of protons and antiprotons together at energies of 270 GeV circulating in opposite directions, giving a centre-of-mass energy of 540 GeV, well above the threshold for which the W and Z particles could be created. The beams could be brought together at various points around the SPS ring where a number of particle detectors were waiting ready to record the tracks of the high-energy particles produced by the annihilation of the protons and antiprotons.

An experimental team of 135 workers from at least 11 research institutions and led by the Italian physicist Carlo Rubbia (1934–) began to search for these particles using a detector called UA1 (Underground Area 1) and another detector situated further round the synchrotron ring called UA2 (Underground Area 2). UA1 and UA2 were large gas-filled chambers containing wires that pick up electrical pulses caused by the ionisation that the charged particles leave behind, as well as containing many other components such as calorimeters and magnets. The central detector in UA1 is an example of a *drift chamber*, which we discussed in Chapter 4 and is essentially an 'electronic bubble chamber'. The detector was surrounded by a strong magnetic field that curved the trajectories of charged particles. In 1983 Rubbia's team triumphantly announced the discovery of the first W particles.

The experimenters were looking for the decays

$$Z^0 \rightarrow e^+ + e^- \qquad Z^0 \rightarrow \mu^+ + \mu^-$$
$$W^+ \rightarrow e^+ + \nu_e \qquad W^+ \rightarrow \mu^+ + \nu_\mu$$
$$W^- \rightarrow e^- + \overline{\nu}_e \qquad W^+ \rightarrow e^+ + \nu_\mu$$

The W particles were found first and then the Z particle in the UA1 detector in May 1984. The decay of the Z^0 can produce an e^+ and e^- or μ^+ and μ^- at large angles, and well clear of other particles produced in the collision. Electrons and positrons are easily identified and, by measuring the curvature of their trajectories in the magnetic field and by finding that the total energy of the electron and positron (or alternatively the μ^+ and μ^-) in the electromagnetic calorimeters is close to the predicted mass of the particle, the experimenters can be sure that their production comes from the decay of the Z^0 particle.

The Z^0 can also decay into a μ^+ and μ^-. Figure 7.2 is a computer display from the UA1 detector that shows the formation of a Z^0 particle from the annihilation of a proton and antiproton inside the central drift chamber (red cylinder). In this case the Z^0 is decaying into a muon and antimuon. The computer has matched hits in the muon counters outside the magnet (red box) and drawn in two tracks (blue slashes) which show a slight curvature in the magnetic field. Measurement of their momentum tells us the total energy of the event, which can be used to infer the creation of a Z particle.

A similar technique was used to find the W particles, although their decay signatures were not so prominent as the Z since the neutrino is not observable and a more subtle analysis had to be made. Nonetheless unambiguous observation of the W and Z has been confirmed in numerous collision events.

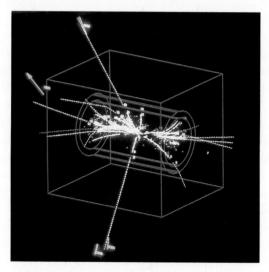

Figure 7.2 The decay of a Z^0 particle into two muons in the UA1 detector

The Eightfold Way

Suppose we plot a diagram on an x–y coordinate system with electric charge on the x-axis and strangeness on the y-axis for all the families of particles for which these values are known. What we find are plots that appear with a definite geometrical symmetry. Recall that there are eight baryons including the neutron and proton that have a spin quantum number of $\frac{1}{2}$, which were listed in Table 5.4, and we list them again in Table 7.2 for our convenience.

Table 7.2 The eight spin-$\frac{1}{2}$ baryons

Particle	Symbol	Mass/MeV/c^2	Charge/e	Strangeness
Proton	p	938.3	+1	0
Neutron	n	939.6	0	0
Lambda-zero	Λ^0	1115.6	0	−1
Sigma-plus	Σ^+	1189.4	+1	−1
Sigma-zero	Σ^0	1192.5	0	−1
Sigma-minus	Σ^-	1189.4	−1	−1
Xi-zero	Ξ^0	1314.9	0	−2
Xi-minus	Ξ^-	1321.3	−1	−2

If we plot the strangeness of these baryons against their electric charge using a sloping axis for the strangeness quantum number, then a diagram with a hexagonal symmetry emerges that looks like Figure 7.3. Of the eight baryons, six form a hexagon, with the remaining two at the centre.

A similar plot (Figure 7.4) can be made for the nine spin-0 mesons, which were listed in Table 5.5 and whose properties are listed again in Table 7.3 overleaf.

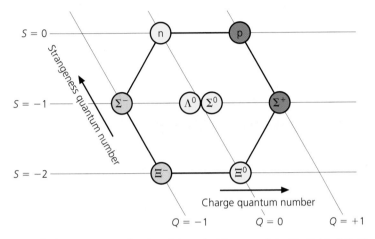

Figure 7.3 The Eightfold Way pattern for the eight spin-$\frac{1}{2}$ baryons listed in Table 7.2. The particles are represented on a strangeness–charge plot using a sloping axis for the strangeness quantum number

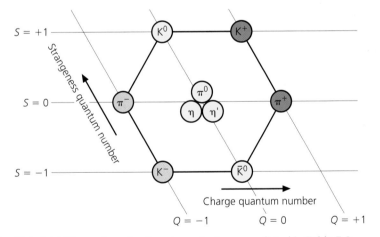

Figure 7.4 The Eightfold Way pattern for the nine spin-0 mesons listed in Table 7.3

Table 7.3 The nine spin-0 mesons

Particle	Symbol	Mass/MeV/c^2	Charge/e	Strangeness
Neutral π meson	π^0	135.0	0	0
Positive π meson	π^+	139.6	+1	0
Negative π meson	π^-	139.6	−1	0
Positive K meson	K^+	493.7	+1	+1
Negative K meson	K^-	493.7	−1	−1
Neutral K meson	K^0	497.7	0	+1
Anti-neutral K meson	$\overline{K}^0$	497.7	0	−1
Eta	η	547.5	0	0
Eta prime	η'	957.8	0	0

These plots are called **Eightfold Way** patterns, and were proposed independently in the early 1960s by two particle physicists, an American Murray Gell-Mann (1929–), of the California Institute of Technology (Caltech), and an Israeli Yuval Ne'eman (1925–), who was working at Imperial College, London. Gell-Mann and Ne'eman recognised that regular patterns such as these were strong evidence of an orderly structure in the particles. Other geometrical symmetries also emerge. An Eightfold Way plot for ten baryons can be arranged in a pattern rather like ten-pins in a bowling alley (Figure 7.5). When this pattern was first proposed, only nine of the particles had been discovered, but so confident was Gell-Mann in the theory and symmetry of the pattern that he predicted the existence of the tenth particle, which he called the omega-minus, and this was discovered in 1964 (see Box 5.3, page 166).

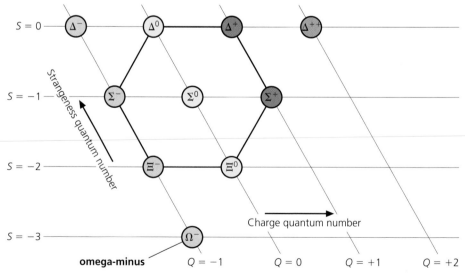

Figure 7.5 The Eightfold Way pattern for ten baryons that predicted the existence of the omega-minus

The important thing to understand about Eightfold Way patterns is that they are to particle physics what the Periodic Table is to chemistry. In each plot, a pattern of organisation emerges in which missing particles stand out in the same way as missing elements were predicted in the early Periodic Table of Mendeleyev. The name 'Eightfold Way' was coined by Gell-Mann, in an allusion to Eastern mysticism, to refer to the eight quantum numbers (not all of which are defined here) that appear in the mathematics predicting the patterns. Just as the Periodic Table suggested to chemists that atoms are not fundamental particles but have an underlying form, the Eightfold Way patterns suggested to particle physicists that baryons and mesons must also have an internal structure and are not truly elementary.

The quark model

Gell-Mann and another particle physicist, George Zweig (1937–) of CERN, showed that the Eightfold Way patterns could be replicated if the mesons and baryons were composed of further elementary particles, which Gell-Mann called **quarks** from the line in James Joyce's enigmatic novel, *Finnegans Wake*, quoted on page 214.

There are six quarks together with six **antiquarks**. We assign to them names called **up** (u), **down** (d), **strange** (s), **charm** (c), **top** (t) and **bottom** (b). Collectively these names are called **quark flavours**. The names have no special significance, but are just whimsical labels thought up by particle physicists. We could just as easily have labelled them after flavours of ice-cream! The quark flavours top and bottom were previously called 'truth' and 'beauty', but these names have now been abandoned.

One of the most striking features of quarks is that they have **fractional charge**. This is a radical departure from what we know about the electric charges on other particles, which are all multiples of the electron charge, either positive or negative. Figures 7.6a and b overleaf show the Eightfold Way patterns for the eight spin-$\frac{1}{2}$ baryons and the nine spin-0 mesons expressed in terms of quarks. The crowning feature of the quark model is that *all mesons and baryons can be understood in terms of appropriate combinations of quarks*, and all matter can now be expressed in terms of quarks and leptons. The quarks, antiquarks and their quantum numbers are listed in Table 7.4. All quarks have spin $\frac{1}{2}$ and are therefore fermions. They are assigned baryon numbers of either $+\frac{1}{3}$ or $-\frac{1}{3}$. All baryons and mesons can be constructed from combinations of the quarks and antiquarks listed.

The quark structure of baryons

Each baryon is made up of three quarks. Consider first the proton, which has a charge $+1e$, baryon number $B = +1$ and spin $\frac{1}{2}$. The proton is made up of a quark composition of uud (two up and one down quark). Checking for charge:

$$(uud) = (+\tfrac{2}{3}e) + (+\tfrac{2}{3}e) + (-\tfrac{1}{3}e) = +1e$$

Since the baryon numbers are all $\frac{1}{3}$, then this means the total $B = +1$. As for spin, each quark has spin $\frac{1}{2}$, so we could have two of the quark spins parallel and one antiparallel, which leads to a net spin of $\frac{1}{2}$.

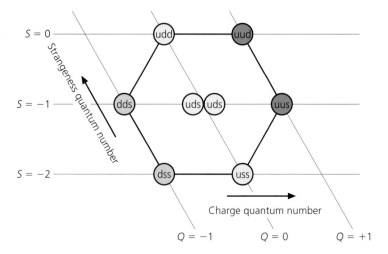

Figure 7.6(a) The quark composition of the eight spin-$\frac{1}{2}$ baryons shown in Table 7.2 and Figure 7.3. Although the two central baryons show the same quark structure, the Σ^0 is an excited state of the Λ^0, which decays into the Λ^0 by emitting a gamma ray photon

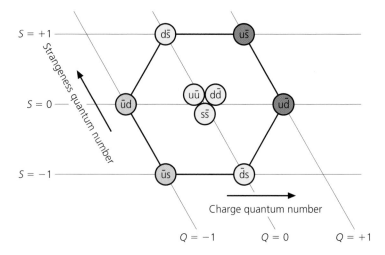

Figure 7.6(b) The quark composition of the nine spin-0 mesons shown in Table 7.3 and Figure 7.4

The neutron is made up of one up and two down quarks (udd). Similarly, checking for charge:

$$(\text{udd}) = (+\tfrac{2}{3}e) + (-\tfrac{1}{3}e) + (-\tfrac{1}{3}e) = 0e$$

The baryon numbers and spins are as before, for the proton.

Table 7.4 The properties of the six quarks and their antiquarks

Flavour	Spin/ $\hbar$	Charge/ e	Baryon number, B	Strangeness, S	Mass*/GeV/c^2
Quarks					
u (up)	$\frac{1}{2}$	$+\frac{2}{3}$	$+\frac{1}{3}$	0	0.004
d (down)	$\frac{1}{2}$	$-\frac{1}{3}$	$+\frac{1}{3}$	0	0.008
s (strange)	$\frac{1}{2}$	$-\frac{1}{3}$	$+\frac{1}{3}$	-1	0.15
c (charm)	$\frac{1}{2}$	$+\frac{2}{3}$	$+\frac{1}{3}$	0	1.5
t (top)	$\frac{1}{2}$	$+\frac{2}{3}$	$+\frac{1}{3}$	0	176
b (bottom)	$\frac{1}{2}$	$-\frac{1}{3}$	$+\frac{1}{3}$	0	4.7
Antiquarks					
$\bar{u}$ (anti-up)	$\frac{1}{2}$	$-\frac{2}{3}$	$-\frac{1}{3}$	0	0.004
$\bar{d}$ (anti-down)	$\frac{1}{2}$	$+\frac{1}{3}$	$-\frac{1}{3}$	0	0.008
$\bar{s}$ (anti-strange)	$\frac{1}{2}$	$+\frac{1}{3}$	$-\frac{1}{3}$	$+1$	0.15
$\bar{c}$ (anti-charm)	$\frac{1}{2}$	$-\frac{2}{3}$	$-\frac{1}{3}$	0	1.5
$\bar{t}$ (anti-top)	$\frac{1}{2}$	$-\frac{2}{3}$	$-\frac{1}{3}$	0	176
$\bar{b}$ (anti-bottom)	$\frac{1}{2}$	$+\frac{1}{3}$	$-\frac{1}{3}$	0	4.7

* The quark masses are approximate; see page 235 for an explanation of how the masses are measured.

WORKED EXAMPLE 7.1

The Σ^- particle is a baryon with charge $-e$, baryon number $+1$ and strangeness -1. Confirm that it has the quark structure dds.

The down (d) and strange (s) quarks have $B = \frac{1}{3}$. Therefore the total baryon number B of the dds combination is

$$B(\text{dds}) = (+\tfrac{1}{3}) + (\tfrac{1}{3}) + (\tfrac{1}{3}) = +1$$

The charge Q is

$$Q(\text{dds}) = (-\tfrac{1}{3}e) + (-\tfrac{1}{3}e) + (-\tfrac{1}{3}e) = -1e$$

The strangeness S works out as

$$S(\text{dds}) = (0) + (0) + (-1) = -1$$

Therefore the sigma-minus consists of two down and one strange quark.

WORKED EXAMPLE 7.2

The Ξ^- particle has spin $\frac{1}{2}$, charge $-e$ and strangeness -2. It is thought to be made of three quarks involving only up, down and strange flavours. What must the combination be?

Since the Ξ^- has $S = -2$ it must contain two strange quarks each of $S = -1$. The third quark must be either an up or a down quark, both with $S = 0$. The two strange quarks have a total charge of $(-\frac{1}{3}e) + (-\frac{1}{3}e) = -\frac{2}{3}e$. However, we need a charge of $-1e$ for the Ξ^-, so we need to add a third quark whose charge is $-\frac{1}{3}e$, that is, a down quark, so that

$$Q(\text{dss}) = (-\frac{1}{3}e) + (-\frac{1}{3}e) + (-\frac{1}{3}e) = -1e$$

and

$$S(\text{dss}) = (0) + (-1) + (-1) = -2$$

The spins of the quarks can be arranged parallel and antiparallel as before.

The quark structure of mesons

All mesons are made up of quark–antiquark *pairs*. We can see that this is so by seeing that the spins of all the mesons in Table 7.3 are zero. In addition, all mesons have baryon number $B = 0$. The baryon number for a quark is $+\frac{1}{3}$ and for an antiquark $-\frac{1}{3}$. For these two conditions to be satisfied, the quarks and antiquarks must spin in the opposite sense to each other (one parallel and the other antiparallel), and the sum of their baryon numbers must be zero. A combination of quark and antiquark would therefore meet these quantum number requirements of the meson.

WORKED EXAMPLE 7.3

Show that the π^+ meson, which has $Q = 1e$, $B = 0$ and $S = 0$, is made up of an up and an anti-down quark pair.

From Table 7.3, consider in turn charge requirement, baryon number and strangeness:

$$Q(u\bar{d}) = (+\frac{2}{3}e) + (+\frac{1}{3}e) = 1e$$
$$B(u\bar{d}) = (+\frac{1}{3}) + (-\frac{1}{3}) = 0$$
$$S(u\bar{d}) = (0) + (0) = 0$$

Experimental evidence for quarks

Scattering experiments

That the proton and nucleon have an internal structure can be inferred by scattering experiments where nucleons are bombarded with beams of electrons, muons or neutrinos with energies from 15 to 200 GeV using particle accelerators. These **deep inelastic scattering** experiments scatter the particles in a way not unlike that of the Rutherford scattering of alpha particles (Chapter 2). In the same way that the presence of a nucleus was deduced within an atom, so is the presence of smaller particles inside a nucleon.

The first of these deep inelastic scattering experiments were done at the Stanford Linear Accelerator Center (SLAC) in the early 1970s. The maximum electron momentum available from the linear accelerator at SLAC was 22 GeV/c, which gives a de Broglie wavelength of

$$\lambda_p = \frac{h}{p} = \frac{(6.63 \times 10^{-34}\,\text{J s})}{(22 \times 10^9 \times 1.6 \times 10^{-19}\,\text{J})/(3.0 \times 10^8\,\text{m s}^{-1})} = 5.7 \times 10^{-17}\,\text{m}$$

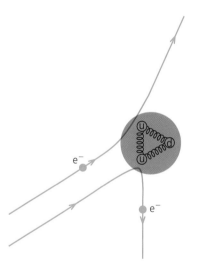

Figure 7.7 An electron scattering off a proton can feel the quarks inside

This wavelength determines the smallest detail that can be resolved by the electron beam, which is a scale of 1/100th the radius of the proton (10^{-15} m). In this way the high-energy electrons can 'feel' the quarks in a similar way that you might feel sweets inside a bag and guess how many there are (Figure 7.7).

The first Eightfold Way patterns for baryons and mesons required the existence of three types of quarks, the up (u), down (d) and strange (s). However, certain experimental discrepancies in decay rates of particles, together with other Eightfold Way patterns, led to the prediction of three more quarks: the charm (c), top (t) and bottom (b).

In 1974 a meson called the **J/ψ** (**J/psi**) was discovered that had a mass of 3.1 GeV/c², over three times the mass of the proton (Box 7.2). The lifetime of this particle could only be explained by introducing another type of quark with the quantum property that particle physicists call **charm**. The J/psi itself has zero charm since it is made up of a charm–anticharm quark pair (c$\bar{\text{c}}$). Charm is believed to be conserved in strong and electromagnetic interactions.

Box 7.2 The discovery of the J/psi particle

In 1974 two teams of particle physicists in the USA independently discovered a very massive meson with a mass of more than 3097 MeV/c^2.

At SLAC the electron–positron collider SPEAR was capable of storing particles of energies between 1.3 and 2.4 GeV per beam. At SPEAR was a detector called the MARK 1, which was a large spark chamber surrounded by an electromagnet. The MARK 1 could show the reconstruction of tracks of charged particles passing through it. One weekend in November 1974, the MARK 1 revealed the existence of the J/ψ particle as a resonance lasting just 10^{-20} s. Although this is a very short time, it is 1000 times longer than would be expected for a meson of this mass, which theory shows should decay much more quickly into lighter particles. Even more surprisingly, the MARK 1 found another meson dubbed the Φ′ at slightly higher energies, which lived for even longer than expected. Figure 7.8 shows the reconstructed track of the J/ψ in the MARK 1 detector.

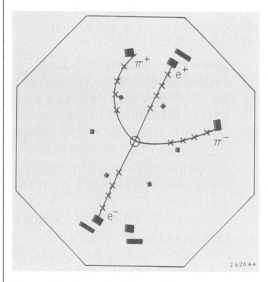

Figure 7.8 J/ψ production and decay were identified in this ψ-shaped track

To explain these long-lived mesons, particle physicists introduced a new property called **charm**, which prohibited the rapid decay of these massive mesons.

Meanwhile at the Brookhaven National Laboratory (BNL), a team led by Samuel Chao Chung Ting (1936–) accelerated protons to energies of 30 GeV and collided them with light nuclei. Using a spark chamber, the reconstructed tracks showed a heavy meson with exactly the same properties as that discovered at SLAC.

Both teams had discovered the same particle at almost the same time! The team at SLAC was headed by Burton Richter (1931–) and initially named the particle the 'psi' because the tracks revealed by the MARK 1 looked like the Greek letter ψ, whereas Ting named it the 'J' after the Chinese character for Ting. To this day, particle physicists have not been able to decide which name to use, so it got stuck at 'J/ψ' (the compromise name 'gypsy' has been suggested, but this has not caught on).

Richter and Ting received the Nobel Prize for Physics in 1976 for the discovery of the J/psi. The J/ψ and Φ′ proved to have exactly the properties of a meson made up of a charm and anti-charm quark pair, which provided particle physicists with strong indications that the quark model was in agreement with the experimental evidence, and stimulated much development of the quark theory as an explanation for the properties of matter at the most elementary level.

The lightest mesons that contain a charmed quark are the D^0 and D^+ with masses of about $1.8\,GeV/c^2$. One baryon containing a charmed quark is the Δ_c^+ with a quark structure udc.

You may notice from Table 7.4 (page 223) that some quarks are extremely massive, which means that particle accelerators need to go to higher and higher energies to produce them. In 1977 a team at Fermilab led by Leon Max Lederman (1922–) found evidence for the bottom quark when they discovered the **upsilon (Υ)** meson using the proton synchrotron, which had recently come into operation accelerating protons to $500\,GeV$. The upsilon was predicted to have the quark–antiquark combination b$\bar{\text{b}}$, and using high-energy proton beams they found this short-lived meson with a mass of $9.46\,GeV/c^2$, zero charge and spin 1.

The upsilon's mass of $9.4\,GeV/c^2$ was too massive to be produced at SPEAR, but particle physicists at DESY modified the DORIS electron–positron collider to reach the energy region of the upsilon. In 1978, they saw the first signs of the upsilon and began to study the way it decayed. This meson's decay was again slowed to $\sim10^{-20}\,$s and decayed into electrons and positrons rather than into other mesons. To account for the slowed decay, the bottom quark has a quantum property called 'bottomness', and new particles containing the bottom quark such as the **B meson** (b$\bar{\text{u}}$) with a mass $0.5\,GeV/c^2$ were soon discovered. In 1978 the DORIS electron–positron collider at Hamburg began operating with sufficient energy to produce bottom quarks in large numbers, and this accelerator, together with the CESR electron storage ring at Cornell, soon confirmed the original discoveries at Fermilab and added several more mesons to the family containing bottom quarks.

It is generally difficult to observe particles containing bottom quarks individually. Particle physicists have learnt about them by analysing a number of events thought to involve particles containing bottom quarks and looking at the angles and energies of the particles produced in their decay. Such events regularly produce tracks in the form of **jets**, which are showers of particles all moving in the same general direction, produced by a quark or antiquark that has been knocked out of a nucleon or parent particle (Figure 7.9, page 228). For reasons discussed below, quarks cannot exist on their own, and will form new groupings if attempts are made to isolate them.

However, in 1986, at CERN, an experiment codenamed WA75 used a beam of π mesons directed into a stack of nuclear emulsion. Electronic detectors surrounding the stack were able to pinpoint the precise position where specific interactions occurred in the nuclear photographic emulsion. Using this information, experimenters were able to find one event where a negative B meson and a neutral B meson were created together, forming a 'vee' associated with the decay of two bottom particles. From this, particle physicists at CERN were able to announce that they were the first to see 'naked bottom'!

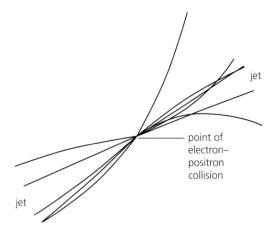

Figure 7.9 A two-jet event from the collision of an electron and a positron

Since the total momentum of the e⁻ and e⁺ is zero, the momentum of the quark–antiquark pair created in the collision must be zero, which means that the velocities of the quarks must also be zero. In order to conserve momentum, the shower of particles produced in the collision must be collinear but in opposite directions. The tracks are curved due to the presence of an external perpendicular magnetic field

The most massive quark, the top quark, was found in 1995 using the 1 TeV Tevatron proton synchrotron accelerator at Fermilab, which collided protons and antiprotons together each with energies of 0.9 TeV (9×10^{11} eV). Very occasionally, the colliding protons will produce a top–anti-top pair (t$\bar{\text{t}}$), which rapidly transforms into particles of lower energy. The jets produced in the decay are carefully analysed to infer the existence of particles containing top quarks. Box 7.3 shows the tracks produced by the decay of a top–anti-top quark pair.

Quarkonium

States of matter that consist of quarks and antiquarks bound together are called **quarkonium**. Mesons, which are quark–antiquark pairs, are examples of quarkonium. The bound state of a charm quark plus anti-charm quark (c$\bar{\text{c}}$) is called **charmonium**. Such a particle has zero charm overall, but the quarks do not get close enough to each other to annihilate. Similarly a strange and anti-strange quark pair (s$\bar{\text{s}}$) is called **strangeonium** and corresponds to a meson called **Φ**(1020).

The quarks in a q$\bar{\text{q}}$ pair will move around each other rather like an electron and proton in the hydrogen atom. Such dual systems of particles have an energy level structure similar to that of the electron energy levels in atoms. If the q$\bar{\text{q}}$ pair orbits at high energy levels, then it constitutes a comparatively massive particle, since, from the mass–energy relation, mass is equivalent to energy. If the q$\bar{\text{q}}$ pair moves to a state of lower energy, then the particle becomes lighter and in the process the excess energy is emitted in the form of other particles such as pions, muons, electrons and

Box 7.3 The discovery of the top quark

In March 1995, particle physicists gathered at a hastily called meeting at Fermilab to announce the discovery of a new particle – the top quark. This confirmed what they knew all along – that this was the last of six quarks from which all hadrons are made. The top quark was found using the CDF detector attached to the Tevatron (Figure 4.12).

A top and anti-top, once produced, change flavour almost instantly. Unlike the up and down quarks, which are stable, the top quark has a lifetime of about 10^{-24} s. According to the Standard Model, the top quark will transform most of the time into a W and a bottom quark. So if a top and anti-top are created in the accelerator, they should generate two W particles, a bottom quark and an anti-bottom quark. However, neither the Ws nor the bottom quarks can be directly observed. The lifetime of the W is also about 10^{-24} s and the bottom quark is unstable. In addition, quarks cannot exist in isolation since the strong force always ensures that quarks are stuck together with other quarks or antiquarks. Instead, what the experimenters see is a jet of mesons or baryons made up of quarks and antiquarks, which have roughly the same direction of motion as the original quark.

The W can also decay into a lepton, such as an electron, positron or muon, which can be detected quite easily, but if a neutrino is created then this passes through the detector unobserved. However, it will carry off momentum, so its presence can be inferred when all the momenta of the detected particles are added up. The missing momentum is assumed to come from the neutrino.

Figure 7.10 shows a classic 'top quark event' as seen in the CDF. A proton and antiproton travelling in opposite directions along the Tevatron beam-pipe collide head-on. A top quark and an anti-top quark are produced, which decay into four jets of particles (red and yellow blocks). These multiple jets are the characteristic signature of a top quark.

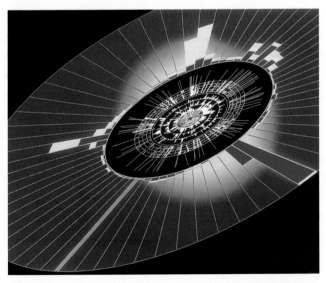

Figure 7.10 A top quark event reconstructed by computer, using data from the CDF detector

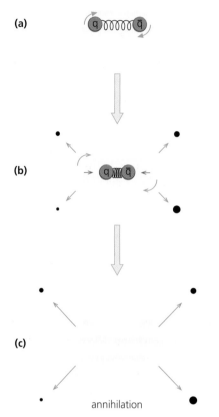

(a)

(b)

(c)

annihilation

Figure 7.11(a) Quarkonium is a state of matter in which quarks and antiquarks are bound together
(b) As the quarks move closer together, the particle (meson) becomes lighter and moves to a state of lower energy, radiating excess energy in the form of other particles
(c) A final burst of particles is emitted as the quarks in the meson annihilate

photons (Figure 7.11). In their lowest energy state, the 'quarkonium ground state', they can no longer radiate particles but instead attract each other and mutually annihilate, producing a final burst of lighter particles. In the case of charmonium, the J/ψ is the lowest energy state possible for the quarks, and the Φ′ the second lowest. It is the energy level structure of the quarkonium states of mesons that gives rise to the numbers of mesons in a particular family. However, the higher the energy state of the $q\bar{q}$ pair, then the higher the energy of the accelerator needed to produce it.

Another look at beta decay

Let's go back to beta decay, a weak process, and see how we can now interpret it in terms of the quark model. An example is β^- decay, where a neutron changes into a proton, emitting an electron and an electron antineutrino,

$$n \rightarrow p + e^- + \bar{\nu}_e$$

In terms of quarks, this is interpreted as the neutron (udd) changing into a proton (uud) by changing one of its down quarks into a up quark, that is,

$$\begin{pmatrix} u \\ d \\ d \end{pmatrix} \rightarrow \begin{pmatrix} u \\ u \\ d \end{pmatrix} + e^- + \bar{\nu}_e$$

Notice how a d quark changes into a u quark. Unlike the strong interaction which does not change the flavour of quarks, processes involving the weak interaction can change quark flavour by the emission and absorption of a W boson. In beta decay

$$d \rightarrow u + W^- \quad \text{followed by} \quad W^- \rightarrow e^- + \bar{\nu}_e$$

While strong interaction processes rearrange quarks and can materialise quark–antiquark pairs, if a quark has changed flavour we can be sure that a weak interaction process is at work.

Quantum chromodynamics (QCD)

Colour

Leptons, such as the electron, possess the property of electric charge, which comes in two types, positive and negative, and allows them to feel the electromagnetic force. Quarks have fractional electric charge and feel the electromagnetic force. Bound as baryons in a nucleus, quarks, unlike leptons, also feel the strong force. This implies that quarks must have a property that enables them to feel the strong force and also binds them together in threes in nucleons or as pairs in mesons. This strong force analogy of electric charge is called **colour**. The quark colours come in three shades, 'red' (r), 'green' (g) and 'blue' (b). This of course has nothing to do with real colour – it is yet another one of those whimsical labels that particle physicists have dreamt up to describe a property of quarks that determines the strength of their interactions with each other through the strong force.

Any stable hadron (baryon or meson) must have an overall 'colour charge' that is *colourless*, rather like combinations of the red, green and blue primary colours in light mix together to make white (Figure 7.12). Baryons such as protons must have a red, green and blue quark, for example u(r), u(g) and d(b), and it doesn't matter what combination you have as long as there is only one of each present. Antiquarks have their corresponding **anti-colour**, 'anti-red' ($\bar{r}$), 'anti-green' ($\bar{g}$) and 'anti-blue' ($\bar{b}$), which cancel out in pairs. For example, mesons are made up of quark–antiquark pairs (q$\bar{q}$), so that a π^+ meson, which is composed of an up and an anti-down pair, would have the colour combination u(r)$\bar{d}$($\bar{r}$) and is overall colourless, since the red is 'cancelled out' by the anti-red.

Colour is a central concept in the theory of the force that holds quarks together and acts as the source of this force in the same way as mass is the source of gravity, and electric charge the source of the electromagnetic force. The colour force between quarks is described by a theory called **quantum chromodynamics** or **QCD**.

proton

up(red) quark + up(green) quark + down(blue) quark
red + green + blue = white (colourless)

π^+ meson

up(red) quark + anti-down(anti-red) quark
red + anti-red cancel out, leaving the
π meson colourless

Figure 7.12 Quantum chromodynamics (QCD). Quarks carry a type of 'charge' called *colour*, which is to the strong force what electric charge is to the electromagnetic force. However, unlike electric charge, the colour charge comes in six variants – three colours and three anti-colours. Any stable baryon or meson must have an overall colour charge that is *colourless*

('Chromo' comes from the Greek word *khroma* meaning colour.) Like QED, QCD is a gauge theory which describes an exchange force involving particles of the **colour field** that carry the force from one quark to another. These carrier particles are called **gluons** (pronounced 'glue-ons') and have zero mass and spin 1. The colour field is analogous to the electromagnetic field and the gluon is analogous to the photon in QED.

All the fundamental forces listed in Chapter 6 *decrease* with distance, but the colour force is different, as it remains constant and even *increases* as the distance between quarks increases. One way to imagine the colour force is to think of two balls connected by a rubber band (Figure 7.13). When the balls are fairly close to each other, they can move freely, but when they are further away, the band stretches and the force between the balls becomes stronger. Stretch the balls any further apart and the rubber band snaps.

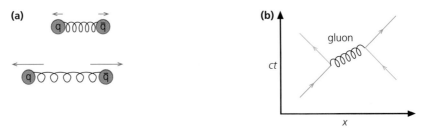

Figure 7.13 The particles that mediate the colour force between quarks are called 'gluons'. Gluons carry a colour and an anti-colour
(a) The force between quarks is rather like two balls connected by a rubber band – as the balls get further apart, the force between them increases
(b) A Feynman diagram for a quark interaction involving the exchange of a gluon. The gluon is represented by a coil. Notice that the gluon generates a colour change for the quarks

The strong interaction is now known to be composed of two components. The first is the fundamental or **colour interaction**, which is the force acting between the quarks and mediated by gluons. The second is called the **residual strong interaction** and is the force acting between colour-neutral nucleons such as the proton and neutron. This is what we originally understood as the strong nuclear force binding nucleons together in the atomic nucleus. What we now see is that the strong force, which nuclear physicists observe as binding the nucleus together and is mediated by the exchange of mesons, actually has a more fundamental origin in the colour force that binds the quarks together. This force is due to the residual or remainder of the strong interactions between the colour 'charged' quarks that make up the nucleons. It can be seen to be mediated by the exchange of *mesons*, which we saw earlier are the π mesons (pions) predicted by Yukawa. Being mesons, π mesons are composite bosons and, at the level of nucleon interactions, the colour force reveals itself as the strong force that binds nuclei together. However, the strong force that binds atomic nuclei involves the exchange of composite particles (mesons), whereas the colour force between quarks involves the exchange of individual bosons (gluons).

Quark confinement

Quarks have never been observed in isolation. It appears impossible to liberate them from hadrons since, if the force increases indefinitely as the distance between them increases, then an infinite amount of energy would be needed to extract one. Attempts have been made to find free quarks, perhaps produced by high-energy cosmic rays on Earth, using apparatus similar to that used in Millikan's oil-drop experiment but looking for multiples of a fractional electronic charge instead. However, the quark model with its concept of 'colourless particles' tells us that particles cannot exist with a residual colour, so that an isolated quark, which would have a colour red, green or blue (or a corresponding anti-colour), would be forbidden.

In a high-energy collision, a quark may become separated from its fellow quarks by such a distance that the energy involved is enough to create a quark–antiquark pair. In this case a meson will emerge from the event, but not a single quark. Coming back to our rubber band analogy, we can think of this event as the snapping of the band attaching the quark to its companions and a new q$\bar{\text{q}}$ pair being formed, each one being attached to a new piece of rubber band (Figure 7.14).

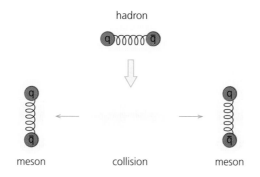

hadron

meson   collision   meson

quark–antiquark pairs are formed but never a single quark

Figure 7.14 Quark confinement. If a quark becomes sufficiently separated during a high-energy collision, then a new quark grouping can be formed, but never a single quark on its own, because this would mean that the overall colour charge *would not* be colourless

Figure 7.15 overleaf shows a two-jet event observed at the DESY accelerator involving an electron–positron collision. The energy released in the annihilation is enough to produce several pairs of q$\bar{\text{q}}$ pairs joined by the colour force. Since the total momentum of the colliding electron–positron pair is zero, by momentum conservation, so must be the momentum of the q$\bar{\text{q}}$ pairs, giving rise to two oppositely directed jets of particles that emerge from the collision.

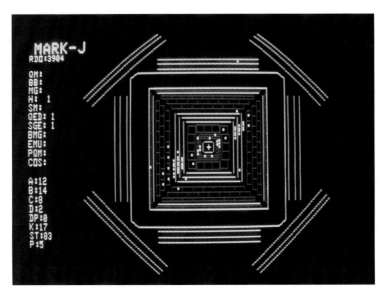

Figure 7.15 A two-jet event at DESY involving an electron–positron collision

The Standard Model

If we combine quantum electrodynamics, the electroweak unification, the quark model and quantum chromodynamics together, we get something that particle physicists call the **Standard Model**:

QED + Electroweak Theory + quark model + QCD = Standard Model

Both quarks and leptons are point-like particles which, as far as we can tell, do not appear to have any deeper structure. This leads us to make another fundamental statement about matter. Instead of having leptons, baryons and mesons, we can now say that:

All matter can be thought of as being composed of combinations of six types of quark and six types of lepton.

What we are saying here is that leptons and quarks are truly elementary. While some baryons and mesons may be more fundamental than others, such as protons and neutrons, they are all composed of quarks. In terms of structure, it seems that the basic building blocks that nature uses to construct matter are quarks and leptons.

Generations of matter

'Everyday matter' is composed of atoms consisting of protons and neutrons, which are composed of u and d quarks. Atoms also contain electrons e and, in radioactive decay, we find that electron neutrinos or antineutrinos (v_e or $\overline{v}_e$) are emitted, so that we can associate these into a pair of leptons and a pair of quarks:

$$(e, v_e) \qquad (u, d)$$

High-energy accelerator experiments show that muons μ and muon neutrinos ν_μ are associated with charmed and strange quarks, so that we have another grouping:

$$(\mu, \nu_\mu) \qquad (c, s)$$

Finally, at even higher energies, we find another pair of leptons, the tau and the tau neutrino, associated with the top and bottom quarks:

$$(\tau, \nu_\tau) \qquad (t, b)$$

These three sets of elementary particles increase in mass so that (e, ν_e), (u, d) are the lightest leptons and quarks, and (τ, ν_τ), (t, b) are the heaviest.

Together with this lepton–quark correspondence, there are the field particles that are exchanged in the electromagnetic interaction (the photon), in the weak interaction (the Z and the $W^\pm$) and in the strong interaction or colour force (the gluon). We can summarise the Standard Model in terms of **three generations of matter** in the diagram shown in Figure 7.16.

Note that the masses of the quarks listed in Table 7.4 are approximate. Since quarks only exist inside hadrons, where they are confined by the colour force, we cannot measure their mass by isolating them. The mass of a hadron includes contributions from the quarks' kinetic energy and potential energy (due to the quarks interacting among themselves). So the quark masses make up only a small part of the hadron mass. In a proton with mass $0.938\,\text{GeV}/c^2$, for example, the three constituent quarks (uud) are thought to make up only $0.02\,\text{GeV}/c^2$ or about 2% of the proton mass.

To attempt to measure quark masses, particle physicists can do calculations to see what will happen to a quark inside a hadron when it is struck by other particles. What we call the 'quark mass' will control its acceleration when an external force is applied to a hadron in a high-energy collision. Total momentum calculations, in

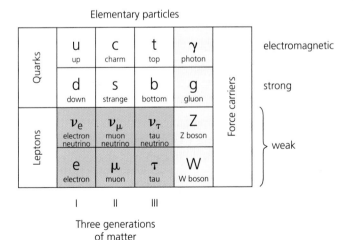

Figure 7.16 The Standard Model of particle physics, illustrating the three generations of matter. Gravity is not accounted for in the Standard Model

which particles such as electrons are scattered off hadrons, also help in understanding how momentum is distributed among the quarks. Neither of these methods can precisely determine quark masses, but these masses are fixed to give the best agreement between theory and experiment.

If we can't isolate a quark, then how can we be sure they are real? The answer is that calculations that presuppose their existence give the right answers when we do experiments. When electrons are scattered off protons and neutrons (in a manner similar to Rutherford scattering), then the pattern of scattering angles is consistent with point-like objects within the nucleus, rather like you can feel the shape of chocolate bars within their wrapper. The jets produced as a result of the creation of quark–antiquark pairs give characteristic predictions, all of which are confirmed experimentally. The accumulation of many such results, where experiments match predictions, is very strong evidence that quarks really do exist.

Is it possible that there may be more quarks and leptons waiting to be discovered at even higher energies? The answer to this question is that particle physicists believe there aren't. All the particles discovered in accelerator experiments to date can fit into the Standard Model with its scheme of six quarks and six leptons in three generations. In addition, according to our current theories of the origin of the universe, its subsequent evolution would have turned out differently if there had been more than three kinds of neutrinos; we will discuss this further in Chapter 8. However, why nature should repeat itself in this hierarchy of quarks and leptons is completely unknown.

What about gravity?

A glaring omission from our discussion of the Standard Model and unifying the interactions is the gravitational interaction. The Standard Model is a unification of sorts and has been spectacularly successful in accounting for the properties of the elementary particles. But where does gravity fit in to this scheme? To answer this question, we need to take a look at our universe of matter at the largest scales, which is the subject of our final chapter, dealing with the relationship between particle physics and the origin of the universe.

Summary

◆ One of the principal objectives of particle physicists is to unify all the forces of nature into a **Theory of Everything** (TOE), which can explain all known phenomena in quantum terms. The first step was the unification of electricity and magnetism into electromagnetism as described by **Maxwell's equations**.

◆ The next step towards unification was taken with the development of a gauge theory called the **Electroweak Theory**, which theorised a single **electroweak force** mediated by the **W** and **Z particles**. This was dramatically confirmed when these particles were discovered at CERN in the early 1980s with masses very close to those predicted by the Electroweak Theory.

◆ If electric charge and strangeness are plotted on a set of coordinate axes for all baryons and mesons, then a geometrical symmetry appears called an Eightfold Way pattern. The Eightfold Way patterns revealed an underlying structure in the families of particles, which can be explained by more elementary particles called **quarks**. The quarks come in six **flavours** called **up**, **down**, **strange**, **charm**, **top** and **bottom**. Each quark has its corresponding **antiquark** and all have **fractional charge**. In the **quark model** all mesons and baryons can be understood in terms of appropriate combinations of quarks, and/or antiquarks. All matter can now be expressed in terms of quarks and leptons.

◆ All baryons are made up of combinations of **three quarks**. All mesons are made up of a **quark–antiquark pair**. Quarks cannot be observed directly but through **deep inelastic scattering** experiments and the ways in which they are expected to decay into observable particles. **Charm** is a quantum property like strangeness, which was introduced to account for some types of particle decays involving the strong and weak interactions. The **J/ψ** (J/psi) **meson** was discovered that had the properties expected of a charm–anti-charm quark pair. States of matter that consist of quarks and antiquarks bound together are called **quarkonium**.

◆ Strong interactions do not change the flavours of quarks. Instead, rearrangements of quarks inside particles can occur, including the materialisation of a quark–antiquark pair. Weak interactions can change the flavour of a quark, by the emission and absorption of a W boson.

◆ **Quantum chromodynamics** is a gauge theory that explains the **colour force** that operates between quarks. The colour force has a carrier particle called a **gluon**, which carries the colour force from one quark to another, and increases with distance. **Colour** is to the strong interaction what electric charge is to the electromagnetic interaction, and each quark is assigned a colour which is either **red**, **green** or **blue**. For all hadrons, the **quark colours** must add up such that they are **colourless**. This is why quarks cannot exist on their own, a state called **quark confinement**. The strong interaction is now thought of as a **colour interaction** between quarks, and a **residual strong interaction**, which is what we observe as the force acting between nucleons in the nucleus.

◆ Particle physicists have come up with the **Standard Model**, in which all matter can be thought of as being composed of combinations of six types of quark and six types of lepton. These can be summarised as **three generations of matter** together with their force-carrying particles. The Standard Model is remarkably successful in accounting for the properties of the fundamental particles, but it lacks the ability to incorporate gravity. A further development in unification is therefore needed.

Questions

1 a Protons and neutrons consist of two types of quark in different combinations. What is the quark combination of:
i) a proton, ii) a neutron?
  b If the 'down' quarks are 2.6 fm apart inside the neutron, what is the magnitude of the electrostatic force between them? [1 fm $= 10^{-15}$ m]

2 a The Feynman diagrams below show interactions which are *not* observed. For each diagram give two reasons why the interaction shown is *not* possible.

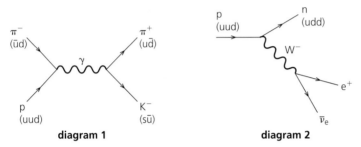

diagram 1	**diagram 2**

Edexcel, January 2000 (part)

  b The Feynman diagram represents the β^- decay process.

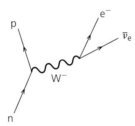

With reference to the diagram describe each stage of the decay. Then redraw the diagram so that it shows the quark transformation that occurs in β^- decay.

ULEAC, Specimen (part)

3 a The K^+ is a meson with strangeness $+1$. One of its common decay modes is

$$K^- \rightarrow \pi^- + \pi^0$$

π-mesons are not strange particles.
i) Name the type of interaction responsible for $K^- \rightarrow \pi^- + \pi^0$ decay.
ii) The following table gives the properties of the relevant quarks. Deduce the possible quark content of the K^-, the π^- and the π^0.

Type of quark	Charge	Baryon number	Strangeness
u	$+\frac{2}{3}$	$\frac{1}{3}$	0
d	$-\frac{1}{3}$	$\frac{1}{3}$	0
s	$-\frac{1}{3}$	$\frac{1}{3}$	-1

iii) The rest mass of a proton is $938\,\text{MeV/c}^2$. The K^- rest mass $= 0.53$ proton masses and the π^- rest mass $= \pi^0$ rest mass $= 0.15$ proton masses.

Assuming that the K^- is stationary when it decays, show that the total energy of each pion produced in the decay is $249\,\text{MeV}$.

Given that $E^2 = m_0^2 c^4 + p^2 c^2$, calculate the momentum of each pion. Express your answer in units of MeV/c.

ULEAC, Specimen (part)

b W^+, W^- and Z^0 exchange particles are produced during annihilation reactions between electrons and positrons at CERN.

i) Explain why more energy is available for creating new particles in a colliding beam experiment than in a fixed target experiment. The Z^0 has a mass of $91.2\,\text{GeV/c}^2$. What is the minimum beam energy required to produce a Z^0?

ii) The Z^0 may decay into a quark/antiquark pair. Draw a Feynman diagram representing the production and subsequent decay of a Z^0.

iii) Although the mass of the W particles is only $80.4\,\text{GeV/c}^2$, explain why they must be created in pairs. State the minimum beam energy required.

Edexcel, January 2001 (part)

4 a A bubble chamber was frequently used as a particle detector until the mid-1980's. A bubble chamber consists of a large quantity of liquid hydrogen. A strong magnetic field acts perpendicular to the liquid surface.

When a single positive particle enters this detector at right angles to the magnetic field, the tracks shown on the diagram below are observed.

Account for the tracks produced at points A and B.

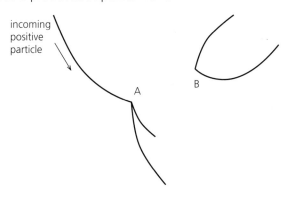

incoming positive particle

A

B

b Two particles D^+ and Δ^{++} can decay in the following ways:

$$D^+ \rightarrow K^- + \pi^+ + \pi^+ \qquad \text{lifetime} \sim 10^{-12}\,s$$
$$\Delta^{++} \rightarrow p + \pi^+ \qquad \text{lifetime} \sim 10^{-23}\,s$$

(note the $++$ means that the Δ^{++} has double the electron charge)
D, K and π are mesons. The D^+ is a charmed meson; the K^- is a strange meson; all the other particles have zero charm and strangeness.
Use the information in the table below to deduce the quark structures of D^+, π^+ K^- and Δ^{++} particles.

Quark	Symbol	Charge	Strangeness	Charm
up	u	$+\frac{2}{3}$	0	0
down	d	$-\frac{1}{3}$	0	0
charm	c	$+\frac{2}{3}$	0	+1
strange	s	$-\frac{1}{3}$	−1	0

Show that the difference between the number of quarks and the number of antiquarks $(q - \bar{q})$ is constant for both decays.
Which fundamental interaction is involved in the Δ^{++} decay?
The D^+ decays by the weak interaction. Give *two* reasons in support of this statement.

Edexcel, January 1999 (part)

5 In 1962 the existence of a particle with strangeness -3 was predicted. The particle Ω^- was identified in 1964 in an experiment involving a strong interaction between a K^- meson of strangeness -1 and a proton in a hydrogen bubble chamber. The interaction involved was

$$K^- + p \rightarrow \Omega^- + K^+ + K^0$$

a Is the Ω^- particle a baryon or a meson? Give two reasons for your answer.
b Using the information given in the table below, deduce the quark composition of all the particles involved.

Type of quark	Charge	Strangeness
u	$+\frac{2}{3}$	0
d	$-\frac{1}{3}$	0
s	$-\frac{1}{3}$	−1

c The Ω^- (lifetime $8.2 \times 10^{-11}\,s$) subsequently decayed in a three stage process to a proton and a number of π-mesons. π-mesons have zero strangeness. What fundamental interaction must be involved at some stage in this process? Give two reasons for your answer.
What exchange particle(s) could be mediating this process?

ULEAC, January 1996 (part)

6 The diagram below is from a bubble chamber and shows a collision between a negative pi-meson and a stationary proton at A.

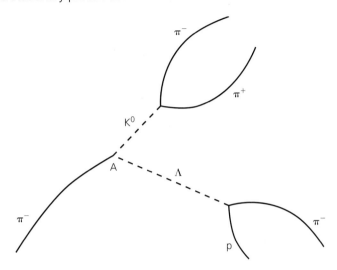

Write down the interaction at A.

The K^0 meson decays into two pi mesons π^- and π^+. The K^0 has strangeness $+1$ and is the only strange particle in the decay.

Use the data in the table below to deduce the quark structure of these three mesons.

Quark	Charge	Strangeness
u	$+\frac{2}{3}$	0
d	$-\frac{1}{3}$	0
s	$-\frac{1}{3}$	-1

By which fundamental interaction does the decay of the K^0 occur? Give a reason for your answer.

Suggest a value for the lifetime of the K^0.

Edexcel, June 1999 (part)

7 Which of the following statements are true?

A Hadrons are particles that feel the strong interaction.

B Charge and baryon number are conserved in all interactions.

C Strangeness is conserved in strong and electromagnetic interactions but not in weak interactions.

D Every hadron has a value of charge Q, baryon number B and strangeness S assigned to it.

E The photon is not a fundamental particle.

F Because the W^+ is observed to decay into a μ^+ and ν_μ we can conclude correctly that the W^+ contains a μ^+ and ν_μ.

Particle physics and the early universe

'I ask you to look both ways for the road to a knowledge of the stars leads through the atom; and important knowledge of the atom has been reached through the stars.'

Sir Arthur Eddington (1882–1944), British astrophysicist

Evidence for the origin of the universe

Where did all the matter and energy in the universe come from? To answer this question, we have to explain why the universe exists as it does. We need a theory that can account for both the large-scale and the small-scale structure of the universe and also tell us what may happen to it in the future.

The branch of science that deals with these questions is called **cosmology**, and scientists who study such things are called **cosmologists**. The task of the cosmologist is to understand how different phenomena of nature, from small elementary particles and fundamental forces, right up to very large-scale structures in the universe such as clusters of galaxies, all fit together. The main question to ask is whether the universe in which we live is ordered or chaotic. Can our experience of the universe, with its many diverse phenomena, be understood in a meaningful and logical way? Particle physics has shown us that an underlying symmetry exists in nature, which is simple and obeys certain laws, yet the Standard Model, despite all its successes, is still quite complicated and tells us nothing about gravity. Particle physicists are working on theories that may provide the next and possibly final step in unification. But before we look at the progress that has been made, let's take a look at the universe on a larger scale, and in particular at three key pieces of evidence that point to its creation at some time in the distant past.

The expanding universe

When you look up at the sky on a clear night, you can probably see many thousands of stars, each appearing as a bright point of light. Some stars appear brighter than others and, if you look carefully, they may even show colours (they certainly do in photographs). If we look through a telescope, then we see a myriad of other stars that are too faint to be seen with the naked eye.

Olbers' Paradox

But why is the sky dark at night? On the face of it, this may seem a silly question to ask, but consider this: If the universe were infinite and contained stars more or less homogeneously, then everywhere we looked we would eventually intercept light coming from some star, and so the sky would appear uniformly bright (Figure 8.1). This question was first discussed in a paper written in 1823 by the German amateur astronomer Heinrich Wilhelm Olbers (1758–1840).

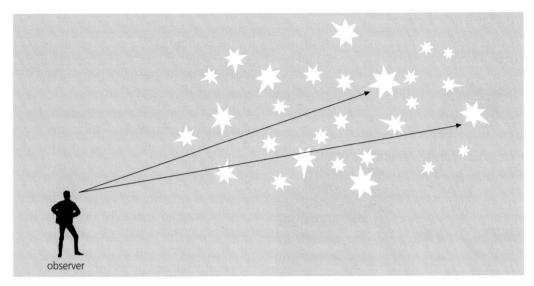

Figure 8.1 Olbers' Paradox. If the universe was infinite and isotropic, then any line of sight drawn from an observer must eventually end at the surface of a star. So the sky should be uniformly bright at night

Olbers' discussion is quite involved, but the essential feature of the argument is that, if the universe is infinite in space and the distribution of stars is isotropic and homogeneous, then any line of sight drawn from an observer looking out from the Earth, in any direction, must eventually end at the surface of a star. We conclude that, by looking in all directions, the night sky should appear uniformly bright. But this is not what we observe. This apparent contradiction is known as **Olbers' Paradox**, and we will see that its resolution lies in the facts that the universe is *expanding* and the speed of light is *finite*. When we look out into space, we look into the universe's past because light from that era has taken a certain time to arrive at Earth. We get no light from the most distant parts of the universe because the universe simply isn't old enough yet for such light to reach us.

Olbers' Paradox is powerful evidence that the universe is finite, and at some point it must have had a beginning, since, if it was infinitely old, then the sky would be bright at night. So next time you go outside at night and look at the starry sky, remember that the darkness tells us this – the universe once had a beginning.

Hubble's discoveries

During the 1920s the American astronomer Edwin Hubble (1899–1953) (Figure 8.2) made a very important discovery. Hubble, together with others, had noticed that the spectral lines of distant galaxies were **redshifted** (Box 8.1), indicating that the galaxies were moving away from us. He found that the extent to which the spectral lines are shifted was proportional to how far away the galaxy is from us. Hubble plotted a graph of radial recessional velocity (v) versus distance (r) and obtained a straight line, which showed that the radial velocity was directly proportional to the distance. This result is known as **Hubble's Law** and if v is measured in $km\,s^{-1}$ and distance r in Mpc (the **megaparsec** (Mpc) is a unit of astronomical distance equal to $3.1 \times 10^{19}\,km$), we get the relation

Figure 8.2 Edwin Hubble (1899–1953)

$$v = H_0 \times r$$

where H_0 is known as the *Hubble constant*, with unit of $km\,s^{-1}\,Mpc^{-1}$.

The value of the Hubble 'constant' is the subject of much debate, and its determination depends on how well we can measure the distances to galaxies by independent methods. The current value is thought to lie between 50 and $100\,km\,s^{-1}\,Mpc^{-1}$, and in this book we will adopt the currently accepted value of $H_0 = 75\,km\,s^{-1}\,Mpc^{-1}$.

Hubble's Law is evidence that the universe is not static but *expanding*, and if this is the case then in the past it must have been smaller than we see it today.

The abundance of the elements

By analysing the light from stars and galaxies, and looking at the spectral lines they contain, astronomers have determined that the universe as we see it now contains 70–80% hydrogen and 20–30% helium by mass. All other heavier elements form a small minority, and most stars are broadly similar in their chemical composition. Astrophysicists have determined that atoms that are heavier than helium are 'cooked' inside stars by nuclear reactions in a process called **stellar nucleosynthesis**, which includes the formation of very heavy elements in supernova explosions. Stars shine by converting hydrogen into helium in a nuclear fusion process called **hydrogen burning**. This is the initial stage of stellar nucleosynthesis, where hydrogen is converted into helium, and then other heavier elements are formed as the star progresses through its life-cycle.

Box 8.1 The redshift and Hubble's Law

You may have noticed when standing at a station that, if a train is coming towards you sounding its horn, the note is higher and then drops in frequency as it passes by and starts to recede. This is an example of the **Doppler Effect** named after the Austrian physicist Christian Doppler (1803–1853). The reason why this happens is that, when the train is approaching, more sound waves per second are reaching your ears than if the train is stationary, and the wavelength becomes shortened as a result. Since, for a given wave speed, the frequency is inversely proportional to the wavelength, the frequency of the note is higher. As the train recedes from you, there are fewer sound waves per second reaching your ears, the wavelength appears longer and so the frequency of the note is lower.

The same thing happens with electromagnetic radiation. With light, depending on the motion of the object with respect to the observer, the colour of the light is affected. An approaching light source is a little more blue (shorter wavelength) than it would otherwise be, and one that is moving away a little redder (longer wavelength). The Doppler Effect for light can be used to estimate the speed at which distant galaxies are receding from the Earth. This is given by the formula

$$\frac{v}{c} = \frac{\lambda - \lambda_0}{\lambda_0}$$

where λ_0 is the wavelength of the light measured in the laboratory and λ is the wavelength emitted from the moving source. This may be written as

$$z = \frac{v}{c} = \frac{\Delta\lambda}{\lambda_0}$$

where $\Delta\lambda = \lambda - \lambda_0$ is the difference between the wavelengths λ and λ_0. If λ is greater than λ_0, then the galaxy is moving *away* from the Earth and v/c is positive. We say that the light has been **redshifted** towards the red end of the electromagnetic spectrum. If λ is less than λ_0, then the galaxy is moving *towards* us and its light has been shifted towards the blue end or **blueshifted**. In this case v/c is negative.

By comparing the wavelengths in the spectral lines of a galaxy with those in a laboratory on Earth, we can measure the difference in wavelength $\Delta\lambda$ and calculate the **redshift** z, from which we can work out the **recessional velocity**. Hubble's Law (Figure 8.3, overleaf) relates the recessional velocity to distance. Galaxies with very high values of z are moving with faster velocities. The fact that we see redshifts in (nearly) all galaxies shows that the universe is *expanding*, and the Hubble constant H_0 is a measure of the rate at which this expansion is taking place.

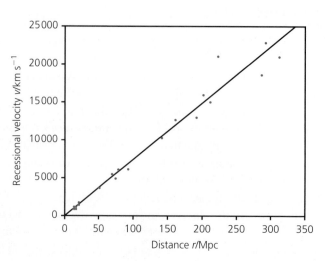

Figure 8.3 Hubble's Law. The radial recessional velocity v of distant galaxies is proportional to their distance r from us

However, the helium abundance in stars cannot all have been formed by stellar nucleosynthesis. If this were so, then we would expect older stars to contain less helium than younger ones, as they would have used up more of it in nuclear fusion processes, but this is not what we find by analysing the starlight. In the history of the universe, the production of helium by stellar nucleosynthesis has still had no more than a modest effect on the abundance of helium in the universe. In addition, the thermonuclear fusion reactions in stars convert hydrogen into helium at a rate that releases large amounts of energy, but, in order to obtain the helium abundances we see today, the stars would have to fuse helium from hydrogen at a much faster rate, making them shine more brightly. We conclude from this that some other processes must have been at work to account for the hydrogen and helium abundances we measure today, and that these occurred before the oldest stars in the universe were formed.

Cosmic microwave background (CMB) radiation

As well as Hubble's Law and elemental abundances, there is another important piece of evidence that supports the idea of an expanding universe and also points to a *hot creation* as being its origin. In 1964 Arno Penzias (1933–) and Robert Wilson (1936–), two scientists at Bell Telephone Laboratories in New Jersey, were carrying out experiments using a microwave antenna for early satellite communications (Figure 8.4). As they pointed the antenna towards the sky, their receiver registered a faint 'hiss' that would not go away. The hiss seemed to be coming from all directions, was constant with time and could be detected at any time of the day or year. The type of radiation that Penzias and Wilson had detected is characteristic of **blackbody radiation** (Box 8.2).

Figure 8.4 Arno Penzias (1933–) and Robert Wilson (1936–) and their microwave antenna

Blackbody radiation is electromagnetic radiation whose emission depends only on the temperature of an object. As we will see later, cosmologists have theorised that, when the universe was very young at a time of about 1 s, it would have been extremely hot. As it expanded, it cooled, and the wavelength of the radiation that it emitted was redshifted down to the microwave region of the electromagnetic spectrum, with a blackbody temperature of about 3 K. One way of thinking of the cosmic microwave background is to imagine looking at the dying embers of a bonfire, from which you can deduce that, at some time in the past, the bonfire must have been very hot and burning brightly.

This theoretically predicted value of 3 K is in close agreement with the blackbody radiation temperature measured by Penzias and Wilson. In 1989, a space astronomy satellite called the **Cosmic Background Explorer** (COBE) was launched, which measured the distribution of microwave radiation in more detail and confirmed a blackbody curve with a peak wavelength corresponding to a temperature of about 2.7 K (Figure 8.6).

Box 8.2 Blackbody radiation

A **blackbody** is an object that absorbs all electromagnetic radiation that falls upon it and reflects and transmits none. If heated, a blackbody will produce **blackbody radiation** whose intensity follows a continuous distribution called a **Planck curve** or **blackbody spectrum** (Figure 8.5). It is important to understand that when you see a blackbody curve you know that the processes that give rise to the emission of the radiation depend *only* on temperature and not on any other property such as the chemical composition of the object. A blackbody is also in **thermal equilibrium**, which means that its components all share a common temperature.

A blackbody emits electromagnetic radiation over a wide range of wavelengths, but there will be one wavelength, called the **peak wavelength**, where the emission of radiation has its *maximum* intensity. The German physicist Wilhelm Wien (1864–1928) discovered a simple relationship called **Wien's Law** between the absolute temperature T of a blackbody and the peak wavelength λ_{max} at which the radiated energy reaches its maximum intensity:

$$\lambda_{max}/m = \frac{2.90 \times 10^{-3}}{T/K}$$

So, for example, blackbody radiation that has a wavelength of 0.1 cm (which lies in the microwave region of the electromagnetic spectrum) has a temperature of

$$T = \frac{2.90 \times 10^{-3}}{0.1 \times 10^{-2}} \approx 3\,K$$

The 3 K radiation that Penzias and Wilson discovered showed that the entire universe was 'warm' (although cold by our standards!). The COBE satellite confirmed the blackbody Planck curve of the universe, showing that it was filled with thermal radiation with a peak wavelength corresponding to a temperature of 2.7 K. The **cosmic microwave background**, as it is commonly known, is a relic of the hot origin of the universe. At a very early time in its history, radiation and matter were at the same temperature and the universe behaved like a blackbody. As the universe expanded, this radiation redshifted so that every wavelength in the original blackbody curve was increased by a common factor with the peak of the spectrum occurring at about 3 K, which is what we observe today.

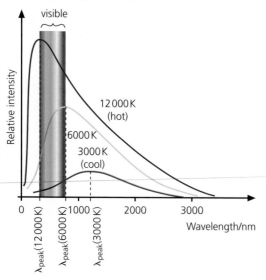

Figure 8.5 Blackbody curves. The peak wavelength for a given temperature is found from Wien's Law. (Relative intensity scale compressed for clarity)

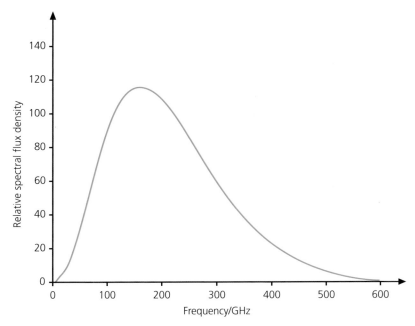

Figure 8.6 The blackbody curve for the cosmic microwave background radiation, as obtained by the COBE satellite. The peak of the curve corresponds to a temperature of 2.73 K

The intensity of the microwave background is virtually constant for any direction in which you measure it. This property, known as its **isotropy**, provides strong support for the **Cosmological Principle**. This principle asserts that the universe at the largest scales is **homogeneous**, which means that it has the same density everywhere, and **isotropic**, which means it 'looks the same' from whatever direction you view it (Figure 8.7).

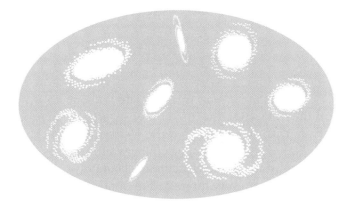

Figure 8.7 The Cosmological Principle. At the largest scales (clusters of galaxies and beyond), the universe is homogeneous and isotropic

The Big Bang and the age of the universe

The three observational features of the universe that we have looked at in the previous section lead cosmologists to conclude the following:

1 The universe was much smaller than it is now.
2 The lighter elements such as hydrogen and helium were created very early on in its history
3 The universe was very much hotter in the past.

These conclusions can be combined into a theory of the universe's origin called the **Big Bang**. The Big Bang asserts that at some point in the distant past all the matter and radiation in the universe were concentrated together in a state of high density at a single point, where some kind of gigantic explosion must have happened to cause it to start expanding. But when did this occur?

To arrive at an estimate of the age of the universe, consider a galaxy moving with constant velocity v. The distance D travelled in time t is

$$D = v \times t$$

and the travel time is

$$t = \frac{D}{v}$$

From Hubble's Law we know that

$$v = H_0 \times D$$

so that

$$t = \frac{D}{H_0 \times D} = \frac{1}{H_0}$$

If we use the value for H_0 of $75 \, \text{km s}^{-1} \text{Mpc}^{-1}$, remembering that $1 \, \text{Mpc} = 3.1 \times 10^{19} \, \text{km}$, we get

$$t = 1 \times \frac{3.1 \times 10^{19} \, \text{km}}{75 \, \text{km s}^{-1}} = 4.13 \times 10^{17} \, \text{s}$$

Since there are $3.15 \times 10^7 \, \text{s}$ in one year, this estimate is about 1.3×10^{10} years. Therefore, this puts the time when the universe first started to expand at some 13 000 million years ago. From this, we can see that the reciprocal of the Hubble constant tells us the age of the universe, and its value changes with time as the universe expands. The Hubble 'constant' is therefore more accurately called the **Hubble parameter**, and cosmologists are very interested in obtaining a reliable estimate of it.

The importance of Hubble's Law is that it implies that the universe is *finite*, and this is what we would expect if Olbers' Paradox is to be resolved. We cannot see any light from galaxies more than 13 000 million light years away, and this marks the limit of the *observable* universe. If the universe is infinite, then light from galaxies beyond this limit simply hasn't had enough time to reach us yet.

The fate of the universe

Bounded or unbounded?

Will the universe go on expanding for ever? To answer this question, you need to know the average density of matter it contains. As the universe expands, gravity will continuously act to slow down and decelerate the expansion. If the gravitational strength generated by this matter is too weak, then the universe will go on expanding and the galaxies will continue to move away from each other. In this case, cosmologists say that the universe is **unbounded**. If the average density of matter is high enough, then gravity will halt the expansion and the universe will at some time in the future reach a maximum size. Gravity will then start pulling all the galaxies back towards each other, and the universe will contract. In this situation, the universe is said to be **bounded**.

Critical density

Between an unbounded and bounded universe is the case where the average density of matter is just high enough so that the galaxies come to a halt in an infinite time when they are infinitely far apart, and cosmologists call this value the **critical density** ρ_c. An important quantity that determines which of these scenarios will happen is the **density parameter** Ω_0, which is defined as

$$\Omega_0 = \frac{\text{density of matter in the universe}}{\rho_c}$$

By 'density', what we mean here is the **mass–energy density** of the universe. This is a consequence of the equivalence of mass and energy. The critical density must account for the total sum of all matter and radiation in the universe that exerts a contribution to ρ_c.

Figure 8.8 shows what happens for different values of Ω_0. If the universe contains no matter at all, then $\Omega_0 = 0$ and the universe expands, since there is no gravity to slow the expansion. For $0 < \Omega_0 < 1$, the density of matter in the universe is less than

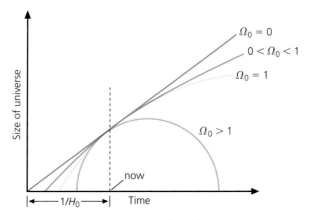

Figure 8.8 The density parameter Ω_0 determines the fate of the universe

the critical value, and it will expand for ever. For $\Omega_0 > 1$, the total mass–energy will pull the universe in again, and at some time in the future it will collapse. However, if $\Omega_0 = 1$, the universe is 'balanced' between open and closed, between being unbounded and bounded.

From statistical surveys of stars and galaxies, astronomers have determined that the value of Ω_0 is remarkably close to 1. In fact, detailed calculations show that if Ω_0 differed from unity by one part in 10^{55} at a very early time in the universe's expansion, then either the universe would have collapsed by now or the mass–energy density would be considerably less than we measure today. For some reason the universe likes to be close to the critical density, and any theory of the universe's origins must take account of this. We can use Hubble's Law to obtain an estimate of how old the universe is, but how do we account for the origin of the lighter elements such as hydrogen and helium, since we know that hydrogen is not made in stars by stellar nucleosynthesis? The abundance of hydrogen that already exists in the universe cannot have come from stars, so it must have come from somewhere else. This is where particle physics plays an important role in explaining the origin of matter and energy in the universe.

Modelling the universe using particle physics

Pair production

Earlier, we mentioned that cosmologists refer to the average mass–energy of the universe due to the equivalence of energy and matter through $\Delta E = \Delta mc^2$. The average density of matter in the universe today is estimated to be about $10^{-26}\,\text{kg}\,\text{m}^{-3}$. Radiation is produced by stars and galaxies in a concentrated form, but these only occupy a small fraction of the entire universe. By far the most dominant source is the cosmic microwave background, which has a mass density equivalent of about $5 \times 10^{-31}\,\text{kg}\,\text{m}^{-3}$ and is some 100 000 times smaller than the density due to matter alone. This suggests that, at the present time, we live in a universe that is dominated by matter. However, cosmologists believe that in the universe's early history it was *radiation* that dominated. In order to understand why, we need to look at a concept that we discussed earlier in this book, that of *pair production*.

We can write pair production as:

$$\text{photon} + \text{photon} \rightarrow \text{particle} + \text{antiparticle}$$

The reverse process is also possible. A particle and antiparticle can collide and annihilate each other, producing two high-energy gamma ray photons:

$$\text{particle} + \text{antiparticle} \rightarrow \text{photon} + \text{photon}$$

For pair production to occur two conditions must be satisfied:

1 The energy of the photons must be greater than the mass–energy of the created particles.
2 Pair production must obey the laws of conservation of energy and of momentum.

Cosmologists model the early universe as a hot, high-energy photon gas, and the energy the photons have depends on their *temperature*. Creation and annihilation of particles occur in equal numbers, so that as many particles are made per second as are being destroyed, and the universe is said to be in a state of **thermal equilibrium**, with an overall symmetry existing between matter and antimatter.

The average energy E of photons in a gas can be approximated by the equation $E = \frac{3}{2}kT$, where k is Boltzmann's constant and T is the temperature of the gas. We can use this equation to work out the temperature at which the gas has enough energy for two photons to create a particle–antiparticle pair consisting of an electron and a positron. The equivalent energy of an electron at rest is $m_e c^2 = 9.1 \times 10^{-31}\,\text{kg} \times (3 \times 10^8\,\text{m s}^{-1})^2 = 8.2 \times 10^{-14}\,\text{J}$ (or 0.51 MeV) and, since a positron is created as well, the total energy required is $2 \times 0.51\,\text{MeV} = 1.02\,\text{MeV}$, or $1.6 \times 10^{-13}\,\text{J}$. Using $E = \frac{3}{2}kT$, the temperature is

$$T = \frac{2E}{3k} = \frac{2 \times 1.6 \times 10^{-13}\,\text{J}}{3 \times 1.4 \times 10^{-23}\,\text{J K}^{-1}} \approx 10^{10}\,\text{K}$$

Protons and antiprotons need higher temperatures to create them because of their higher mass. The important point to understand here is that the creation of different kinds of particles by pair production is dependent on the *temperature* of the photons.

The four eras in the history of the universe

The Big Bang model requires that the universe started from a primeval fireball of infinitely high temperature and density. For reasons explained later, we will outline the development of the universe from when its temperature was $10^{33}\,\text{K}$ and its age was greater than $10^{-43}\,\text{s}$. We model the universe initially as a high-energy photon gas, which, like any gas, cools when it expands. Cosmologists divide the history of the universe into four periods or *eras* of time, each corresponding to a particular range of temperatures:

- a *heavy-particle* era
- a *light-particle* era
- a *radiation* era
- a *matter* era

Heavy-particle era

Temperature of universe $< 10^{33}\,\text{K}$, time after Big Bang $> 10^{-43}\,\text{s}$

During this period, the universe is hot enough for all massive elementary particles to be created by pair production. The universe contains quarks, antiquarks, leptons and antileptons in equal numbers and is in thermal equilibrium. The universe expands rapidly, with the first stable protons and neutrons being formed at about $10^{-6}\,\text{s}$. The nucleon rest energy is about 940 MeV so, at this time, the temperature of the photons that produce nucleons by the reactions

$$\gamma + \gamma \rightarrow p + \bar{p}$$

and

$$\gamma + \gamma \rightarrow n + \bar{n}$$

is

$$T = \frac{2E}{3k} = \frac{2 \times 940 \times 10^6 \times 1.6 \times 10^{-19} \text{J}}{3 \times 1.4 \times 10^{-23} \text{J K}^{-1}} \approx 10^{13} \text{K}$$

At this temperature, the universe contains a 'soup' of p, $\bar{\text{p}}$, n, $\bar{\text{n}}$, e^-, e^+, μ^-, μ^+, π^0, π^-, π^+ plus other particles including photons, neutrinos and antineutrinos.

Light-particle era

Temperature of universe $< 10^{12}$ K, time after Big Bang $> 10^{-4}$ s
The temperature of the universe is no longer hot enough for massive particles to be made. Lighter particles such as electrons and positrons can still be produced, and protons and electrons combine to make neutrons. We mentioned earlier that we live in a matter-dominated universe and, for this to be the case, there must have been an excess of particles over antiparticles during the transition from the heavy-particle to light-particle era, and the symmetry between particles and antiparticles is broken.

Radiation era

Temperature of universe $< 10^{10}$ K, time after Big Bang > 10 s
In this era, nuclei are being built. Neutrons and protons left over from the heavy-particle and light-particle eras interact to form the first stable nuclei. The most important nucleus to form is that of deuterium $^2_1$H and, from this, stable nuclei of helium $^4_2$He and light helium $^3_2$He as well as beryllium $^7_4$Be and lithium $^7_3$Li can be made. Deuterium can be formed by the reaction

$$^1_0\text{n} + ^1_1\text{H} \rightarrow ^2_1\text{H} + \gamma$$

and the other elements by

$$^2_1\text{H} + ^1_1\text{H} \rightarrow ^3_2\text{He} + \gamma$$

$$^2_1\text{H} + ^2_1\text{H} \rightarrow ^4_2\text{He} + \gamma$$

$$^3_2\text{He} + ^2_1\text{H} \rightarrow ^4_2\text{He} + ^1_0\text{n} + ^0_1\text{e}$$

$$^4_2\text{He} + ^3_2\text{He} \rightarrow ^7_4\text{Be} + \gamma$$

$$^7_4\text{Be} + ^0_{-1}\text{e} \rightarrow ^7_3\text{Li} + \nu$$

The manufacture of light elements from neutrons and protons formed in the heavy-particle and light-particle eras is called **primordial nucleosynthesis** (don't get this confused with *stellar nucleosynthesis*, which is the formation of heavier elements in the interior of stars). About 25–30% of this baryonic matter by mass is helium, the rest being mainly hydrogen, with trace amounts of beryllium and lithium. The Big Bang model predicts that the universe should contain a helium abundance of at least these percentages, with more being created by nuclear reactions in stars. Observations of the chemical composition of various celestial objects tend to support this.

Matter era

Temperature of universe $< 3000\,K$, time after Big Bang $> 10^6$ years (10^{13} s)

In previous eras matter and radiation still interact with each other and are locked together. The radiation cannot escape and the universe is *opaque* to radiation. However, when the temperature of the universe has dropped to 3000 K, the first hydrogen and helium *atoms* form, and matter and radiation are no longer coupled together. The universe becomes *transparent* to radiation, and cosmologists call this event **decoupling**. The decoupling of photons and baryons marks the stage at which the radiation expands with the universe, leaving the matter to interact with itself via gravity.

Owing to local variations in density, the baryonic matter starts to clump together and material condensation occurs, leading cosmologists to believe that it was from the beginning of this era that large-scale structures such as galaxies could first start to form. The radiation that spread out after decoupling now comes to us as the 3 K cosmic microwave background, which, because of its smoothness, suggests that, before this event, matter and radiation must have been very uniformly distributed. Exactly how galaxies formed after decoupling is far from clear. It is hoped that instruments such as the Hubble Space Telescope will enable astronomers to peer further back in time so that we can see galaxies in the early stages of their formation, which will help us understand how this happened.

At 10^{18} s we have reached the present. The universe continues to expand, with the galaxies receding from each other. The Hubble parameter decreases and the temperature of the universe continues to drop, reaching absolute zero in an infinite amount of time.

You might wonder at this point if Hubble's Law violates the Cosmological Principle, since it seems to pick out a unique point in space from where the universe began and a unique 'edge' to all the expanding mater. It is important to understand that the Big Bang *did not* happen at some point in space in an otherwise empty universe. The Big Bang *is* space itself expanding, and the galaxies are moving with it. We are 'inside' the Big Bang, as it were. In this way, the universe cannot be said to have an edge, and its properties of homogeneity and isotropy as required by the Cosmological Principle remain intact. Figure 8.9 shows a chart of the history of the universe.

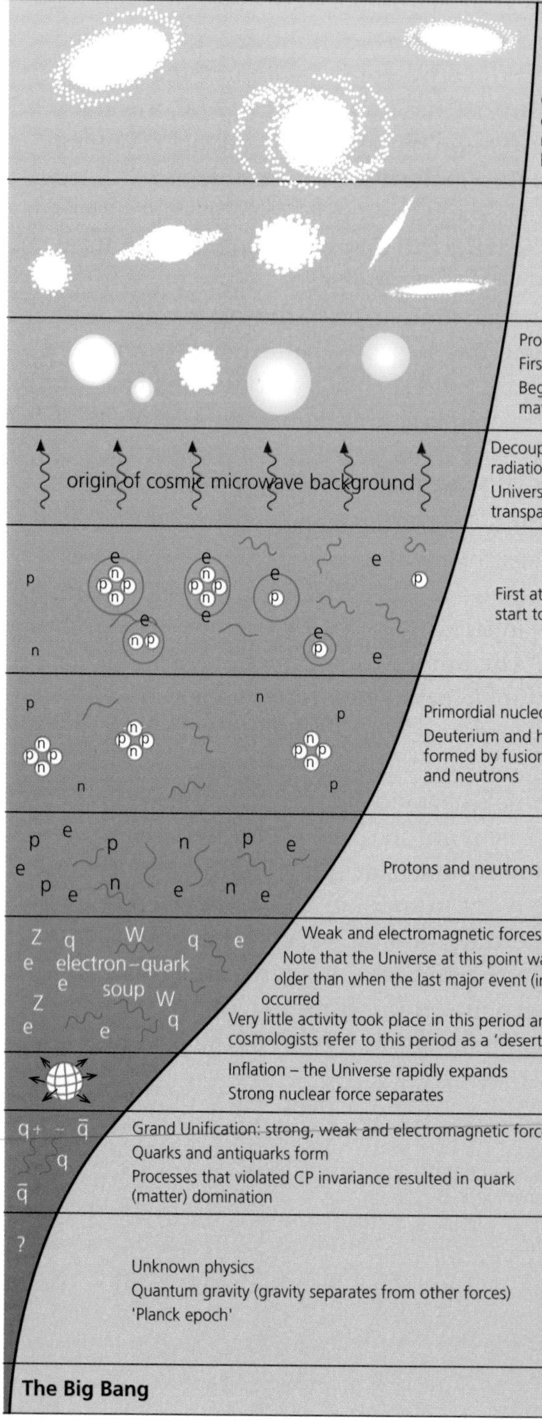

	Time after Big Bang	Temperature	Particle energies	Density /kg m^{-3}
COBE detects cosmic microwave background	**NOW** (10^{18} s)	3 K	>eV	10^{-26}
Galaxy discs	10^{16} s	10^{-3} K	>eV	10^{-25}
Protogalaxies First stars Beginning of matter era	10^{15} s	10 K		3×10^{-25}
Decoupling of radiation from matter Universe becomes transparent	10^{11} s	10^4 K	<1 keV	10^{-19}
First atoms start to form	10^4 s	10^5 K	1 MeV–1 keV	10^{-3}
Primordial nucleosynthesis Deuterium and helium nuclei formed by fusion of protons and neutrons	10^2–10^3 s	10^9 K	~MeV	10^4
Protons and neutrons created	10^{-6} s	10^{13} K	<1 GeV	10^{17}
Weak and electromagnetic forces separate Note that the Universe at this point was 10^{20} times older than when the last major event (inflation) occurred Very little activity took place in this period and cosmologists refer to this period as a 'desert'	10^{-12} s	10^{15} K	>1 GeV	10^{20}
Inflation – the Universe rapidly expands Strong nuclear force separates	10^{-32} s	10^{25} K	10^{18} GeV	$<10^{75}$
Grand Unification: strong, weak and electromagnetic forces unified Quarks and antiquarks form Processes that violated CP invariance resulted in quark (matter) domination	10^{-35} s	10^{27} K	$>10^{18}$ GeV	10^{75}
Unknown physics Quantum gravity (gravity separates from other forces) 'Planck epoch'	10^{-45}–10^{-42} s	$>10^{30}$ K	$>10^{18}$ GeV	10^{97}
The Big Bang	0	∞	–	∞

Figure 8.9 A chart depicting the history of the universe

The origins of the fundamental interactions

Grand Unified Theories

All of what we have described so far involves particles interacting with each other at different stages of their development. At our present time or *epoch*, we observe four distinct interactions (three if you count the electroweak as a single one), but what were the interactions like in the early history of the universe?

The Standard Model is a gauge theory that explains all the known particles and their interactions as far as the energies of our accelerators will allow, but it is incomplete as it only manages to partially unify the forces of nature together. Unification of the four fundamental interactions of nature is a central objective of particle physicists. A classical theory, that of James Clerk Maxwell, started this off, with the unification of electricity and magnetism through his equations of electromagnetism. Weinberg, Glashow and Salam performed a similar feat with electromagnetism and the weak interaction, unifying them into a single electroweak force, the confirmation of which was the discovery of the W and Z particles. The success of the electroweak theory relied on the idea of broken symmetry to highlight the differences in the electromagnetic and weak forces. Particle physicists have naturally wondered whether unification could be extended to embrace the remaining strong and gravitational interactions. Quantum chromodynamics, which is modelled on QED, explains the forces acting between quarks. Attempts are being made to unify the electroweak theory and QCD into a single theory that embraces electricity, magnetism, the weak and the strong force. Such a theory is called a **Grand Unified Theory (GUT)**.

There are many different GUTs, but the main idea behind them all is that the fundamental interactions become *indistinguishable at very high energies* and the broken symmetry between them is restored. To understand why, consider the W and Z particles that mediate the weak force. These particles all have mass. In order to carry the weak interaction from one particle to another, they have to be created from the uncertainty in the energy in the vacuum (see Box 8.3, page 266) while they are carrying the force, which is then repaid without violating the Uncertainty Principle. Since they can only do this for a short interval of time, they can only travel a small distance before they vanish, and so the weak force has only a short range. The electromagnetic force is mediated by photons, which, being massless, do not need to borrow energy and so have infinite lifetimes, which means that electromagnetic forces in principle have an infinite range. At significantly higher energies above 100 GeV (e.g. 1000 GeV), the W and Z particles can be created without having to borrow energy from the vacuum. They become real particles and can exist for considerably longer times than allowed by the Uncertainty Principle. Since their lifetimes are extended, the range of the weak force becomes greater, and the weak and the electromagnetic interactions become like each other. This implies that, when the universe was hot enough with more than enough energy to create W and Z particles, electromagnetism and the weak force were on an equal footing as a single electroweak force. As the universe cooled, and W and Z particles could no longer be made outside the confines of the Uncertainty Principle, the electromagnetic force 'froze out' from the weak force, giving us the two distinct interactions we see today.

Grand Unified Theories attempt to extend this concept to all the interactions. In the case of QCD, it is much harder to unify it with QED and the electroweak theory. In QED there is only a single particle – the photon. In the electroweak theory there are three – the W^+, W^- and Z^0. In QCD there are *eight* types of gluon that mediate the force between the quarks, and it all starts getting rather complicated. A GUT that united QCD with the electroweak theory would have to treat quarks (which are hadrons, and feel the strong force) and leptons (which do not) as some kind of single equivalent particle. There is tantalising evidence that quarks and leptons belong to a 'super-family' of particles, since, as we saw in Chapter 7, there is a correspondence of pairs of leptons with pairs of quarks in three generations at successively higher masses. This suggests that, at high enough energies, quarks could change into leptons and vice versa. An exchange particle dubbed an **X boson** (remember that all exchange particles associated with forces are bosons) that could transform quarks into leptons (and vice versa) would have to be very massive, around $10^{15}\,GeV/c^2$ – well beyond the energy range of any accelerator that could be built with existing technology. GUTs that unite the strong with the weak and electromagnetic interactions require 12 exchange X bosons; the number of bosons needed increases as the forces become more united!

The decay of the proton

Despite the large energies that are needed to test these theories, GUTs do make one prediction that could be tested experimentally. Very rarely, a quark inside a nucleon might be able to borrow enough energy from the vacuum to manufacture an X boson. In the case of a proton (uud) an up quark emits a virtual X boson and transforms itself into an antiquark. The X boson is absorbed by a down quark, which becomes a positron, and the remaining quark–antiquark pair converts itself into a π meson.

The likelihood of an event is such that the average lifetime of a proton would be over 10^{30} years, some 10^{20} times greater than the estimated age of the universe. However, although it is rare, proton decay is a *random* process, and if you had a large number of protons, statistically, you might expect at least one of them to decay each year. One way of thinking about it is to imagine an actuary who calculates the expected lifetime of people for insurance purposes. The actuary can be certain that at least 50% of us will be dead by the time we reach 80. However, there will be some who will live to 100 and others who will die young.

In a proton decay experiment it is the protons that 'die young' in which we are interested. According to GUTs, proton decay can occur when a u quark emits an X boson and turns into a $\bar{u}$ quark, which transforms into a positron, the X boson mediating the transformation of a quark into a lepton, that is,

$$p \rightarrow \pi^0 + e^+ \qquad \text{(Figure 8.10)}$$

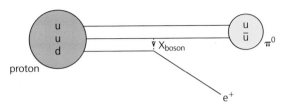

Figure 8.10 The decay of a proton into $\pi^0 + e^+$ via the emission of an X boson by a u quark

If we note that the neutral π meson π^0 decays into two photons and that the positron e^+ could annihilate with any nearby electron, then, if proton decay exists, it leads us to the alarming conclusion that eventually all the matter in the universe will convert itself into electromagnetic radiation, running down like some cosmic torchlight battery!

In some GUTs the proton lifetime is thought to be about 10^{31} years, and 100 tonnes of matter would contain about 5×10^{31} nucleons, from which you might expect to see five decays in a year. The universe is reckoned to be about 10^{10} years old, so that proton decay does not pose any immediate threat to our existence. Experiments are being carried out to look for proton decay, but have so far proved unsuccessful. This does not mean that the GUT is necessarily wrong, merely that the proton lifetime is greater than 10^{31} years!

Matter–antimatter asymmetry

It is clear that we live in a matter-dominated rather than antimatter-dominated universe. But how has this come about? In the early universe, matter and antimatter were created in equal amounts. Pair production created pairs of electrons and positrons, quarks and antiquarks, which all should have eventually annihilated. So the universe should not have favoured matter over antimatter, but this is exactly what seems to have happened. We have seen that on a fundamental level things in the universe tend to be symmetric, so some kind of asymmetry must have been at work to ensure that nature produced the matter-dominated universe we see today.

In Chapter 6 we have already seen an instance of particle asymmetry. Recall that, under the weak interaction, the decay of the K^0_L meson violates **CP** conservation. The K^0_L displayed a preference for decaying into positrons rather than electrons. Such a phenomenon proves that nature sometimes demonstrates a capacity for fundamental lopsidedness in particle interactions. Particle physicists think that something like this was responsible for the preponderance of matter over antimatter in the early universe.

While **CP** violation has been observed in the decay of the K^0_L meson, electroweak theory predicts that **CP** violation should be more easily observable in the decays of **B mesons**, which are similar to, but much heavier than, K mesons. At SLAC, an experiment called **BaBar** that uses the PEP-II electron–positron collider has been set up to find out more about **CP** violation. B mesons can be created from the annihilation of electrons and positrons. The experiment started running in 1999 and, by mid-2001, over 32 million B and anti-B pairs had been recorded and their subsequent decays studied. BaBar will continue to accumulate data for many years, allowing a precise and comprehensive test of the theory of **CP** violation.

One idea is that, when the universe was less than 10^{-4}s old, there existed a baryon–antibaryon soup. X bosons, which can transform quarks into leptons, could have had a sizeable **CP** violation. Perhaps, in the early universe, an excess of baryons was created from the decays of the $\underline{X}$ boson as compared to the creation of antibaryons from the decays of the $\overline{X}$ boson. If this was so, then **CP** violation is the reason behind the excess of matter over antimatter that we see in the universe today.

All this is highly speculative. The best evidence we could have that X bosons exist would be the detection of proton decay, but so far no such event has been observed, although the search goes on.

Magnetic monopoles

If you cut a magnet in half, you get two magnets, each with north and south poles, or a **dipole**. Do the same again, and you get smaller magnets, again with north and south poles. However often you do this, you always get dipoles. Even on the atomic and nuclear level, matter always comes in magnetic dipoles. No one has ever seen an individual north or south pole, otherwise called a **monopole**. However, some GUTs predict that at very high energies some particles may exist that carry a single unit of 'magnetic charge', in other words a single north or south pole. These particles would be very massive, corresponding to a mass of $10^{16}\,\text{GeV}/c^2$. This is equivalent to the mass of a small bacterium!

Cosmic rays, which are nature's own particle accelerators, can produce particles with energies up to at least $10^{11}\,\text{GeV}$, and could possibly reach the energies needed to create monopoles. Experiments have been set up to search for magnetic monopoles. They rely on the fact that a magnetic charge passing through a coil of superconducting wire would generate an electric current, which, due to the superconductivity, would persist and be measurable. However, to date, there is no firm evidence that magnetic monopoles have been experimentally observed.

Neutrinos with mass

A neutrino is an elementary particle and is a lepton with zero charge and spin $\frac{1}{2}$. In Chapter 7 we saw that they come in three kinds, the electron neutrino, the muon neutrino and the tau neutrino. Neutrinos left over from the Big Bang fill the universe in enormous quantities, but while they do not appear to have mass, some GUTs suggest that this may not be true. The neutrino could have a tiny mass, and experiments designed to look for this have now shown conclusive evidence that this is so. Neutrinos have been observed to change into other kinds of neutrinos, a phenomenon called **neutrino oscillations**, and only possible if neutrinos have mass.

New evidence for massive neutrinos

In a specially carved out cavity in an old zinc mine 600 m under Mount Ikena near Kamioka in the Japanese Alps is the Super-Kamiokande neutrino detector. This giant detector consists of a 50 000 tonne double-layered tank of ultra-pure water surrounded by 11 146 photomultiplier tubes (Figure 8.11).

Figure 8.11 The Super-Kamiokande experiment

Neutrinos are produced in many circumstances. The ones detected by Super-Kamiokande come from the interaction of cosmic rays with the Earth's atmosphere. The primary cosmic rays make a shower of secondary particles, all travelling in generally the same direction at nearly the speed of light. Most of these secondaries are π and K mesons, as well as muons, which decay producing neutrinos. The majority of these neutrinos go straight through the Earth without interacting with it. Occasionally, though, one will interact with a quark in the nucleus of an oxygen atom, water being the molecule H_2O. A single electron or a single muon is produced, which travels close to the speed of light and exceeds the speed of light in water, which is about 75% of its speed in a vacuum. This results in the optical equivalent of a sonic boom which, as we saw in Chapter 4, is called *Cerenkov radiation* (see Figures 4.19a, b). A flash of blue light is emitted in a 42° half-angle cone trailing the particle and detected by the photomultipliers surrounding the tank. The muons travel in relatively straight lines and produce clean ring images. The electrons are scattered more and produce fuzzier images.

Particle physicists have determined that the production of neutrinos in the atmosphere by cosmic rays is such that there should be twice as many muon neutrinos as electron neutrinos. However, earlier observations made more than ten years ago with smaller-scale neutrino detectors showed that the ratio of muon neutrino interactions to electron neutrino interactions was closer to one not two. This became known as the 'atmospheric neutrino anomaly' and a number of explanations were proposed. One explanation was that the muon neutrino was oscillating into another type of neutrino on its way down to Earth. However, most particle physicists believed that a simpler explanation such as deficiencies in the detector or errors in their calculations might be responsible. There were no compelling arguments that a neutrino oscillation might be at work.

Since the beginning of its operation in 1996, the Super-Kamiokande detector has been the most sensitive in the world for monitoring neutrinos from various sources. It is able to detect much higher-energy muon neutrinos and in much greater numbers than any previous neutrino detector. On 5th June 1998 it was announced at a physics conference in Takayama, Japan, that Super-Kamiokande has been able to determine unambiguously that the muon deficit is not due to experimental or theoretical errors but is the result of the muon neutrino oscillating into another kind of neutrino. Oscillations of neutrinos require them to have mass, which means that a revision of the Standard Model of particle physics is necessary, as it currently requires the mass of all neutrinos to be set to zero.

Particle physicists have now assigned tentative upper limits to neutrino masses. The electron neutrino is thought to be $< 2.2 \, eV/c^2$, the muon neutrino $< 170 \, keV/c^2$ and the tau neutrino $< 28 \, MeV/c^2$. Some cosmological models tell us that the neutrino mass cannot be more than $100 \, eV/c^2$ or the universe would have contracted already from gravitational attraction, but these models may have to be revised. However, oscillations mean that the neutrinos certainly have mass and they have now become very strong candidates for *dark matter*.

Dark matter

From observations of the motions of clusters of galaxies and the way in which individual galaxies rotate, astronomers have deduced that there is much more mass in the universe than can be directly detected from the emission of electromagnetic radiation at different wavelengths. Some of this matter, called **dark matter**, may be baryonic in form. For reasons that we will discuss later, cosmologists believe the universe should be 'closed', that is, there should be enough matter in the universe so that its present expansion will be brought to a halt. However, the amount of matter in the universe in baryonic form, such as the amount of helium produced in the Big Bang and in old stars, is not enough to close the universe. At least ten times as much matter must exist for this to happen.

If neutrinos have mass, then they could contribute to this missing matter, but other more massive exotic forms of matter never detected on Earth could also exist. As much as two-thirds of the missing mass may exist as **WIMPs** (weakly interacting massive particles), which were created in large quantities during the Big Bang. The name WIMP *does not* mean they take part in the weak interaction (in fact, a WIMP would not interact with baryonic matter except through gravity or, for example, by direct collisions with the nucleons in an atom). The name is simply yet another one of those whimsical fancies to which particle physicists are inclined.

A WIMP is a consequence of a GUT that embraces a theory called **supersymmetry**, which holds that there is a pairing of all bosons and fermions. In other words, at very high energies such as existed in the Big Bang, there is no longer any distinction to be made between particles of different spin. Each boson and fermion that we see today must have a **supersymmetric partner**, which we cannot yet detect. For every known boson, supersymmetry predicts the existence of a 'new' kind of *fermion*, and for every known fermion, a 'new' kind of *boson*. The new supersymmetric bosons are given names by taking the name of the equivalent fermion and adding an 's' to the front. So, for example, the electron (which is a fermion) has a supersymmetric counterpart that is a boson called the 'selectron'. The new supersymmetric fermions are named by adding 'ino' after the name, so a photon (which is a boson) has a supersymmetric counterpart (which is a fermion) called the 'photino'. We get other supersymmetric names. Quarks become 'squarks' and, collectively, leptons become 'sleptons'. However, particle physicists have yet to observe any of these particles experimentally.

Dark matter could also exist in the form of undeveloped stars such as **brown dwarfs**. These are objects that were not massive enough to start nuclear reactions and shine. Other candidates could be **black holes**, which are the remains of very massive stars that have collapsed to points of infinite density and are already cold non-luminous compact objects. However, as we have seen above, a promising new candidate is neutrinos with mass, since they exist in copious amounts in the universe.

The superforce

As we stated earlier, central to GUTs is the notion that all forces were unified as a single 'superforce' when the universe was at a temperature that exceeded 10^{32} K. As the universe cooled, gravity, and then the strong force, followed by the weak and the electromagnetic forces, froze out to the four interactions we know today (Figure 8.12). The nature of this single force is beyond our current knowledge of physics. However, quantum mechanics still enables us to speculate as to what the Big Bang emerged from. To this end, we must take a closer look at what we mean by a vacuum.

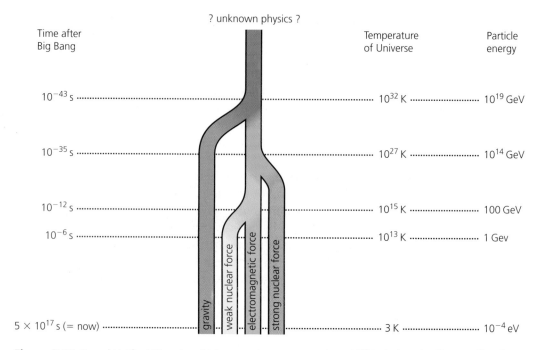

Figure 8.12 Grand Unified Theories. At temperatures greater than 10^{32} K during the first 10^{-43} s after the Big Bang, the forces of nature that we know today were unified as a single force

Quantum cosmology

Something from nothing

In our description of the Big Bang model, we traced the development of the universe from a time of 10^{-43} s after the initial explosion. So what happened before then?

If the history of the universe is wound back, then 13 000 million years ago it must have been in a superdense superhot state. In the first few minutes of the universe, primordial protons, neutrons and electrons were formed into a mixture of roughly 75% hydrogen, 25% helium and traces of light elements such as deuterium and lithium. In theory, rewinding the universe implies that it emerged from a point of zero volume and infinite density. Cosmologists call such a state of matter a **singularity**.

By using the Uncertainty Principle, cosmologists are able to arrive at an interval after which space and time become measurable, called the **Planck time** t_P, which is made up of the fundamental constants G (gravitational constant), h (Planck constant) and c (speed of light in vacuum) as

$$t_P = \sqrt{\frac{Gh}{c^5}} = \sqrt{\frac{(6.67 \times 10^{-11} \, \text{N} \, \text{m}^2 \, \text{kg}^{-2}) \times (6.63 \times 10^{-34} \, \text{J} \, \text{s})}{(3.00 \times 10^8 \, \text{m} \, \text{s}^{-1})^5}} = 1.35 \times 10^{-43} \, \text{s}$$

In quantum terms, the Planck time is the smallest unit of time that can exist and, because of quantum uncertainty, intervals smaller than this have no meaning. However, quantum uncertainty tells us something else. A **vacuum** is not a region devoid of everything but seethes and foams with activity. QED tells us that **virtual pairs** of particles are constantly appearing and disappearing in accordance with the Uncertainty Principle (Figure 8.13). In these **quantum fluctuations**, energy is borrowed from the vacuum and repaid almost at once. This sea of virtual particles can be seen to exert a force, as demonstrated by the Casimir Effect (Box 8.3).

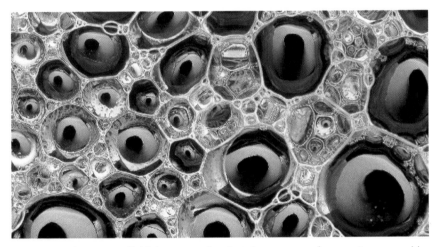

Figure 8.13 QED and quantum field theory predict that the vacuum of space–time resembles a foam of erupting and collapsing virtual particles, rather like bubbles forming and collapsing in soapy water. These so-called 'vacuum fluctuations' are evidenced by the Casimir Effect

In the conditions of infinite density in the Big Bang, you might wonder how the universe could expand at all. The intense gravitational field would crush it out of existence. However, it is precisely this large reservoir of gravitational energy that can provide a virtual pair with enough energy to materialise and become real particles in the real world. In fact, the vast gravitational energy associated with the early moments of the Big Bang provides enough energy to fill the universe completely with all kinds of particles and antiparticles after the Planck time.

Inflation

One of the problems associated with the Big Bang model is how to account for the apparent 'flatness' of the universe. Einstein's General Theory of Relativity (see below) tells us that space–time is curved in the presence of mass. The overall geometry of the universe is determined by the average distribution of all the matter it contains. Cosmological models show that the density parameter corresponding to $\Omega_0 = 1$ means that we live in a flat universe with the geometrical properties of the plane. As we have seen earlier, we have reason to believe that Ω_0 is, in fact, close to or equal to 1.

Box 8.3 The Casimir Effect

According to quantum theory, the space inside a vacuum is not empty but is filled with vast amounts of fluctuating energy. This is a consequence of the Uncertainty Principle, which implies that there is always a degree of uncertainty in the energy of a particle due to **quantum fluctuations**. Even a particle that was at absolute zero would still jiggle randomly about its resting point with a small amount of residual or **zero-point energy**. The adjective 'zero-point' is used because, even at a temperature of absolute zero where a particle has no kinetic energy, there is still a motion due to quantum uncertainty. This is why you can never cool anything down to a temperature of 0 K. The quantum uncertainty in energy will always ensure that there is a temperature – although it is very small. The quantum fluctuations can be intense enough to enable virtual pairs of particles to form spontaneously from the vacuum and then disappear again, providing they don't violate the Uncertainty Principle.

The zero-point fluctuations of electromagnetic energy are the easiest to detect. This is because the photons are easy to make, since they have no rest mass. In the 1940s the Dutch physicist Hendrik Brugt Gerhard Casimir (1909–2000) suggested a way in which these quantum fluctuations could be detected. Think of virtual photons as electromagnetic or travelling waves. Then there is an infinite number of possible modes in all frequencies and directions of propagation, which, when added all together, add up to a very large amount of energy (Figure 8.14a). At first one might suppose that all this energy should be readily observable, but we cannot see its effects in everyday life because it averages out uniformly across space and tends to cancel itself out.

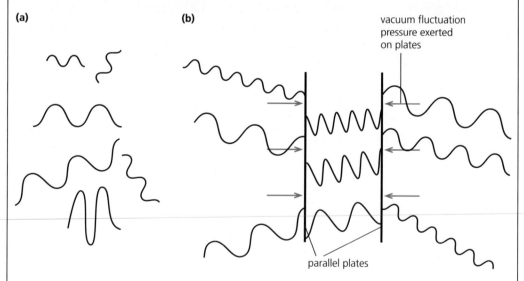

(a)

(b)

vacuum fluctuation pressure exerted on plates

parallel plates

Figure 8.14 The Casimir Effect
(a) In the vacuum, virtual photons can exist in an infinite number of oscillating modes in all frequencies and directions
(b) An integral number of wavelengths only is allowed between the plates, so there are fewer photons than outside

However, suppose we place two highly reflective metal plates close together, as shown in Figure 8.14b. Inside the space between the plates, photons will be reflected from the walls and only vacuum fluctuations with certain wavelengths will 'fit' within the gap; all other modes will be excluded. You can imagine it rather like standing waves on a string set up between two fixed points. Only certain wavelengths can exist between the points, and the string can only vibrate in certain modes.

The number of vibrational modes of the electromagnetic waves (all of which carry energy and momentum) is now greater *outside* the plates than between them. In other words, there are fewer photons existing between the plates than outside them, and this imbalance means the plates have a net force or an excess 'vacuum fluctuation pressure' exerted on them externally, which pushes them together. The vacuum pressure is very small, but experiments have been done using very flat and highly reflective plates separated by a distance of a few nanometres which have shown that the Casimir Effect exists and does indeed show that quantum fluctuations are a real phenomenon.

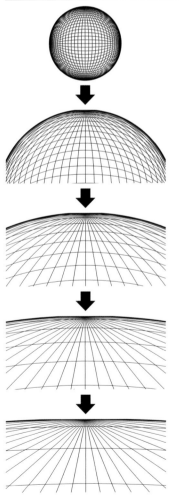

Cosmologists think that, when the universe was between 10^{-43} and 10^{-30} s old, it expanded extremely rapidly. Calculations show that the universe doubled in size every 10^{-35} s. A quantum fluctuation of size 10^{20} times smaller than a proton would have inflated into a sphere about 10 cm across in just 1.5×10^{-32} s. This period of **inflation**, called the **inflationary epoch**, was over in a period of 10^{-36} s to 10^{-33} s. It had the effect of moving much of the material that was close together in the first instants of the universe very much further away, so that as the universe expanded the distribution of matter at large scales became smooth and uniform. In the same sense that we cannot see the curvature of the Earth from the scale of a school playing field, and so it subsequently appears flat and level, inflation flattens the universe rather like the surface of a sphere gets flatter as the sphere expands (Figure 8.15).

Inflation can also explain why the cosmic microwave background is isotropic. To understand why, consider cosmic microwave radiation coming towards us from opposite parts of the sky. This radiation has been travelling for at least 13 000 million years, and the size of the known universe is at least 10^{25} m across. So how did these two unrelated parts of the universe come to

Figure 8.15 The rapid inflation of the early universe would flatten out space in the same way as the surface of a sphere becomes 'flatter' as it expands

be at the same temperature of 2.7 K? The answer is that, during the inflationary epoch, these two parts of the universe were originally very close to each other, and this common origin is why they have the same temperature *after* inflation has pushed them far apart in the universe's very early history.

Inflation is not inconsistent with Einstein's postulate that the speed of light is constant to all observers, since the rapid expansion of the universe during the inflationary epoch was the *expansion of space itself* and did *not* involve the *motion* of objects through it. Inflation was when the distances between particles suddenly increased by a large amount. The particles didn't move – it was the space between them that expanded!

Gravity and mass
The General Theory of Relativity
In Chapter 1, we briefly discussed how Einstein's General Theory of Relativity dealt with the more general case of bodies undergoing acceleration in a non-inertial frame of reference, as opposed to the Special Theory, which only considered the motion of bodies in an inertial frame. The equations of motion in General Relativity rely on a geometrical description of space and time in which bodies move along paths called **geodesics** where space has been 'curved' by the presence of mass (in geometry, a geodesic is the shortest distance between two points). The larger the mass, the greater the curvature and deflection of the body's motion (Figure 1.19, page 19).

As we saw in Chapter 1, the Principle of Equivalence seems to show that the effects of acceleration are indistinguishable from the effects of a uniform gravitational field. The Equivalence Principle can be stated in another way. From Newton's 2nd Law a mass m can be defined in relation to a force F acting on it and the acceleration a it experiences. Mass defined in this way is called **inertial mass** and we write it as $F = ma$. Alternatively, we can define mass using Newton's Law of Universal Gravitation as $F = GMm/r^2$, where r is the distance between the two masses M and m. When defined like this, it is called **gravitational mass**. The mass m that appears in the equation for inertial mass is, said Einstein, the same m that appears in the one for gravitational mass – the two are equivalent, even though they are defined differently! In recent years, experiments have confirmed this to be true to at least 1 part in 10^{11}, which is a very high level of agreement.

Flat and curved space–time
Newtonian mechanics demands that the geometry of the universe is flat and that masses interact gravitationally by means of forces, with space and time being independent quantities. Einstein, however, said that the geometry of the universe can have any shape depending on the distribution of matter in it, and space and time are linked together. Objects move along the curvature of space–time caused by the presence of mass. Newton's mechanics is not, however, invalidated as it turns out to be just a special case of the General Theory. We can sum up Einstein's mechanics as: 'space–time tells matter how to move' and 'matter tells space–time how to curve'.

Gravitational radiation

Another prediction of General Relativity is that of **gravitational radiation**. General Relativity predicts the existence of **gravity waves**, which are generated when there is a change in the strength of a gravitational field. One of the consequences of Maxwell's Equations is that a charged particle like an electron emits electromagnetic waves when accelerated. This is how radio waves are emitted from aerials, which are simply conductors with electrons accelerating inside them. In a similar way, a **gravitational wave** is emitted when a mass accelerates and should produce small 'ripples' or distortions in the fabric of space–time through which it passes. However, because gravity is some 10^{36} times weaker than electromagnetism, these distortions in matter are expected to be very small, in fact smaller than the diameter of an atomic nucleus. Some astronomical objects that might produce gravity waves of sufficient strength that could be detectable on Earth are the merger of a binary star system, or the collapse of a supernova into a black hole. The technology now exists to measure the effects of gravitational waves, and gravitational wave observatories have recently been built to try to detect them (Box 8.4).

Quantum gravity

However successful the General Theory of Relativity is in describing gravity, it appears to have no connection with quantum theory. Electromagnetism and the weak and strong nuclear forces can all be described in quantum terms, but gravity seems the 'odd man out' – in fact, the General Theory does away with the concept of force altogether! So is it possible to construct a quantum theory of gravity? A Theory of Everything that incorporates a quantum theory of gravity would have to combine all the existing features of both Special and General Relativity and be compatible with the existing quantum field theories of the other interactions.

One approach is to model gravity in a similar manner to quantum electrodynamics, with the gravitational force between masses being carried by an exchange particle called the **graviton**. Whereas in electromagnetism 'like' charges repel and 'unlike' attract, there is only one kind of 'gravitational charge', i.e. 'mass', and it always attracts. Calculations show that the graviton would have a spin of 2, but would also interact with itself and other gravitons. This last property makes it impossible to renormalise quantum gravity to get rid of the infinities that appear in the equations, and a complete theory of quantum gravity continues to elude particle physicists.

What is mass?

You are probably used to thinking of mass as the 'amount of matter' an object has. We can define mass in terms of Newton's 2nd Law as a resistance to an applied force (inertial mass) or in terms of Newton's Law of Universal Gravitation as a measure of the strength with which one body attracts another (gravitational mass). Mass and energy are interchangeable in accordance with Einstein's mass–energy relation $E = mc^2$. But none of these definitions really tell us what mass is and, more importantly, how an object *acquires* mass.

Box 8.4 Detecting gravitational radiation

Gravitational radiation is a prediction of Einstein's General Theory of Relativity and is associated with acceleration and orbital motion. Objects with mass that accelerate will emit gravitational radiation that travels at the speed of light. Since the strength of the gravitational interaction is so small (only 10^{-40} times as strong as the electromagnetic), then the radiation will be very weak as it spreads out as 'ripples' in space–time rather like a stone being dropped in a pond. Only objects that have strong gravitational fields are likely to emit gravitational radiation that is strong enough to be detected. There are two sources of radiation that may be able to produce ripples big enough to be detected by gravitational wave detectors. One of these is a **supernova** explosion. These occur fairly frequently in our galaxy (about once every 25 years) and are caused by the sudden collapse of a massive star to form a **neutron star** or **black hole**. If close enough, such an event should produce a huge burst of gravitational radiation that could be easily detected for a few microseconds of time.

Another way would be to detect the gravitational radiation produced by two neutron stars in orbit around each other. Such an arrangement is called a **binary pulsar** (Figure 8.16). In 1974, the American radio astronomer Russell Hulse and his research student Joseph Taylor discovered a very unusual binary star system in our galaxy. The system consisted of two **neutron stars** in orbit around each other. Neutron stars emit radio waves rather like a lighthouse with a definite period, and measurements of the periodic Doppler shift of the pulsars' radio emissions showed that the orbit was decaying or slowly shrinking in size. According to General Relativity, this could be accounted for if the binary pulsar was losing energy in the form of gravitational radiation. Furthermore, the rate of decay of the orbit was exactly what General Relativity predicted even if the gravity waves themselves could not be detected. Taylor and Hulse's binary pulsar provides very strong evidence for the existence of gravity waves, and they were awarded the 1993 Nobel Prize for Physics for their discovery.

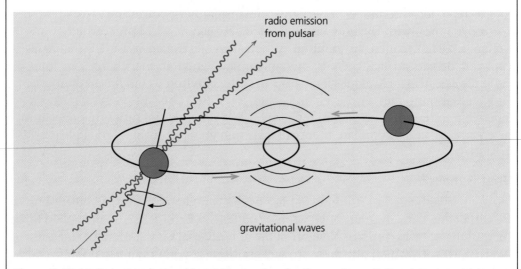

Figure 8.16 Neutron stars (pulsars) in orbit around each other emit gravitational waves, which may be inferred from the decay of their orbit

A new approach to detecting gravitational radiation is **LIGO**. LIGO stands for Laser Interferometer Gravitational-wave Observatory and is a giant laser interferometer consisting of mirrors suspended at each of the corners of a giant L-shaped vacuum system measuring 4 km along one length. Precision laser beams in the interferometers will sense small distortions of the mirrors when a gravitational wave passes through them, and scientists can measure changes in the interference fringes produced. The system is so accurate that it can detect distortions in the mirror spacings by about one-thousandth of a fermi (10^{-18} m). Two LIGO installations have been built, one at Livingston, Louisiana, and the other at Hanford, Washington State, USA (Figure 8.17).

Figure 8.17 The LIGO observatory at Hanford

We saw that in QED the carriers of the electromagnetic force, because of its long range, require the existence of the photon, which has zero mass. In the early development of the Electroweak Theory, the W and Z particles were also required to have zero mass, which would be inconsistent with the range of the weak interaction. The W and Z needed to be massive so that the weak interaction did not extend beyond its observable limits, but how could the W and Z acquire mass?

A solution was put forward by the British physicist Peter Higgs (1929–). He proposed the existence of an otherwise undetectable quantum field called the **Higgs field** with which is associated a previously unknown particle called the **Higgs boson**. The Higgs boson allows any previously massless particle to become massive by absorbing a Higgs boson by means of the **Higgs mechanism**. The Higgs field, if it exists, is different from other quantum fields as it is a **scalar field**. This means that it has no preferred direction in space and is the same strength everywhere, which is why we cannot detect it, unlike an electromagnetic field, which has a definite direction in space depending on the distribution of charge. Calculations show that the Higgs boson is predicted to have zero spin and a mass anywhere between 4 and $1000\,\text{GeV}/c^2$.

The Higgs mechanism was a crucial part of the Electroweak Theory in explaining how the W and Z particles get mass. The fact that the Electroweak Theory has proved so successful is strong evidence that Higgs bosons exist, and accelerator energies are now reaching the stage where the Higgs boson may be created as a by-product of Z^0 production.

In this way, the Higgs mechanism may explain the origin of mass. Higgs bosons created in the Big Bang are absorbed by some particles and become massive. There may even be more than one type of Higgs boson. Only their discovery in accelerator experiments will enable us to investigate their properties and understand why some particles have mass and others don't.

Strings

Supersymmetry

We saw in Chapter 6 that the key concept that appears again and again is that of symmetry. The fact that we see symmetry so often suggests to particle physicists that there may be an underlying fundamental symmetry or supersymmetry that permeates all of nature and could form the basis of a Theory of Everything (TOE).

Supersymmetry, or **SUSY** as it is commonly known, is a GUT that restores the symmetry between bosons and fermions. Recall that fermions are the particles that make up the material world, such as electrons and protons, whereas bosons are the exchange particles associated with the transmission of forces. SUSY adds another four dimensions to those of ordinary four-dimensional space–time. These extra four dimensions are abstract mathematical ones that enable fermions to be transformed into bosons and vice versa. A consequence of this eight-dimensional geometry, called **superspace**, is that every boson and fermion has a **supersymmetric partner**, which doubles the number of particles that exist. The fact that these supersymmetric partners are nowhere to be seen suggests that they may be much more massive than their counterparts that we observe today and could only be created at very high energies.

One version of SUSY called **N = 8 supergravity** (since it involves eight dimensions) incorporates gravity in such a way that it can explain everything – forces, particles, space–time, . . . the lot! Even the infinities that crop up in the quantum field equations of gravity cancel out. It is also possible that these supersymmetric partners may be the candidates for the dark matter necessary to close the universe. $N = 8$ supergravity is a GUT that seems to renormalise itself, and advocates of SUSY regard it as the most promising GUT yet. Searches for these supersymmetric partners are being carried out, but despite its mathematical elegance, SUSY remains unproven as a TOE.

String-driven things

In the 1970s a new concept for describing particles was developed. Particle physicists were accustomed to treating elementary particles like quarks and leptons as mathematical points. When the quark model was first proposed, the quarks inside hadrons were described as being held together by the exchange of gluons, and in Chapter 7 we saw how this could be modelled as two quarks held together by a piece of elastic. The energy stored in the 'elastic' is so strong that a quark pair could be thought of as behaving like a stretched piece of string. This led to the idea, put forward by the Japanese physicist Yoichiro Nambu (1921–), that all particles might be regarded not as points but as one-dimensional entities called **strings**. Nambu proposed that elementary particles were actually tiny strings that vibrated at certain energies. The lengths of these strings were extremely small, only 10^{-15} m long, but the quantum properties of particles, like their masses, electric charges and spin, corresponded to different modes of vibration, just like violin strings vibrating at different frequencies produce different musical notes. Strings can interact with each other, forming different kinds of strings representing different particles (Figure 8.18).

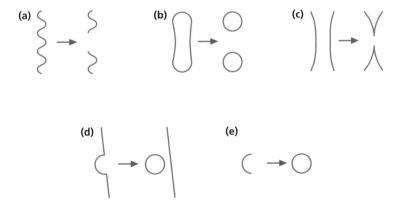

Figure 8.18 Particles as strings. Five types of string interactions are represented here:
(a) A single string splits and creates two smaller strings
(b) A closed string pinches and produces two smaller closed strings
(c) Two strings collide and form two new strings
(d) A single open string re-forms and creates an open and closed string
(e) The ends of an open string touch and create a closed string

String theory remained little more than a mathematical curiosity until, in 1984, Michael Green (1946–) of Queen Mary College, London, and John Schwarz (1941–) of the California Institute of Technology (Caltech) combined the idea of supersymmetry with strings to make an improved version called **superstring theory**. Superstring theory is able to explain gravity in quantum terms. Each point in space is regarded as a six-dimensional ball about the size of the **Planck length**. The Planck length is the smallest measurement of length that has any meaning. Its value can be calculated from the gravitational constant G, the speed of light c and the Planck constant (expressed as $\hbar = h/2\pi$) as

$$\text{Planck length} = \sqrt{\frac{G\hbar}{c^3}}$$

$$= \sqrt{\frac{(6.67 \times 10^{-11}\,\text{N}\,\text{m}^2\,\text{kg}^{-2}) \times (1.06 \times 10^{-34}\,\text{J}\,\text{s})}{(3.00 \times 10^8\,\text{m}\,\text{s}^{-1})^3}} = 1.62 \times 10^{-35}\,\text{m}$$

At the Planck length, quantum effects dominate. Space–time itself is quantised into 'lumps' and is not smoothly continuous. In superstring theory, particles only look like one-dimensional strings from a scale greater than a few Planck lengths. Each string has, in fact, many dimensions except one that are all 'rolled up' or *compactified*. Elementary particles are associated with small loops of string barely larger than these balls of space, and at larger scales, superstring theory reduces to the conventional quantum field theory, which we currently use to describe particles and their interactions. While superstrings are mathematically elegant and seem to be able to answer many of the problems associated with unifying gravity with the other forces of nature, it is simply not possible to build particle accelerators that are powerful enough to probe matter at scales comparable to the Planck length.

Membranes

One of the problems with string theory is that there are several different versions, each of which has advantages and disadvantages when it comes to describing particles and their interactions in a TOE. There are in fact five possible string theories, and it can be shown mathematically that these are the only contenders. So how do particle physics theorists know which string theory is the right one?

One way forward is to consider particles not as strings but as two-dimensional sheets or **membranes**. In the mid-1990s this idea, known as **M-theory**, became extremely popular among theoretical particle physicists, because if we describe particles like this, then there is only *one* mathematical package that fits the bill and can provide a TOE. What's more, M-theory incorporates all the existing one-dimensional string theories in a manner reminiscent of the way Einstein's Theory of Relativity incorporates Newtonian mechanics. The kind of energies needed to test some of the predictions of M-theory will be available when the Large Hadron Collider (LHC) comes into operation at CERN, and experimental particle physicists are waiting to see if these ideas of their theoretical counterparts turn out to be true.

At the moment all of this is speculative. However, it should be appreciated that, as well as experimental results, particle physics can also make progress on theoretical grounds alone. The mathematical ingenuity involved poses formidable challenges to the human intellect and how we perceive the world conceptually.

The future of particle physics

What does the future hold for particle physics? The answer to this question undoubtedly lies in a combination of more accelerator experiments and advances in theory. On 2nd November 2000, the LEP accelerator at CERN completed its final run ready to be dismantled for the Large Hadron Collider (LHC) to be built in the same 27 km tunnel. Over the summer of 2000, LEP was tweaked to give collision energies of up to 209 GeV, within range of creating the Higgs boson. In fact, for the past few years, LEP had been running in the energy band where the Higgs boson had been most expected. Each time the energy was increased, particle physicists held their breath, until, when events above 206 GeV were analysed, tantalising evidence was found of the creation of a Higgs boson! In these events, an electron–positron pair could produce a Higgs boson back-to-back with another particle. But these events are right at the edge of LEP's capabilities and are difficult to disentangle from the routine production of W and Z particles. The initial candidates saw four jets of particles, two from the Higgs boson. Four separate detectors around the LEP ring called ALEPH, DELPHI, L3 and OPAL provided evidence that was consistent with a Higgs event.

In 2005, CERN's latest accelerator, the LHC, will start accelerating hadrons, and will be able to collide protons and antiprotons together at energies of up to 14 TeV. This energy is a long way from Grand Unification energies, but the LHC should be able to manufacture the Higgs boson, which will confirm theories of the Higgs mechanism and the origins of mass. Is it possible that some new particles, such as the supersymmetric particles that SUSY predicts, may appear? This would be a major step forward in validating string theory and the mathematical ideas that have been built around it. Confirmation of proton decay would be enormously significant and provide the first verified prediction of Grand Unification. A positive confirmation of neutrino oscillation and mass will possibly help to clear up the mystery of dark matter, and this may well have arrived.

Advances in particle physics may come from a completely different direction. We mentioned earlier the cosmic microwave background (CMB) radiation that has been measured in detail by the COBE satellite. This blackbody radiation is a relic of the Big Bang. The high degree of uniformity of the CMB means that the early universe must have been smooth and homogeneous. But soon after the decoupling of radiation from matter, density fluctuations in this otherwise even distribution must have occurred, from which galaxies and other large-scale structures formed. The COBE satellite detected minute fluctuations in the CMB, which, although corresponding to variations in temperature of only a few millionths of a kelvin, are nonetheless there and echo the point at which matter in the universe first began to

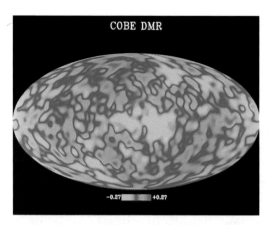

Figure 8.19 A computer-enhanced image showing the very small temperature variations (of the order of 10^{-4} K) in the cosmic microwave background radiation

organise itself (Figure 8.19). On 30th June 2001, NASA's Microwave Anisotropy Probe (MAP) was launched, which is able to make high-resolution measurements of the CMB, as will the European Space Agency's Planck satellite due for launch in 2007. These satellites may be able to tell us more about how the decoupling process occurred and help cosmologists refine their theories of the Big Bang.

In early 2001, the first results from the Sudbury Neutrino Observatory in Canada confirmed that neutrinos are able to 'oscillate' from one kind to another. This is only possible if the three types of neutrino – electron, muon and tau – have mass. This result has important implications for cosmological models and the total mass–energy density of the universe.

One of the mysteries of the universe is that, at the largest scales, it has a **filamental** structure. It is populated by large clusters of galaxies in which there are empty **voids** containing little or no galactic matter (Figure 8.20). We do not know why the universe has this kind of structure, but it must in some way be connected with the early formation of galaxies soon after the Big Bang. Cosmologists believe that the density fluctuations revealed by the COBE satellite may be useful in this regard, since, for this to have happened, the distribution of matter in the early stages of the Big Bang cannot have been as smooth as was first assumed. Some cosmologists have supposed that some form of exotic dark matter may have been responsible, such as WIMPS that would have been able to clump together (assuming they have mass) to form the seeds from which galaxies could form. Cosmologists have tested these ideas using computer models, but at this stage we simply don't know. However, what you should appreciate is that this union of particle physics with astrophysics – the very small explaining the very big – is one of the most remarkable achievements to emerge from late 20th century physics.

Particle physics in the service of humankind

Particle physics experiments cost millions of pounds to run. In 2000, the total budget for running CERN was some £400 million. Couldn't sums like this be better spent in alleviating poverty, advancing medicine or improving the standard of living for the world's population?

This is a very important question. Particle physics involves using technology at the 'cutting edge'. An example is the magnetic resonance imaging (MRI) scanners, which are routinely used in hospitals to image the body. MRI scanners provide remarkably detailed images of the human body that doctors and surgeons can use to

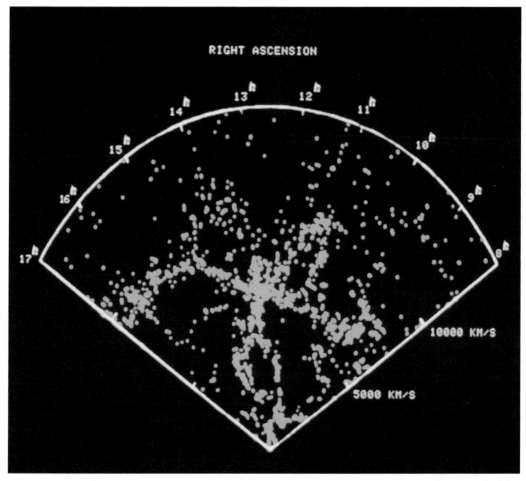

Figure 8.20 A wedge of galaxy distributions showing the filamental or thread-like appearance of the universe, an image nicknamed the 'stickman'

treat disease and sickness. Their operation depends on the use of powerful superconducting magnets that cause protons in the body to emit electromagnetic radiation at radio frequencies. The technology for developing these magnets can be traced directly to the superconducting magnets used in synchrotrons for making particles move in curved paths. Medicine has benefited in this respect from the sums poured in to building accelerators.

We saw in Chapter 4 that linear accelerators using particle physics technology are employed in treating cancer by bombarding tumours with various types of particle. Recently, beams of π mesons have been employed to destroy cancerous cells by producing them in such a way that they deliver all of their energy to the rogue cells without damaging the healthy surrounding tissue. These π mesons act like 'exploding depth charges', giving up all their energy to the cancer cell in the last instants of their existence, so destroying it in the process.

277

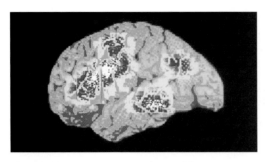

Figure 8.21 A computer-enhanced image obtained from a PET scan of the human brain while the person is occupied in verbal thinking and speaking

Antimatter has been used to probe the workings of the human brain. In a technique called **positron emission tomography (PET)** a patient is given a radioactive positron emitter that attaches itself to glucose in the blood. The glucose concentrates in regions of the brain where there is high metabolic activity. The positrons annihilate with electrons in the brain, producing 'back-to-back' gamma rays, which are detected by a network of scintillator crystals around the patient's head. The signals are fed into a computer, which displays images of the brain; these reveal how the brain functions when fed with external stimuli (Figure 8.21). Such studies are of great value to doctors in treating patients with neurological disorders such as Alzheimer's and Parkinson's diseases and may help us to understand many psychiatric illnesses.

CERN invents the World Wide Web

The Internet has now revolutionised the way we communicate with each other. The 'Web' consists of millions of computers around the world, all linked together in a network forming a 'virtual community' of users, which can exchange information using a common electronic language. The 'www' prefix at the beginning of an Internet address for a 'website' is based on an idea developed by the CERN computer scientist Tim Berners-Lee in 1990. The communications software that defines the Universal Resource Locators (URLs) and the Hypertext Mark-up Language (HTML) that describes the way web pages are drawn were developed by Berners-Lee and a colleague Robert Cailliau. Since CERN is an international institution now made up of 20 member states, a way was needed to send particle physics data directly to national institutes for analysis rather than being processed at CERN. Using the World Wide Web, scientists in their own countries can access information from CERN in a consistent and simple way, as well as from each other. This fostered a *de facto* standard for computer communications, which is now used by both academic and commercial users to exchange information over the Internet.

As a result, e-mail has now become an essential tool for human communication, which is faster than a preprint or letter and less intrusive than a telephone call. So next time you surf the Web and use a www.com address, remember that it was particle physics research at CERN that made it possible!

However, particle physics is not about a select group of scientists performing experiments for their own benefit. It is a shared experience for all. Answers to questions like 'Why do we live in a world like this?', as well as being answered in philosophical and theological terms, can now be investigated from a scientific viewpoint. Science will never be able to answer all questions of existence, but it complements the way we understand the world and contributes useful and practical benefits for us all.

Like all scientists, particle physicists have a duty to explain to the world what they have found and what their discoveries mean. It is in the nature of human beings to be curious. It is an interesting thought that, just as the early Greeks used mathematical form to describe the physical world, so particle physicists today are using symmetry to describe its fundamental structure. In this sense our understanding of the physical world has come full circle. Nature displays a beautiful pattern and order. To know why the world is, is to know ourselves.

Summary

- **Cosmology** is the study of the universe from the very small to the very large. It is concerned with the development of the universe from its origins to its ultimate fate, and it embraces both particle physics and astrophysics.

- The universe is expanding, as evidenced from **Hubble's Law**. This suggests that it must have been much smaller and much denser in the distant past. The amounts of hydrogen and helium in stars indicate that their abundances cannot be accounted for by their formation in stellar interiors through **stellar nucleosynthesis** and must have been present before the oldest stars were formed.

- The universe is filled with a low-temperature **isotropic** radiation called the **cosmic microwave background**, which indicates that the universe must once have been much hotter than it is now. The Hubble expansion, hydrogen and helium abundances and the microwave background are all evidence that lead cosmologists to conclude that the universe had a definite beginning in time called the **Big Bang**. The fate of the universe is determined by the **density parameter** Ω_0. The **critical density** is where the average **mass–energy** of the universe is just high enough so that its expansion comes to a halt in an infinite time. Current estimates of matter in the universe suggest that it contains insufficient mass to halt the expansion.

- Particle physicists have found that matter can be created by **pair production**. The creation of matter involved pair production at very high temperatures early on in the Big Bang. The history of the universe can be divided into four eras of time, in which elementary particles, nuclei and the first atoms were formed. The formation of light elements in the Big Bang is called **primordial nucleosynthesis** and allows us to make estimates of the abundance of helium in the universe. Radiation and matter were initially locked together but became separate after **decoupling**. From then on, matter could start to organise itself via gravity to form the galaxies we see today.

- **Grand Unified Theories (GUTs)** attempt to unify the forces of nature into a single force at high energies. Some consequences of GUTs are **proton decay** and the existence of many kinds of exotic matter. **Supersymmetry (SUSY)** is a GUT that asserts that there is no distinction to be made between bosons and fermions. Each fermion and boson has a **supersymmetric partner** that is very massive and cannot be manufactured by current accelerator technology.

◆ The **Planck time** is an interval after which space and time become measurable and, in quantum terms, because of the Uncertainty Principle, it is the smallest unit of time that can exist. Before the Planck time, the universe itself may have originated from **quantum fluctuations** in the **vacuum**. The existence of quantum fluctuations is demonstrated by the **Casimir Effect**.

◆ From observations of the motions of clusters of galaxies and the way individual galaxies rotate, there is much more mass in the universe than can be directly detected from the emission of electromagnetic radiation at different wavelengths. This hidden matter is called **dark matter** and may exist partly in the form of **neutrinos with mass** and partly as exotic particles predicted by GUTs. Cosmological theories strongly suggest that the universe should be **closed** and dark matter may provide the answer to the 'missing mass'. The geometrical flatness of the universe can be accounted for by the concept of **inflation** by which the universe expanded very rapidly early on in its history. Inflation can also account for the isotropy of the cosmic microwave background radiation.

◆ The origin of mass is thought to come about by means of the **Higgs mechanism**. The universe is permeated by a quantum field called the **Higgs field** by which particles acquire mass by means of the **Higgs boson**. The Higgs mechanism played an important part in the formulation of the electroweak unification, and the Higgs boson should be detected in the next generation of particle accelerators.

◆ Einstein's General Theory of Relativity predicts the existence of **gravitational radiation**. This occurs when a mass undergoes acceleration, and it is very weak. However, gravitational radiation should be detected from objects where gravity is very strong, such as **supernova** explosions and **binary pulsars**.

◆ Particle physicists are trying to develop a **quantum theory of gravity**. Such a theory asserts that the universe began from a **quantum fluctuation** and postulates the existence of a carrier particle of the gravitational force called the **graviton**. A serious difficulty with quantum gravity is that it cannot be renormalised. One version of SUSY called $N = 8$ **supergravity** (since it involves eight dimensions) incorporates gravity in such a way that it can explain everything (forces, particles, space–time) as a true **Theory of Everything (TOE)**.

◆ A different approach is to model particles as one-dimensional objects called **strings**. **Superstrings** combine both string theory and supersymmetry by treating elementary particles as small loops of string at the fundamental length scale called the **Planck length**. String theory can overcome many of the mathematical difficulties associated with a quantum theory of gravity and the union of relativity theory with quantum mechanics. There are many different versions of string theory. **Membranes** or **M-theory** is a two-dimensional version in which particles are visualised as sheets or membranes. There is only one unique M-theory that gives a TOE and incorporates the successes of string theory. Some of the predictions of M-theory may be testable at the level of the strong interaction.

Questions

1 a i) What is meant by the term *redshift*? What is its cosmological significance?

ii) State Hubble's Law and from it deduce the units of the Hubble parameter H_0.

iii) Current estimates of H_0 vary by as much as a factor of 2. What do you think is the problem in obtaining a reliable estimate for H_0?

iv) An approximate age of the universe t_0 can be obtained from the formula $t_0 = 1/H_0$. What assumption has been made in obtaining this formula?

b By looking at the spectrum of the star Alpha Centauri, it is found that one of the calcium absorption lines has a wavelength of 396.820 nm. The same line in the spectrum of the Sun (which is close to the Earth) has a wavelength of 396.849 nm. What is the speed at which Alpha Centauri is receding from us?

c Suppose that, at some distant time in the future, astronomers found that the spectra of light from the stars was shifted towards the blue end of the spectrum; what could be deduced about the general motion of the universe?

2 What evidence do we have which indicates that the Universe is expanding?

Explain why the rate of expansion is slowing down.

Explain the statement: *By observing distant galaxies, we are seeing the Universe as it would have looked when it was much younger.*

Sketch a graph showing how the scale of the Universe would vary with time in an open universe.

What determines whether or not an expanding Universe will be open or closed?

Edexcel, January 2001 (part)

3 a The diagram below shows the intensity of the radiation emitted by two light bulbs at a temperature of 1500 K and 2000 K.

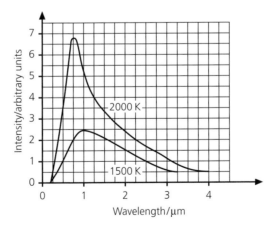

i) What are the peak wavelengths λ_{peak} of the radiation from the light bulbs?

ii) Do the data in the diagram support Wien's Law $\lambda_{peak} T = $ constant, where T is the absolute temperature of the light bulb filaments?

b Given Wien's Law, $\lambda_{peak}T = 2.9 \times 10^{-3}\,\text{m K}$, what is the peak wavelength of the 2.7 K cosmic microwave background blackbody radiation? In what part of the electromagnetic spectrum does it lie?

4 a The early history of the universe was dominated by thermal radiation. Consider the formation of proton–antiproton pairs or neutron–antineutron pairs by the reactions

$$\gamma + \gamma \rightarrow p + \bar{p}$$
$$\gamma + \gamma \rightarrow n + \bar{n}$$

For these reactions to take place, the photons must have an energy at least as great as the nucleon rest energy.
i) What is the nucleon rest energy E? [Take m_0 for a nucleon to be $1.66 \times 10^{-27}\,\text{kg}$.]
ii) If the energy of the photons is given by $E = kT$, where k is Boltzmann's constant, at what temperature must the universe be for the first nucleons to be made?

b Approximately four minutes after the Big Bang the matter in the Universe was almost entirely composed of helium and hydrogen nuclei ($^4_2\text{He}^{2+}$ and $^1_1\text{H}^+$). The total mass of the hydrogen was approximately three times the total mass of the helium. Show that this means that just a few seconds earlier, the ratio of protons to neutrons must have been 7:1.

5 a At one stage in its expansion the Universe resembled a giant nuclear fusion reactor with ^1_1H, ^2_1H and ^3_1H nuclei fusing to form helium nuclei. This era is shaded in on the graph shown below which shows the relationship between temperature and time in the expanding Universe.
Use the graph to estimate the time at which helium formation began, and the temperature of the Universe at this time.

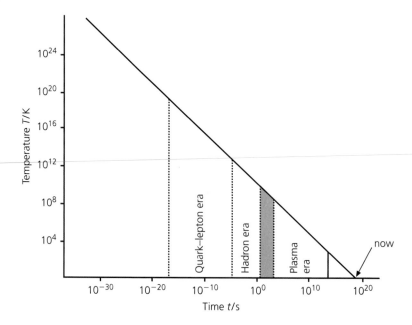

Explain in terms of the appropriate fundamental interactions why the fusion reactions only take place at very high temperatures.

b In an accelerator at CERN, electrons and positrons with energies of the order of 100 GeV are made to collide resulting in the interaction shown below.

In what era of the expanding Universe did such interactions take place?
What exchange particle is involved in this interaction?

Edexcel, June 1998 (part)

6 The value of the critical density parameter is given by

$$\rho_c = \frac{3H_0^2}{8\pi G}$$

a Calculate ρ_c when H_0 is
  i) $30 \, \mathrm{km \, s^{-1} \, Mpc^{-1}}$,
  ii) $50 \, \mathrm{km \, s^{-1} \, Mpc^{-1}}$,
  iii) $75 \, \mathrm{km \, s^{-1} \, Mpc^{-1}}$.

b How is the age of the universe affected by these differing values of H_0?
  [$1 \, \mathrm{Mpc} = 3.1 \times 10^{19} \, \mathrm{km}$]

Tables of particles

Where the most accurate recommended values of physical quantities have been quoted here and in the main text, these have been taken from the following three sources: CODATA *Recommended Values of the Fundamental Physical Constants, 1998; Journal of Physical and Chemical Reference Data, Vol 28, No 6, 1999; Reviews of Modern Physics, Vol 72 No 2, 2000.* The reader is referred to these sources for more precise information on particle preferred values.

Particle rest masses

electron m_e $9.109\,381\,88(72) \times 10^{-31}$ kg   or $5.485\,799\,110(12) \times 10^{-4}$ u
proton m_p $1.672\,621\,58(13) \times 10^{-27}$ kg   or $1.007\,276\,486\,688(13)$ u
neutron m_n $1.674\,927\,16(13) \times 10^{-27}$ kg   or $1.008\,664\,915\,78(55)$ u

Energy equivalences of rest masses
electron $0.510\,998\,902(21)$ MeV
proton $938.271\,998(38)$ MeV
neutron $939.565\,330(38)$ MeV

Particle categories

Fermions   Particles that obey Fermi–Dirac statistics and have half-integer values of spin ($\frac{1}{2}$, $\frac{3}{2}$ and so on). Fermions obey the Exclusion Principle. **Quarks** and **leptons** are fermions.

Bosons   Particles that obey Bose–Einstein statistics and have integer values of spin (1, 2 and so on). Bosons do not obey the Exclusion Principle.

Hadrons   Any particle that feels the strong force. All hadrons are composed of quarks.

Baryons   Any *fermion* that feels the stong force. Baryons are members of the *hadron* family.

Mesons   Particles that are composed of a quark and antiquark. Mesons are members of the *hadron* family.

The quarks

Name	Symbol	Charge/(+e)	Mass/(GeV/c^2)	Stable?
up	u	+2/3	0.33	yes
down	d	−1/3	~0.33	no
charm	c	+2/3	1.58	no
strange	s	−1/3	0.47	no
top	t	+2/3	180	no
bottom	b	−1/3	4.58	no

The leptons

Name	Symbol	Charge/(+e)	Mass/(GeV/c^2)	Lifetime/s
electron	e^-	-1	5.31×10^{-4}	stable
electron neutrino	ν_e	0	$0^\dagger$	stable
muon	μ^-	-1	0.106	2×10^{-6}
muon neutrino	ν_μ	0	$0^\dagger$	stable
tau	τ^-	-1	1.78	3×10^{-13}
tau neutrino	ν_τ	0	$0^\dagger$	stable

$\dagger$ Recent observations of neutrino oscillations have provided very stong evidence that neutrinos have mass. The mass of the electron neutrino is thought to be $2.2 \, eV/c^2$, the muon neutrino $< 179 \, keV/c^2$ and the tau neutrino $< 28 \, MeV/c^2$.

Some baryons

Name	Symbol	Charge/(+e)	Mass/(GeV/c^2)	Lifetime/s	Quark structure	Strangeness
proton	p	$+1$	0.938	$>10^{39}$	uud	0
neutron	n	0	0.940	900	udd	0
lambda	Λ	0	1.115	2.6×10^{-10}	uds	-1
sigma-plus	Σ^+	$+1$	1.189	0.8×10^{-10}	uus	-1
sigma-minus	Σ^-	-1	1.197	1.5×10^{-10}	dds	-1
sigma-zero	Σ^0	0	1.192	6×10^{-20}	uds	-1
xi-minus	Ξ^-	-1	1.321	1.6×10^{-10}	dss	-2
xi-zero	Ξ^0	0	1.315	3×10^{-10}	uss	-2
omega-minus	Ω^-	-1	1.672	0.8×10^{-10}	sss	-3

Some mesons

Name	Symbol	Charge/(+e)	Mass/(GeV/c^2)	Lifetime/s	Quark structure	Strangeness
pi-zero	π^0	0	0.135	0.8×10^{-16}	$u\bar{u}/d\bar{d}$	0
pi-plus	π^+	$+1$	0.140	2.6×10^{-8}	$u\bar{d}$	0
pi-minus	π^-	-1	1.115	2.6×10^{-10}	$d\bar{u}$	0
K-zero	K^0	0	0.498	indeterminate	$d\bar{s}$	$+1$
K-plus	K^+	$+1$	0.494	1.2×10^{-8}	$u\bar{s}$	$+1$
K-minus	K^-	-1	0.494	1.2×10^{-8}	$s\bar{u}$	-1

Mesons containing charmed quarks

Name	Symbol	Charge/(+e)	Mass/(GeV/c^2)	Lifetime/s	Quark structure	Charm
psi	ψ	0	3.1	10^{-20}	$c\bar{c}$	0
D-zero	D^0	0	1.86	4.3×10^{-13}	$c\bar{u}$	$+1$
D-plus	D^+	$+1$	1.87	9.2×10^{-13}	$c\bar{d}$	$+1$

Particle physics timeline

The following is a list of important milestones in the history of particle physics. It is not exhaustive and the reader is referred to other sources in the Bibliography (page 291) for more detail of the historical contexts.

640–540BC   Thales of Miletus postulates that water is the basic substance of the Earth.

430–390BC   Empedocles asserts that all matter is made up of four elements: earth, air, fire and water.

470–380BC   Democritus comes up with the idea of indivisible pieces of matter which he calls *atomos*.

384–322BC   Aristotle formalises the gathering of scientific knowledge incorporating the four elements of Empedocles and thus provides the fundamental basis of science for nearly 2000 years.

1642–1727   Sir Issac Newton uses the scientific method of observation and measurement, and formulates the laws of classical mechanics which describe motion in mathematical terms.

1791–1867   Michael Faraday carries out experiments with electricity and magnetism, and provides evidence that they are related. He discovers electrolysis which suggests that electricity might be particulate in nature. He also introduces the idea of a field.

1803   John Dalton publishes his Law of Partial Pressures and introduces an atomic theory.

1869   Dmitri Mendeleyev publishes his Periodic Table which classifies the properties of the chemical elements according to their atomic weights.

1873   James Clerk Maxwell combines electricity and magnetism together to produce a theory of electromagnetism that describes the propagation of electromagnetic waves.

1883   Cosmic rays are discovered in balloon experiments.

1887   Michelson's and Morley's experiment disproves the existence of the ether.

1896   Henri Becquerel discovers radioactivity.

1897   J.J. Thomson discovers the electron.

1898   Marie and Pierre Curie separate radioactive elements.

1900   Max Planck introduces quantum theory to explain thermal radiation.

1905   Albert Einstein publishes an explanation of the photoelectric effect based on quantum ideas first introduced by Planck. He also introduces the Special Theory of Relativity.

1909   Robert Millikan determines the value of the electronic charge.

1911   Ernest Rutherford infers the existence of the atomic nucleus.

1912   Einstein explains gravity as the curvature of space–time in his General Theory of Relativity.

1913   Niels Bohr constructs a model of the atom based on quantum ideas.

1919   Sir Arthur Eddington and other astronomers measure the gravitational deflection of starlight, so confirming Einstein's General Theory of Relativity.

1924   Louis de Broglie proposes that matter has wave properties.

1925   Wolfgang Pauli formulates his Exclusion Principle to explain the electronic structure of atoms.

1925   Samuel Goudsmit and George Uhlenbeck introduce the idea of quantum spin to explain the splitting of spectral lines.

1926   Erwin Schrödinger introduces quantum mechanics based on wave equations.

1927   Werner Heisenberg formulates the Uncertainty Principle.

1927   The wave behaviour of electrons is experimentally demonstrated by Clinton Davisson and Lester Germer.

1929   Edwin Hubble's observations provide experimental evidence that the universe is expanding.

1930   Wolfgang Pauli suggests the existence of the neutrino to explain the energy spectrum of beta decay.

1931   By combining quantum mechanics with Special Relativity, Paul Dirac predicts the existence of antimatter.

1931   The positron is discovered by Carl Anderson.

1931   The neutron is discovered by James Chadwick.

1932   The first artificially produced nuclear reaction is achieved using a high-voltage accelerator by Cockcroft and Walton.

1932   The cyclotron is developed by Ernest Lawrence.

1934   Enrico Fermi proposes a theory of beta decay based on the weak interaction.

1935   Hideki Yukawa proposes a theory of nuclear interactions based on the idea of 'mesons'.

1937   A 'heavy electron' – the muon – is discovered.

1938   Nuclear fission is observed.

1942   The first controlled fission reactor is created.

1945   The first atomic (fission) bomb is exploded in the New Mexico desert.

1946   'Big Bang' cosmology is proposed by George Gamow.

1947   The first synchrocyclotron comes into operation.

1947   The π meson is discovered.

1948   The first linear proton accelerator (32 MeV, University of California at Berkeley) comes into operation.

1952   The first proton synchrotron (2.3 GeV, Brookhaven National Laboratory) comes into operation.

1952   The first thermonuclear (fusion) bomb is detonated at Eniwetok Atoll in the Pacific.

1953   The strangeness hypothesis is introduced.

1956   The neutrino is detected experimentally.

1956   Parity violation is found in weak interactions.

1964   CP violation is observed in neutral K meson decay.

1964   The three-quark model is introduced by Murray Gell-Man and also independently by George Zweig.

1965   The 3 K cosmic microwave background radiation is discovered by Arno Penzias and Robert Wilson.

1967   Steven Weinberg, Sheldon Glashow and Abdus Salam propose a theory that links the electromagnetic and weak interaction together (the Electroweak Theory).

1970   The property 'charm' is proposed by Glashow.

1970–73   The Standard Model of particle physics is developed.

1970s   String theory is proposed as a way of describing particles as one-dimensional strings rather than single points.

1972   A 500 GeV proton synchrotron comes into operation at Fermilab.

1974   The J/ψ particle is discovered and the existence of the charmed quark confirmed.

1975   The τ lepton is discovered.

1976   The Super Proton Synchrotron begins operating at CERN, initially with a peak energy of 500 GeV.

1977   Evidence for the fifth quark (bottom) is found by Leon Lederman and colleagues.

1980s   Supersymmetry and string theory are combined in a way that removes many theoretical limitations of conventional particle models of matter and points towards a complete Theory of Everything.

1983   A proton–antiproton collider at CERN begins operating at 300 GeV.

1983   The $W^{\pm}$ and Z^0 weak bosons, predicted by the Electroweak Theory, are discovered at CERN.

1990s   Membranes or M-theory is developed as a way of describing particles, and provides a unique theory of Grand Unification of matter and forces.

1992   The COBE satellite finds ripples in the microwave background radiation that are consistent with the Big Bang theory.

1995   The top quark is discovered at Fermilab.

1998   Scientists announce that they have discovered neutrino oscillations at the Super Kamiokande detector, providing strong evidence that neutrinos have mass.

1998   By studying high-redshift supernovas, scientists announce evidence for a repulsive force in the universe that seems to be at work on a cosmic scale, partly neutralising the force of gravity and suggesting that the expansion of the universe may be speeding up rather than slowing down.

2000   Fermilab scientists announce the discovery of the τ neutrino.

2000   The elusive Higgs boson, required by the Electroweak Theory, may have been glimpsed at CERN.

2000   Gravitational wave observatories come into operation to look for gravity waves, predicted by General Relativity.

Physical and mathematical data

Physical constants

	symbol	value
Gravitational constant	G	$6.67 \times 10^{-11}\,\mathrm{N\,m^2\,kg^{-2}}$
Acceleration of free fall	g	$9.81\,\mathrm{m\,s^{-1}}$ (close to the Earth)
Gravitational field strength	g	$9.81\,\mathrm{N\,kg^{-1}}$ (close to the Earth)
Speed of light in a vacuum	c	$2.998 \times 10^8\,\mathrm{m\,s^{-1}}$
Elementary charge	e	$1.602 \times 10^{-19}\,\mathrm{C}$
Boltzmann's constant	k	$1.381 \times 10^{-23}\,\mathrm{J\,K^{-1}}$
Stefan–Boltzmann constant	σ	$5.670 \times 10^{-8}\,\mathrm{W\,m^{-2}\,K^{-4}}$
Wien's Law constant	$\lambda_{max}T$	$2.90 \times 10^{-3}\,\mathrm{m\,K}$
Planck constant	h	$6.626 \times 10^{-34}\,\mathrm{J\,s}$
Avogadro constant	N_A	$6.022 \times 10^{23}\,\mathrm{mol^{-1}}$
Permittivity of free space	ϵ_0	$8.854 \times 10^{-12}\,\mathrm{F\,m^{-1}}$
Permeability of free space	μ_0	$4\pi \times 10^{-7}\,\mathrm{H\,m^{-1}}$

Conversion factors

Electronvolt $1\,\mathrm{eV} = 1.602\,189 \times 10^{-19}\,\mathrm{J}$

Barn $1\,\mathrm{b} = 10^{-28}\,\mathrm{m^2}$

Unified mass constant $1\,\mathrm{u} = 931.502\,\mathrm{MeV}/c^2 = 1.660\,566 \times 10^{-27}\,\mathrm{kg}$

Temperature $\mathrm{K} = {}^{\circ}\mathrm{C} + 273$

Mathematical constants and formulae

$\pi = 3.1416$

Angular measure

$360^{\circ} = 2\pi$ radians

1 radian(rad) $= 57^{\circ}\,17'\,45'' = 206\,264.8''$

$1^{\circ} = 60' = 3600'' = 0.017\,45\,\mathrm{rad}$

Areas and volumes

circumference of circle $= 2\pi$ (radius)

area of circle $= \pi$ (radius)^2

surface area of a sphere $= 4\pi$ (radius)^2

volume of a sphere $= \frac{4}{3}\pi$ (radius)^3

arc length of circle $=$ radius $\times$ angle subtended in radians

Difference of two squares

$a^2 - b^2 = (a + b)(a - b)$

Powers of 10

tera (T)	$= 10^{12}$	kilo (k)	$= 10^3$	deci (d)	$= 10^{-1}$	micro (μ)	$= 10^{-6}$
giga (G)	$= 10^9$	hecto (h)	$= 10^2$	centi (c)	$= 10^{-2}$	nano (n)	$= 10^{-9}$
mega (M)	$= 10^6$	deca (da)	$= 10$	milli (m)	$= 10^{-3}$	pico (p)	$= 10^{-12}$

Inequalities

Sometimes it is not possible to say exactly what the size of a physical quantity is. Instead we state the limit of its upper or lower value by using inequality symbols:

$\geq$ means 'is greater than *or* equal to'
$\leq$ means 'is less than *or* equal to'
$>$ means 'is greater than'
$<$ means 'is less than'
$\approx$ means 'is approximately equal to'

Significant figures, standard form and rounding

Significant figures are the number of meaningful digits in a numerical quantity. In experimental or observational measurements the accuracy of the method employed will determine the number of significant figures used. Some examples are:

0.005 432 1	five significant figures
$5.432\,1 \times 10^{-9}$	five significant figures
5.1×10^{-6}	two significant figures
5.43	three significant figures

Zeros at the beginning or end of the number are not counted. However, zeros in the middle of a number *are* counted so that, for example, 5.04, 50 400 and 0.050 4 all have three significant figures.

In general, when doing calculations the result cannot have more significant figures than the quantity used in the formula having the *least* significant figures. If, for example, one of the quantities in the computation is only accurate to, say, two significant figures (with the rest being known to more than this) then we can express the result only to two significant figures.

In particle physics we often encounter very large and very small numbers. It is useful to write them in **standard form**, that is, as $a \times 10^n$ where a is a number between 1 and 10, and the index n is an integer (positive or negative whole number, or 0). When multiplying numbers together, the numbers are multiplied and the indices added. When dividing, the numbers are divided and the indices subtracted. For example:

$$(3.4 \times 10^4) \times (5.6 \times 10^{-3}) = (3.4 \times 5.6) \times 10^{(4-3)} = 19.04 \times 10^1 = 190.4$$

$$(7.5 \times 10^9) \div (4.8 \times 10^{-6}) = (7.5/4.8) \times 10^{(9-(-6))} = 1.6 \times 10^{15}$$

To **round off** numbers we examine the last digit. If it is greater than or equal to 5 we round up, less than 5 we round down, e.g. 34.56 = 34.6 (round up); 976.854 = 976.85 (round down).

The Greek alphabet

Alpha A α	Eta H η	Nu N ν	Tau T τ
Beta B β	Theta Θ θ	Xi Ξ ξ	Upsilon Υ υ
Gamma Γ γ	Iota I ι	Omicron O o	Phi Φ ϕ, φ
Delta Δ δ	Kappa K κ	Pi Π π	Chi X χ
Epsilon E ϵ	Lambda Λ λ	Rho P ρ	Psi Ψ ψ
Zeta Z ζ	Mu M μ	Sigma Σ σ	Omega Ω ω

The Periodic Table of the chemical elements

period	1	2	3	4	5	6	7	8	9	10	11	12	13	14	15	16	17	18
1	1.0079 1 H																	4.0026 2 He
2	6.941 3 Li	9.012 4 Be											10.81 5 B	12.011 6 C	14.007 7 N	15.999 8 O	18.998 9 F	20.179 10 Ne
3	22.990 11 Na	24.305 12 Mg											26.982 13 Al	28.086 14 Si	30.974 15 P	32.06 16 S	35.453 17 Cl	39.948 18 Ar
4	39.098 19 K	40.08 20 Ca	44.956 21 Sc	47.90 22 Ti	50.942 23 V	51.996 24 Cr	54.938 25 Mn	55.847 26 Fe	58.933 27 Co	58.70 28 Ni	63.546 29 Cu	65.38 30 Zn	69.72 31 Ga	72.59 32 Ge	74.922 33 As	78.96 34 Se	79.904 35 Br	83.80 36 Kr
5	85.468 37 Rb	87.62 38 Sr	88.906 39 Y	91.22 40 Zr	92.906 41 Nb	95.94 42 Mo	(97) 43 Tc	101.07 44 Ru	102.91 45 Rh	106.4 46 Pd	107.87 47 Ag	112.41 48 Cd	114.82 49 In	118.69 50 Sn	121.75 51 Sb	127.60 52 Te	126.90 53 I	131.30 54 Xe
6	132.91 55 Cs	137.33 56 Ba	rare earths 57–71	178.49 72 Hf	180.95 73 Ta	183.85 74 W	186.21 75 Re	190.2 76 Os	192.22 77 Ir	195.09 78 Pt	196.97 79 Au	200.59 80 Hg	204.37 81 Tl	207.2 82 Pb	208.98 83 Bi	(209) 84 Po	(210) 85 At	(222) 86 Rn
7	(223) 87 Fr	(226) 88 Ra	acti- nides 89–103	(257) 104 Rf	(263) 105 Ha	(263) 106 Sg	107 Bh	108 Hs	109 Mt									
group	1	2	3	4	5	6	7	8	9	10	11	12	13	14	15	16	17	18

rare earths (lanthanides)	138.91 57 La	140.12 58 Ce	140.91 59 Pr	144.24 60 Nd	(145) 61 Pm	150.4 62 Sm	151.96 63 Eu	157.25 64 Gd	158.93 65 Tb	162.50 66 Dy	164.93 67 Ho	167.26 68 Er	168.93 69 Tm	173.04 70 Yb	174.97 71 Lu
actinides	(227) 89 Ac	232.04 90 Th	(231) 91 Pa	238.03 92 U	(237) 93 Np	(244) 94 Pu	(243) 95 Am	(247) 96 Cm	(247) 97 Bk	(251) 98 Cf	(254) 99 Es	(257) 100 Fm	(258) 101 Md	(259) 102 No	(260) 103 Lr

In each box, the number at the top is the *atomic mass number*, i.e. the mass (in grams) of one mole or, alternatively, the mass (in u) of one atom. The number at the bottom is the *atomic or proton number*. Numbers in parentheses denote the atomic masses of the most stable or best-known isotope of the element; all other numbers represent the average masses of a mixture of several isotopes as found in naturally occurring samples of the element. The names for elements 104–109 are those recommended by the American Chemical Society.

Bibliography

Internet

An abridged version of this book can be found at www.ph.surrey.ac.uk/starbase/pp, a website maintained and run by the Physics Department at the University of Surrey. The website contains many hot links to particle physics sites around the world, including national accelerator laboratories.

Books

Q is for Quantum: Particle Physics from A to Z, John Gribbin. Phoenix Giant, Orion Books (1999)
An excellent one-stop glossary for particle physics that explains many of the key ideas and the jargon used in particle physics and the quantum world.

QED: The Strange Theory of Light and Matter, Richard Feynman. Penguin (1990)
The story of QED as told by one of its principal architects. Without using mathematical equations, Feynman takes the reader through the important concepts and introduces the idea of phasors.

The Character of Physical Law, Richard Feynman. Penguin (1995; originally published in 1965)
Based on a series of lectures given at Cornell University and originally televised on BBC2, Feynman discusses the principles and nature of physical laws and gathers their common features into one broad principle of invariance.

Fearful Symmetry: The Search for Beauty in Modern Physics, A. Zee. Princeton University Press (1999)
Zee's book is an excellent and readable account of how today's theoretical physicists are discovering the deep symmetries and simplicities inherent in nature.

In Search of SUSY, John Gribbin. Penguin (1998)
A very readable account of supersymmetry and the quest for a Theory of Everything.

Quarks, Leptons and the Big Bang, Jonathan Allday. Institute of Physics Publishing (1998)
A clear and readable introduction to particle physics and related areas of cosmology.

Nucleus: A Trip into the Heart of Matter, R. Mackintosh, J. Al-Khalili, B. Jonson, T. Pena. Canopus Publishing (2001)
An excellent and well illustrated introduction to the properties of the atomic nucleus for the general reader, which includes many relevant sections on particle physics and cosmology.

Astrophysics, Christopher Bishop. John Murray Publishers (2000)
A companion volume to *Particle Physics* which includes a chapter on cosmology.

Answers to numerical questions

Chapter 1

2 a $B = \dfrac{4\pi \times 10^{-7}\,\text{NA}^{-2} \times 2\,\text{A}}{2 \times \pi \times 1\,\text{m}} = 4 \times 10^{-7}\,\text{T}$

b $B = \dfrac{4\pi \times 10^{-7}\,\text{NA}^{-2} \times 2\,\text{A}}{2 \times \pi \times 2\,\text{m}} = 2 \times 10^{-7}\,\text{T}$

c $B = \dfrac{4\pi \times 10^{-7}\,\text{NA}^{-2} \times 2\,\text{A}}{2 \times \pi \times 3\,\text{m}} = \dfrac{4}{3} \times 10^{-7}\,\text{T}$

The graph should follow a 1/r curve.

3 a No. of electrons $= \dfrac{1\,\text{C}}{1.602 \times 10^{-19}\,\text{C}} = 6.2 \times 10^{18}$

b $F = \dfrac{1}{4\pi \times 8.85 \times 10^{-12}\,\text{Fm}^{-1}} \times$

$\dfrac{(+)25.0 \times 10^{-9}\,\text{C} \times (-)7.20 \times 10^{-9}\,\text{C}}{(6 \times 10^{-2}\,\text{m})^2}$

which has magnitude $4.5 \times 10^{-4}\,\text{N}$. Since they are charges with opposite polarity, the force is an attractive one.

c Rearranging Coulomb's Law,

$q_2 = \dfrac{F \times 4\pi\epsilon_0 r^2}{q_1} =$

$\dfrac{0.18\,\text{N} \times 4\pi \times 8.85 \times 10^{-12}\,\text{Fm}^{-1} \times (12 \times 10^{-2}\,\text{m})^2}{(-)30.0 \times 10^{-8}\,\text{C}}$

$= 9.6 \times 10^{-7}\,\text{C}$

For the force to be attractive, the polarity of the charge must be positive.

4 a $F_{\text{grav}} = 6.67 \times 10^{-11}\,\text{Nm}^2\text{kg}^{-2} \times$

$\dfrac{1.67 \times 10^{-27}\,\text{kg} \times 9.11 \times 10^{-31}\,\text{kg}}{(0.5 \times 10^{-10}\,\text{m})^2}$

$= 4.1 \times 10^{-47}\,\text{N}$

b $F_{\text{Coul}} = \dfrac{1}{4\pi \times 8.85 \times 10^{-12}\,\text{Fm}^{-1}} \times$

$\dfrac{(-)1.6 \times 10^{-19}\,\text{C} \times (+)1.6 \times 10^{-19}\,\text{C}}{(0.5 \times 10^{-10}\,\text{m})^2}$

$= 9.2 \times 10^{-8}\,\text{N}$

c The gravitational force on the sub-atomic scale is extremely feeble. In a hydrogen atom, it is weaker than the electrostatic force by a factor of 10^{39}.

5 $c = \dfrac{1}{\sqrt{4\pi \times 10^{-7}\,\text{NA}^{-2} \times 8.854 \times 10^{-12}\,\text{C}^2\text{N}^{-1}\text{m}^{-2}}}$

$= 2.998 \times 10^8\,\text{ms}^{-1} \approx 3.0 \times 10^8\,\text{ms}^{-1}$

In order to show homogeneity, first of all express μ_0 and ϵ_0 in SI base units.

$\mu_0 = 4\pi \times 10^{-7}\,\text{kg}\,\text{ms}^{-2}\,\text{A}^{-2}$
$\epsilon_0 = \text{A}^2\text{s}^2\text{kg}^{-1}\text{m}^{-1}\text{s}^2\text{m}^{-2} = \text{A}^2\text{kg}^{-1}\text{s}^4\text{m}^{-3}$

To make things simpler, square both sides of the equation to get rid of the square root sign:

$c^2 = \dfrac{1}{\mu_0\epsilon_0}$

and using the notation L = length (m), T = time (s), M = mass (kg) and A = amperes:
left-hand side of equation has dimensions L^2T^{-2}
right-hand side of equation has dimensions

$\dfrac{1}{[\text{MLT}^{-2}\text{A}^{-2}][\text{A}^2\,\text{M}^{-1}\text{T}^4\text{L}^{-3}]} = \dfrac{1}{\text{L}^{-2}\text{T}^2} = \text{L}^2\text{T}^{-2}$

Both sides of the equation have the same dimensions, therefore the equation is homogenous with respect to units.

6 a Using $E = hf = hc/\lambda$, energy of an individual photon =

$\dfrac{6.63 \times 10^{-34}\,\text{Js} \times 3.0 \times 10^8\,\text{ms}^{-1}}{630 \times 10^{-9}\,\text{m}} = 3.2 \times 10^{-19}\,\text{J}$

Now $60\,\text{W} = 60\,\text{Js}^{-1}$ so the number emitted per second =

$\dfrac{60\,\text{Js}^{-1}}{3.2 \times 10^{-19}\,\text{J}} = 1.9 \times 10^{20}\,\text{photons s}^{-1}$

b Using Einstein's photoelectric equation $\text{KE}_{\text{max}} = hf - \Phi$, we can first calculate the work function Φ given the max. KE of the electrons and the corresponding frequency of the radiation, i.e.

$\Phi = hf - \text{KE}_{\text{max}} = 6.6 \times 10^{-34}\,\text{Js} \times 7.5 \times 10^{14}\,\text{Hz}$
$- 1.6 \times 10^{-19}\,\text{J} = 3.4 \times 10^{-19}\,\text{J}$

Now the *minimum* frequency of radiation for which electrons will be emitted is when they do so with zero kinetic energy, i.e. when the energy of the radiation is just sufficient to eject them onto the metal surface. For this to happen the energy of the radiation must be equal to the work function, $hf = \Phi$.

$f_{\text{min}} = \dfrac{\Phi}{h} = \dfrac{3.4 \times 10^{-19}\,\text{J}}{6.6 \times 10^{-34}\,\text{Js}} = 5.2 \times 10^{14}\,\text{Hz}$

c Using $E = mc^2$ the energy equivalent of $1\,\mu g =$
$10^{-9}\,kg \times (3.0 \times 10^8)^2\,m^2\,s^{-2} = 9 \times 10^7\,J$
This amount of energy is released in one hour, so
the energy release per second is

$$\frac{9 \times 10^7\,J}{3600\,s} = 25 \times 10^3\,W \text{ or } 0.025\,MW$$

Chapter 2

1 Since the oil drop is stationary, the weight of the
drop W acting downwards is balanced by the force
due to the electric field E between the plates. If the
drop has a charge q, then $qE = W$. The field strength
$E = V/d$ where V is the potential and d the distance
between the plates.

$$\therefore q = \frac{W}{E} = \frac{Wd}{V} =$$

$$\frac{4.90 \times 10^{-15}\,kg \times 9.81\,m\,s^{-2} \times 15 \times 10^{-3}\,m}{1500\,V}$$

$$= 4.8 \times 10^{-19}\,C$$

The charge on one electron is $1.6 \times 10^{-19}\,C$ so

$$\text{no. of electrons attached to drop} = \frac{4.8 \times 10^{-19}\,C}{1.6 \times 10^{-19}\,C} = 3$$

2 b i) If the α-particle travelled along the same path
PQ with twice the initial kinetic energy then at
point X the magnitude of the electrostatic force
would not change. The magnitude of the force
depends only on the respective charges of the
α-particle and the nucleus, and the distance
between them. However the electrostatic force on
the α-particle would act for a shorter time so the
impulse given to the α-particle would be less.
ii) The magnitude of the electrostatic force is given
by Coulomb's Law:

$$F = \frac{1}{4\pi\epsilon_0}\frac{q_1 q_2}{r^2}$$

Because of their different proton numbers, we
have

$$\frac{F_{silver}}{F_{gold}} = \frac{47}{79} = 0.6$$

(All other factors in Coulomb's Law cancel.) The
magnitude of the electrostatic force for silver
nuclei would be 0.6 times that for gold.
c i) The energy of the α-particle at all points in its
motion is constant and is the sum of its electric
potential energy V plus its kinetic energy K, i.e.
$V + K = $ constant. At the distance of closest
approach, $K = 0$, so that all the energy is in the
form of electric potential energy.

$$V = 8.0 \times 10^{-13}\,J = \frac{1}{4\pi\epsilon_0} \times \frac{79 \times 2e^2}{r}$$

where r is the distance of closest approach.
(Remember the α-particle has a magnitude of two
electron charges.) So

$$r = \frac{1}{4\pi \times 8.85 \times 10^{-12}\,F\,m^{-1}} \times$$

$$\frac{158 \times (1.6 \times 10^{-19}\,C)^2}{8.0 \times 10^{-13}\,J} = 4.5 \times 10^{-14}\,m$$

ii) The recoil velocity of the gold nucleus is
negligible. This is a reasonable assumption to make
since the mass of the gold nucleus is nearly 50
times as great as that of the α-particle.

3 The energy of E of the emitted photon is given by
$E = hf = hc/\lambda$. Therefore $\lambda = hc/E$. So for radiation to
be emitted with the longest wavelength we want the
energy difference in the transition to be as small as
possible. This happens in the transition between the
2nd excited state and the 1st excited state, i.e.

$$-1.51\,eV - (-3.39\,eV) = 1.88\,eV =$$

$$1.88 \times 1.6 \times 10^{-19}\,J = 3.01 \times 10^{-19}\,J$$

$$\therefore \lambda = \frac{6.63 \times 10^{-34}\,J\,s \times 3.0 \times 10^8\,m\,s^{-1}}{3.01 \times 10^{-19}\,J} = 660\,nm$$

which is in the near infrared range of the
electromagnetic spectrum.

4 We use the relation for nuclear radii $R = R_0 A^{\frac{1}{3}}$ given on
page 43. The hydrogen nucleus ($A = 1$) has a radius
of $R_0 = 1.2\,fm$. From the Periodic Table on page 290,
we see that the nucleus with the highest atomic
number is that of Lawrencium with $A = 260$. The
radius of the Lawrencium nucleus is

$$1.2 \times 10^{-15}\,m \times (260)^{\frac{1}{3}} = 7.6 \times 10^{-15}\,m \text{ or } 7.6\,fm$$

Lawrencium is the most massive nucleus known but is
less than 8 times the radius of the hydrogen nucleus,
so the teacher's assertion is essentially correct.

5 Using the expression for electric potential energy

$$V = \frac{1}{4\pi\epsilon_0}\frac{q_1 q_2}{r}$$

the electric potential energy of a proton in contact
with a lithium nucleus is

$$\frac{1}{4\pi \times 8.85 \times 10^{-12}\,F\,m^{-1}} \times \frac{7 \times 2 \times (1.6 \times 10^{-19}\,C)^2}{(2.3 + 1.2) \times 10^{-15}\,m}$$

$$= 9.2 \times 10^{-13}\,J \text{ or } \frac{9.2 \times 10^{-13}}{1.6 \times 10^{-19}}\,eV = 5.8\,MeV$$

(Note that we consider the nucleus and the α-particle as point charges with the effective distance between them equal to the sum of their radii.)

Our answer indicates that it is necessary to accelerate protons to energies of millions of electronvolts in order to get them anywhere near a nucleus. Lithium has a low atomic number so to place a proton near a heavier nucleus with more positive charge would require energies even larger than this. For this reason, particle accelerators need to be able to produce energies well in excess of a few MeV in order to probe the nucleus. Compared to 'everyday' energies of a few eV for light photons, for example, this is very high energy.

6 a $^2_1H + ^3_1H \rightarrow ^4_2HE + ^1_0n$

(Note that A and Z numbers have to balance on each side of the nuclear equation.)

b On the left-hand side of the equation:

$2.0141\,u + 3.0161\,u = 5.0302\,u$

On the right-hand side of the equation:

$4.0028\,u + 1.0087\,u = 5.0115\,u$

Difference $= (5.0302 - 5.0115)\,u = 0.0187\,u =$

$0.0187 \times 931\,MeV = 17.4\,MeV$ or

$17.4 \times 10^6 \times 1.6 \times 10^{-19} = 2.8 \times 10^{-12}\,J$

If we consider the molar mass of deuterium then 2 g of ^2_1H contains 6.023×10^{23} nuclei. So 1 kg of ^2_1H contains

$\dfrac{6.023 \times 10^{23}}{2 \times 10^{-3}} = 3.0 \times 10^{26}$ nuclei

If we assume that the fusion of a single nucleus of deuterium releases 17.4 MeV, then this corresponds to an energy release of $3.0 \times 10^{26} \times 2.8 \times 10^{-12} = 8.4 \times 10^{14}\,J$. Using the relationship that energy = power $\times$ time,

$50\% \times 8.4 \times 10^{14}\,J = 1.0 \times 10^6\,W \times t$.

$\therefore t = \dfrac{4.2 \times 10^{14}\,J}{1.0 \times 10^6\,W} = 4.2 \times 10^8\,s$

or $\dfrac{4.2 \times 10^8}{24 \times 60 \times 60} = 4861$ days

We can see then that the fusion of 1 kg of deuterium in a nuclear fusion reactor could generate 1 MW of electrical power continuously for about 13 years! No wonder there is a concerted effort by nuclear scientists and engineers to build such a device.

7 Alpha decay is taking place.

The nuclear equation is of the form $X \rightarrow Y + ?$

From the grid we see that Z: $62 \rightarrow 60$

N: $85 \rightarrow 83$

$A = N + Z$: $147 \rightarrow 143$

So the nuclear equation is $^{147}_{62}X \rightarrow ^{143}_{60}Y + ^4_2He$.

If Y subsequently decays into an isotope W of X by β^- emission, then the equation is $^{143}_{60}Y \rightarrow ^{143}_{61}W + ^{\,\,0}_{-1}\beta$. Thus the point P has $N = 82$ and $Z = 61$.

Chapter 3

1 c We use the length contraction formula

$$x = x_0\sqrt{1 - \dfrac{v^2}{c^2}} = \dfrac{x_0}{\gamma}$$

We first calculate the Lorentz factor γ.

$$\gamma = \dfrac{1}{\sqrt{1 - \dfrac{(2.650 \times 10^8\,ms^{-1})^2}{(3 \times 10^8\,ms^{-1})^2}}} = 2.13$$

The length of the UFO as seen in its own frame of reference is $x_0 = 300$ m. As observed from Earth it will have a contracted length

$$x = \dfrac{x_0}{\gamma} = \dfrac{300}{2.13} = 140.8\,m$$

d We use the relativistic mass equation

$$m = \dfrac{m_0}{\sqrt{1 - \dfrac{v^2}{c^2}}} = \gamma m_0$$

As before, we first calculate γ.

$$\gamma = \dfrac{1}{\sqrt{1 - \dfrac{(2.450 \times 10^8\,ms^{-1})^2}{(3 \times 10^8\,ms^{-1})^2}}} = 1.73$$

m_0 is the mass of an alpha particle as measured by an observer travelling *with* the particles, i.e. 6.65×10^{-27} kg. Their mass as viewed by an observer at rest in a laboratory is therefore 6.65×10^{-27} kg $\times 1.73 = 1.15 \times 10^{-26}$ kg. The observer in the laboratory frame observes the alpha particles to have *increased* in mass.

e i) The rest-mass energy is given by $E = m_0c^2$ where m_0 is the mass of the proton as measured when at rest with respect to the observer, i.e. its rest mass. So

$E = 1.66 \times 10^{-27}\,kg \times (3.0 \times 10^8)^2\,m^2\,s^{-2} =$

$1.5 \times 10^{-10}\,J$

ii) When the proton is moving at an appreciable fraction of the speed of light we need to use the relativistic total energy equation

$$E = \dfrac{m_0c^2}{\sqrt{1 - \dfrac{v^2}{c^2}}} = \gamma m_0c^2$$

Calculating the Lorentz factor γ gives

$$\gamma = \frac{1}{\sqrt{1 - \frac{(2.50 \times 10^8\,m\,s^{-1})^2}{(3 \times 10^8\,m\,s^{-1})^2}}} = 1.81$$

Therefore the total energy is
$1.81 \times 1.5 \times 10^{-10}\,J = 2.7 \times 10^{-10}\,J$ or $1.68\,GeV$

2 a This question involves using the time dilation formula

$$\Delta t = \frac{\Delta t_0}{\sqrt{1 - \frac{v^2}{c^2}}}$$

In the kaon's own inertial frame (i.e. if you were travelling with the kaon) the distance travelled would be simply its speed $(0.990c) \times$ its *proper time* $(0.123\,7\,\mu s)$, i.e.

$0.990 \times 3.0 \times 10^8\,m\,s^{-1} \times 0.1237 \times 10^{-6}\,s = 36.7\,m$

However, when the half-life of the K^+ is measured from a laboratory frame we find that it has been dilated by a factor γ where

$$\gamma = \frac{1}{\sqrt{1 - \frac{v^2}{c^2}}} = \frac{1}{\sqrt{1 - \frac{(0.99c\,m\,s^{-1})^2}{(c\,m\,s^{-1})^2}}} = 7.09$$

So Δt, the lifetime of the K^+ in the laboratory frame, is $7.09 \times 0.123\,7\,\mu s = 7.09 \times 0.123\,7 \times 10^{-6}\,s = 8.77 \times 10^{-7}\,s$. In the laboratory frame the K^+ seems to live about 7 times longer than it does in its own frame. This means that in the laboratory frame it will travel a distance of

$0.990 \times 3.0 \times 10^8\,m\,s^{-1} \times 8.77 \times 10^{-7}\,s = 260.5\,m$

Many experiments have been done using particle accelerators which confirm that particles seem to 'live longer' when travelling at significant fractions of the speed of light.

b This problem is about the relativity of time and the meaning of 'proper time', and leads to a very extraordinary conclusion. Recall that on page 65 we stated that the proper time is an interval of time measured in the *same* location and in the *same* inertial reference frame. Aboard your starship's inertial reference frame, you measure an interval of 10 years to get to the star system and 10 years to get back, making a total of 20 years travelling in proper time. However, mission control on Earth does not measure proper time. In the Earth's inertial reference frame the time interval going out is $\Delta t = \gamma \Delta t_0$ where

$$\gamma = \frac{1}{\sqrt{1 - \frac{(0.999c\,m\,s^{-1})^2}{(c\,m\,s^{-1})^2}}} = 22.37$$

so that the travel time out is 22.37×10 years $= 224$ years. On the journey back to Earth we have a symmetrical situation so the round trip time of the starship as measured from Earth's inertial reference frame is $2 \times 224 = 448$ years. Thus astronauts in the starship would find that their voyage has lasted only 20 years while to mission control on Earth it would have lasted 448 years. So Special Relativity suggests that we can travel into the future by travelling at near light velocities to adjust the rate at which time passes!

3 Mass: we will consider eV/c^2 (since MeV/c^2 is simply a multiple). Using the usual notation for dimensional analysis, e is measured in coulombs and has dimensions AT, volt V has dimensions $ML^2T^{-3}A^{-1}$, and c^2 has dimensions L^2T^{-2}. Therefore eV/c^2 has dimensions

$$\frac{ATML^2T^{-3}A^{-1}}{L^2T^{-2}} = M$$

which is the dimension of mass.
Similarly for momentum, we can express eV/c dimensionally as

$$\frac{ATML^2T^{-3}A^{-1}}{LT^{-1}} = MLT^{-1}$$

which is the dimension of momentum.

a We need to use

$$E = \frac{m_0c^2}{\sqrt{1 - \frac{v^2}{c^2}}} = \gamma m_0 c^2$$

where $\gamma = \dfrac{1}{\sqrt{1 - \frac{(0.8c\,m\,s^{-1})}{(c\,m\,s^{-1})^2}}} = 1.67$

Rest energy $= 0.511\,MeV = 0.511 \times 10^6 \times 1.6 \times 10^{-19}\,J = 8.2 \times 10^{-14}\,J = m_0c^2$
$E = \gamma m_0 c_2 = 8.2 \times 10^{-14}\,J \times 1.67 = 1.4 \times 10^{-13}\,J$ or $0.875\,MeV$

b Using the formula for kinetic energy

$$E_K = m_0c^2(\gamma - 1) = 8.2 \times 10^{-14}\,J \times (1.67 - 1)$$
$$= 5.5 \times 10^{-14}\,J$$

c Using $E^2 = (pc)^2 + (m_0c)^2$ we have
$(0.875\,MeV)^2 = (pc)^2 + (0.511\,MeV)^2$. Then

$$pc = \sqrt{(0.875\,MeV)^2 - (0.511\,MeV)^2} = 0.71\,MeV$$

and so, giving the momentum in units of energy divided by c,

$p = 0.71\,MeV/c$

4 b The relativistic equation for momentum is $p = \gamma m_0 v$ but at 5% of the speed of light γ is very close to 1 so we can use the usual non-relativistic expression for momentum $p = m_0 v$.

$$p = 9.1 \times 10^{-31}\,kg \times 0.05 \times 3.0 \times 10^8\,ms^{-1}$$
$$= 1.4 \times 10^{-23}\,kg\,m\,s^{-1}$$

$$\lambda_p = \frac{h}{p} = \frac{6.63 \times 10^{-34}\,Js}{1.4 \times 10^{-23}\,kg\,m\,s^{-1}} = 4.7 \times 10^{-11}\,m$$

The electron wavelength is of the same order of magnitude as the spacing between molecules, so electron waves can be diffracted by them, enabling their structure to be studied.

5 a Charged π meson:

$$\Delta E = \frac{\hbar}{\Delta t} = \frac{1.05 \times 10^{-34}\,Js}{26 \times 10^{-9}\,s} = 4.0 \times 10^{-27}\,J$$
$$\text{or } 2.5 \times 10^{-14}\,MeV$$

$$\frac{\Delta E}{E} = \frac{2.5 \times 10^{-14}\,MeV}{140\,MeV} = 1.8 \times 10^{-16}$$

b Uncharged π meson:

$$\Delta E = \frac{1.05 \times 10^{-34}\,Js}{8.3 \times 10^{-17}\,s} = 1.3 \times 10^{-18}\,J$$
$$\text{or } 8.1 \times 10^{-6}\,MeV$$

$$\frac{\Delta E}{E} = \frac{8.1 \times 10^{-6}\,MeV}{135\,MeV} = 6 \times 10^{-8}$$

c Rho meson:

$$\Delta E = \frac{1.05 \times 10^{-34}\,Js}{4.4 \times 10^{-24}\,s} = 2.4 \times 10^{-11}\,J \text{ or } 149\,MeV$$

$$\frac{\Delta E}{E} = \frac{149\,MeV}{765\,MeV} = 0.19$$

The precision of our particle detectors is $1/10^6 = 10^{-6}$. Looking at the precision of our knowledge of the energy of the particles, measured by $\Delta E/E$, our detector could make accurate measurements of the rho meson, but would not be precise enough to measure the energy of the π mesons.

6 a Wavelength of the fundamental =
2 × size of atom = $2 \times 2 \times 10^{-10}\,m$
$= 4 \times 10^{-10}\,m$

b Momentum $= m_e v = \dfrac{h}{\lambda_p} = \dfrac{6.63 \times 10^{-34}\,Js}{4 \times 10^{-10}\,m}$

$$= 1.7 \times 10^{-24}\,kg\,m\,s^{-1}$$

c $KE = \dfrac{mv^2}{2} = \dfrac{(mv)^2}{2m} = \dfrac{(1.7 \times 10^{-24}\,kg\,m\,s^{-1})^2}{2 \times 9.1 \times 10^{-31}\,kg}$

$$= 1.6 \times 10^{-18}\,J \text{ or } 10\,eV$$

This is less than the ionisation energy so it can remain in the atom.

d Since $\lambda_p = 2L/n$, $mv = \dfrac{h}{\lambda_p} = \dfrac{nh}{2L}$

$$KE = \frac{mv^2}{2} = \frac{(mv)^2}{2m} = \left(\frac{nh}{2L}\right)^2 \times \frac{1}{2m} = \frac{h^2 n^2}{8mL^2}$$

For $n = 1$,

$$KE = \frac{(6.6 \times 10^{-34}\,Js)^2 \times 1^2}{8 \times 9.1 \times 10^{-31}\,kg \times (4 \times 10^{-10}\,m)^2}$$
$$= 3.7 \times 10^{-19}\,J$$

Similarly for $n = 2$, $KE = 1.5 \times 10^{-18}\,J$, and for $n = 3$, $KE = 3.4 \times 10^{-18}\,J$.

This problem shows how wave ideas can be used to describe the motion of an electron in a hydrogen atom.

Chapter 4

1 a i) Use $T = \dfrac{2\pi m_p}{Bq}$ (see page 108).

$$T = \frac{2 \times \pi \times 1.67 \times 10^{-27}\,kg}{1.250\,T \times 1.6 \times 10^{-19}\,C} = 5.2 \times 10^{-8}\,s$$

ii) The frequency of the applied alternating voltage is equal to the cyclotron frequency:

$$f_{cyc} = \frac{1}{T} = \frac{1}{5.2 \times 10^{-8}\,s} = 19\,MHz$$

iii) The protons emerge from the cyclotron when they have reached an orbital radius of 0.3 m. Using $r = mv/Bq$ (page 108) then

$$v = \frac{Bqr}{m} = \frac{1.250\,T \times 1.6 \times 10^{-19}\,C \times 0.3\,m}{1.67 \times 10^{-27}\,kg}$$
$$= 3.6 \times 10^7\,m\,s^{-1}$$

iv) $KE = \tfrac{1}{2}m_p v^2 =$
$0.5 \times (1.67 \times 10^{-27}\,kg) \times (3.6 \times 10^7\,m\,s^{-1})^2$
$= 1.1 \times 10^{-12}\,J$
or $(1.1 \times 10^{-12})/(1.6 \times 10^{-19}) = 7\,MeV$

b Use $r = mv/Bq$ so that $B = mv/qr$. The momentum mv is

$1\,TeV/c = (10^{12} \times 1.6 \times 10^{-19}\,C)/(3 \times 10^8\,m\,s^{-1}) =$
$5.3 \times 10^{-16}\,kg\,m\,s^{-1}$

So

$$B = \frac{5.3 \times 10^{-16}\,kg\,m\,s^{-1}}{1.6 \times 10^{-19}\,C \times 1000\,m} = 3.3\,T$$

Protons of momentum 1 TeV/c have speeds very close to that of light. The circumference of the Tevatron is $2\pi \times 1000$ m so the time taken for one proton to complete a circuit of the Tevatron is $(2000\pi)/c = 21\,\mu$s.

2 a The energies of linear accelerators are limited by their practical length. A vacuum is necessary to avoid the particles losing energy by collisions with gas molecules in the tube. Two other factors are synchrotron radiation and the strength of the magnets required.

For a charged particle moving in a circular path $mv^2/r = Bqv$ or

$$v = \left(\frac{Bq}{m}\right)r$$

i.e. its velocity is proportional to its radius. The period is

$$T = \frac{2\pi r}{v} = 2\pi \times \left(\frac{r}{v}\right)$$

If the speed of the particle increases, the radius increases in the same proportion, so that the quotient r/v remains the same and the period remains constant.

b From Worked Example 4.3, the KE of a charged particle of mass m moving in a circular path R in uniform magnetic field B is given by

$$KE = \tfrac{1}{2}\left(\frac{q^2 B^2}{m}\right)R^2$$

The KE of the particle is proportional to the square of the radius of the accelerator. So small changes in its circumference will affect the energy of the particle beams.

3 a i) $KE = \tfrac{1}{2}m_p v^2$ so

$$v = \sqrt{\frac{2 \times KE}{m_p}} = \sqrt{\frac{2 \times 8 \times 10^{-12}\,J}{1.67 \times 10^{-27}\,kg}} \approx 10^8\,m\,s^{-1}$$

ii) Circumference of 28 GeV proton synchrotron $= 2\pi \times 100$ m $= 200\pi$ m. Therefore the time to travel round ring is

$$\frac{200\pi\,m}{10^8\,m\,s^{-1}} = 6.3 \times 10^{-6}\,s \approx 6\,\mu s$$

iii) Ignoring relativistic effects,
$p = m_p v = 1.66 \times 10^{-27}\,kg \times 10^8\,m\,s^{-1}$
$= 1.6 \times 10^{-19}\,kg\,m\,s^{-1}$

b i) Use $Q = It$. Total charge in accelerator is $(100 \times 10^{-3}\,A) \times (6 \times 10^{-6}\,s) = 6 \times 10^{-7}\,C$. Number of protons injected = charge in accelerator/charge on proton $= (6 \times 10^{-7}\,C)/(1.6 \times 10^{-19}\,C) = 3.8 \times 10^{12}$ protons.

ii) To avoid the protons losing energy by collisions with gas molecules.

c i) The protons are first 'pulled in' and then 'pushed out' of the electrodes.
ii) Energy gained by proton = charge on proton $\times$ potential difference between electrodes

$$= 1.60 \times 10^{-19}\,C \times 4000\,V = 6.4 \times 10^{-16}\,J \text{ or } 4000\,eV$$

Since there are 14 acceleration points, the energy of one proton increases by $14 \times 4000\,eV = 56\,000\,eV$.
iii) $28\,GeV = 28 \times 10^9\,eV$ so no. of times $= (28 \times 10^9\,eV)/(56\,000\,eV) = 500\,000$
iv) They would be impossibly long to build.

d i) Use $Bev = m_p v^2/r$ so that $B = m_p v/er$ and $B \propto m_p v$ for constant e and r.
ii) From **a i)** 50 MeV protons have a speed of about $10^8\,m\,s^{-1}$.

$$B = \frac{m_p v}{er} = \frac{1.67 \times 10^{-27}\,kg \times 10^8\,m\,s^{-1}}{1.60 \times 10^{-19}\,C \times 100\,m} = 10^{-2}\,T$$

iii) The voltage across the electrodes at the acceleration points must be synchronised with the passage of protons through them so they receive maximum acceleration. Since they are increasing in speed, they pass through the acceleration points more frequently and so the frequency of the acceleration voltage must be increased to compensate.

e i) From **a iii)**, the momentum at injection is $1.6 \times 10^{-19}\,kg\,m\,s^{-1}$. The momentum has thus increased by a factor

$$\frac{1.6 \times 10^{-17}\,kg\,m\,s^{-1}}{1.6 \times 10^{-19}\,kg\,m\,s^{-1}} = 100$$

Since from **d i)** the B-field is proportional to the momentum, the B-field must also increase by 100.
ii) For the purposes of estimation, ignore the relativistic expression for momentum and use $p = m_p v$. Then the increased mass of the proton is given by

$$\frac{\text{momentum at 28 GeV}}{\text{speed}}$$

$$= \frac{1.6 \times 10^{-17}\,kg\,m\,s^{-1}}{3 \times 10^8\,m\,s^{-1}} = 5.3 \times 10^{-26}\,kg$$

The mass of the proton has increased by a factor $(5.3 \times 10^{-26}\,kg)/(1.67 \times 10^{-27}\,kg) = 32$. This is a consequence of Special Relativity.
iii) At 400 GeV the protons are moving very close to the speed of light so their speed as they circulate through the accelerator is constant. However, their *mass* is increasing so their momentum is still increasing. From **d i)** the B-field is proportional to the momentum so this must also increase.

4 b On the right-hand side of the interaction equation

total mass of particles $= (938 + (7 \times 140) +$
$(7 \times 140) + 494 + 1115)\,\text{MeV}/c^2 = 4507\,\text{MeV}/c^2$
or an energy of 4.5 GeV

Assuming that this energy is the rest-mass energy of the particles, the incident protons need to have at least 4.5 GeV/2 ≈ 2.3 GeV each. This is the threshold energy needed to produce the particles. When one of the protons is stationary, then this is a fixed target experiment and the reaction products would have KE as well as rest-mass energy. The incident proton would require more energy from the accelerator to produce them.

c Initial total energy
= KE of incident proton + rest-mass energy of incident proton + rest-mass energy of stationary proton
= 6 GeV + 1.0 GeV + 1.0 GeV = 8 GeV
Momentum p is given by

$$pc = \sqrt{E^2 - m_0^2 c^4}$$

Total energy E of incident proton =
6 GeV + 1.0 GeV = 7 GeV, so momentum of incident proton is

$$p = \sqrt{7^2\,(\text{GeV})^2 - (1.0)^2\,(\text{GeV})^2}/c = \sqrt{48\,(\text{GeV})^2}/c$$
$$= 6.9\,\text{GeV}/c$$

d $K^0 \rightarrow \pi^+ + \pi^-$. The Q-value for the decay is
$(K^0_{\text{rest mass}} - \pi^+_{\text{rest mass}} - \pi^-_{\text{rest mass}}) = (0.498 - 0.140 - 0.140)\,\text{GeV}/c^2 = 0.218\,\text{GeV}/c^2$ or an energy of 0.218 GeV.

This is the KE of the two pions so the KE of each pion is 0.109 GeV. Total energy E of each pion is KE of pion + rest-mass energy of pion

$$= 0.109\,\text{GeV} + 0.140\,\text{GeV} = 0.249\,\text{GeV}$$
$$pc = \sqrt{E^2 - m_0^2 c^4}$$
$$\therefore p = \sqrt{(0.249)^2\,(\text{GeV})^2 - (0.140)^2\,(\text{GeV})^2}/c$$
$$= \sqrt{0.042\,(\text{GeV})^2}/c = 0.2\,\text{GeV}/c$$

5 a The energy loss per proton is 0.4 MeV cm^{-1} so over a track of 2 m in the bubble chamber this represents a total energy loss of 0.4 MeV × 200 = 80 MeV.
The incident proton has an energy of 1000 MeV so the number of hydrogen ions produced by this energy loss is 1000 MeV/80 MeV ≈ 13.

b i) In the scintillator the protons will lose energy of 1 MeV as they pass through the scintillator material. Therefore the energy available to produce photons is 1 GeV − 1 MeV = 999 MeV or 999×10^6 eV. It takes 100 eV to produce a photon, so

$$\text{no. of photons produced} = \frac{999 \times 10^6\,\text{eV}}{100\,\text{eV}}$$
$$= 999 \times 10^4$$

ii) The light collection efficiency is only 10% so the no. of photons produced that are passed into the photomultiplier is 10% × $999 \times 10^4 = 999 \times 10^3$. The detection efficiency is 25% so the number detected by the photomultiplier tube is 25% × $999 \times 10^3 = 2.5 \times 10^5$.

Chapter 5

1 a An elementary particle is one with no apparent internal structure. Leptons are all particles that do not feel the strong force and are elementary. Baryons and mesons are composite particles made up of quarks and feel the strong force.

b $^{23}_{10}\text{Ne} \rightarrow {}^{23}_{11}\text{Na} + {}^{0}_{-1}\beta + \bar{\nu}$
Assuming that the neutrino has zero mass and the Ne nucleus has zero recoil, then the Q-value of this reaction is

22.994 465 u − 22.989 768 u = 0.004 697 u

Now 1 u = 931 MeV so the maximum KE of the beta particle is

0.004 697 × 931 MeV = 4.37 MeV

2 b Use $L = I\omega$.
i) The Earth spins once in 24 hours, or 2π radians in 24 × 3600 = 86 400 s, or $2\pi/86\,400$ rad s^{-1}. Moment of inertia for Earth is

$\frac{2}{5} \times 6 \times 10^{24}\,\text{kg} \times (6.4 \times 10^6\,\text{m})^2 = 9.8 \times 10^{37}\,\text{kg m}^2$

$\therefore L = 9.8 \times 10^{37}\,\text{kg m}^2 \times 2\pi/86\,400\,\text{rad s}^{-1} =$
$7 \times 10^{33}\,\text{N m s}$

ii) $L = 10\,\text{kg m}^2 \times 20\,\text{rad s}^{-1} = 200\,\text{N m s}$
iii) $L = 0.5 \times 1.05 \times 10^{-34}\,\text{J s} = 5.25 \times 10^{-35}\,\text{N m s}$
c Angular momentum at the atomic and sub-nuclear levels is quantised.

3 a See page 150.
b A positron (same mass as electron but with opposite charge). A neutrino is emitted.
c In *electron capture* an atomic electron strays too close to the nucleus and is swallowed, allowing the conversion of a proton into a neutron by the process $p + e^- \rightarrow n$. In this process Z goes down by one unit and N increases by one unit but $A = Z + N$ remains the same.

4 b Surface area of Earth $= 4\pi r^2 = 4 \times \pi \times$
$(6400 \times 10^3 \, m)^2 = 5.1 \times 10^{14} \, m^2$. Therefore the
number hitting the Earth per second is

$$\frac{5.1 \times 10^{14} \, m^2}{1500 \, m^2 s^{-1}} = 3.4 \times 10^{11} \, s^{-1}$$

The charge on a proton $= 1.6 \times 10^{-19} \, C$ and so,
using $I = Q/t$, the current I is
$1.6 \times 10^{-19} \, C \times 3.4 \times 10^{11} \, s^{-1} = 5.4 \times 10^{-8} \, A$

5 a Energy released in an Earth–anti-Earth collision is

$$E = m_{Earth}c^2 + m_{anti-Earth}c^2$$

Rest-mass energy of Earth $= m_{Earth}c^2$
$= 6 \times 10^{24} \, kg \times (3 \times 10^8 \, m s^{-1})^2 = 5.4 \times 10^{41} \, J$

This is the same for the anti-Earth so the total
energy released in the annihilation is
$2 \times 5.4 \times 10^{41} \, J \approx 10^{42} \, J$.

b Mass of proton = mass of antiproton $= 1.67 \times$
$10^{-27} \, kg$. So time taken to produce 1 kg of
antiprotons is

$$\frac{1 \, kg}{1.67 \times 10^{-27} \, kg \times 6 \times 10^{10} \, h^{-1}} \approx 10^{16} \, h \text{ or about } 10^{12} \text{ years}$$

Mass of antiprotons accumulated is
$1.67 \times 10^{-27} \, kg \times 1.2 \times 10^{12} = 2 \times 10^{-15} \, kg$

6 b 1, 8, 18.

7 Fermions are particles which have half-integer spin
and obey Fermi-Dirac statistics. Bosons are particles
which have integer spin and obey Bose-Einstein
statistics.

8 a Any hadron that contains a strange quark is said to
have the property of strangeness. Strangeness is
always conserved in the strong and
electromagnetic interactions.

b i) S is not conserved: $+1 \to 0 + 0$, LHS $\neq$ RHS so it
cannot proceed by strong interaction.
ii) S is conserved: $-1 + 0 \to -1 + 0$, LHS = RHS so
it proceeds via strong interaction.
iii) S is not conserved: $-1 \to 0 + 0$, LHS $\neq$ RHS so it
cannot proceed by strong interaction.
iv) S is conserved: $-1 + 0 \to -1 + 0$, LHS = RHS so
it proceeds via strong interaction.

Chapter 6

1 a A square has four lines of symmetry.
b A rhombus has two lines of symmetry (its diagonals).
c The letter 'A' has only one line of symmetry.
d A circle has an infinite number of lines of
symmetry.

2 The spectral lines of H depend on energy differences
between electron (or positron) orbits. There are no
reasons to think at the current time that these should
be any different in an antihydrogen atom.

3 a A law is invariant if it remains the same under the
action of a transformation.

i) $F = \dfrac{1}{4\pi\epsilon_0} \dfrac{Q_1 Q_2}{(-r)^2} = \dfrac{1}{4\pi\epsilon_0} \dfrac{Q_1 Q_2}{r^2}$

ii) $F = \dfrac{1}{4\pi\epsilon_0} \dfrac{(-)Q_1 \times (-)Q_2}{r^2} = \dfrac{1}{4\pi\epsilon_0} \dfrac{Q_1 Q_2}{r^2}$

(since the product of two minuses is a plus)

b Under space inversion, $x \to -x$, $t \to t$, so $v \to -v$
(since $v = \Delta x/\Delta t$) and so the momentum expression
becomes

$$p = \frac{-m_0 v}{\sqrt{1 - \left(\dfrac{-v}{c}\right)^2}} = -p$$

Space inversion changes the sign of the
momentum expression. Note however that the *law*
of conservation of momentum is invariant under
space inversion. Changing the sign of the velocity
vectors in a system of particles has no effect on the
amount of momentum conserved.

4 a i) Here we have a mix of electron lepton and
muon lepton families so we must check the lepton
number for each.

$$\mu^- \to e^- + \bar{\nu}_e + \nu_\mu$$
$L_e: 0 \to 1 + (-1) + 0 \quad$ LHS = RHS ✓
$L_\mu: 1 \to 0 + 0 + 1 \quad$ LHS = RHS ✓

The reaction conserves lepton number.
ii) Here we are dealing only with the muon lepton
family, so for

$$\bar{\nu}_\mu + p \to \mu^+ + n$$
$L_\mu: -1 + 0 \to -1 + 0 \quad$ LHS = RHS ✓

(Note here that the μ^+ is the antiparticle of the μ^-
so it has $L = -1$.)
iii) This reaction involves both the electron and
muon lepton families.

$$\mu^- \to e^- + \gamma$$
$L_e: 0 \to 1 + 0 \quad$ LHS $\neq$ RHS
$L_\mu: 1 \to 0 + 0 \quad$ LHS $\neq$ RHS

This reaction violates conservation of lepton
number.

b Reaction iii) violates conservation of charge. On the
LHS Q is $+2e$. On the RHS $Q = +3e$.

c i) is forbidden. On the LHS $B = +2$, on the RHS
$B = +3$. B is not conserved.
ii) is OK. Note that a $\bar{p}$ has $B = -1$ so that on the
LHS $B = +2$ and on the RHS $B = +3 - 1 = +2$.
iii) is forbidden. Note that the $\bar{\Lambda}^0$ is an antibaryon
having $B = -1$, π^- is a meson having $B = 0$, so on
the LHS $B = -1$, on the RHS $B = 1 + 0 = +1$.
iv) is forbidden. On the LHS $B = 1$, on the RHS
$B = 0 + 0 = 0$.

d i) This cannot occur because Q is not conserved. The LHS has $Q = -1e + 1e = 0$, the RHS has $Q = +e$.

ii) This cannot occur. Q is conserved but B is not conserved. The Σ^-, n and p are baryons with $B = +1$, and the K^0 is a meson with $B = 0$. So the LHS has $B = +2$, the RHS has $B = 0 + 1 = +1$.

iii) This reaction can occur. For Q, the LHS has $-1e + 1e = 0$, the RHS has $0e + 0e = 0$. Q is conserved.

For B, the K^- and $\overline{K}^0$ are mesons with $B = 0$, and the p and n are baryons with $B = +1$. So the LHS has $0 + 1 = +1$, the RHS has $0 + 1 = +1$. B is conserved.

For S, the K^- and $\overline{K}^0$ have $S = -1$, and the p and n have $S = 0$. So the LHS has -1 and the RHS has -1. S is conserved.

e i) Lepton number.   **iv)** Strangeness.
ii) Strangeness.   **v)** Baryon number.
iii) Baryon number.

5 a The particles involved are all members of the lepton family. We need to check for conservation of both muon and electron lepton number in each case.

$$\mu^+ \rightarrow e^+ + \nu_e$$
$L_\mu: -1 \rightarrow 0 + 0$ LHS $\neq$ RHS therefore L_μ is not conserved.
$L_e: 0 \rightarrow -1 + 1 = 0$ LHS $=$ RHS so L_e is conserved.

But for the decay to occur, both L_μ and L_e have to be conserved.

$$\mu^+ \rightarrow e^+ + \nu_e + \overline{\nu}_\mu$$
$L_\mu: -1 \rightarrow 0 + 0 - 1 = -1$ LHS $=$ RHS so L_μ is conserved.

L_e is conserved as before, so this decay can occur by conservation of lepton number.

b By conservation of Q, X must have a charge of zero since ν_μ has zero charge. This decay involves both muon and electron-type leptons, the lepton numbers of which must both be conserved. Examining L_μ we see that $1 \rightarrow 1$ so L_μ is conserved. L_e is also conserved if X has an electron lepton number of -1, i.e. L_e is conserved if $0 \rightarrow 1 + (-1) + 0$ which implies that X is an electron-type lepton with zero charge and a lepton number of -1. The particle that fits this description is an anti-electron neutrino $\overline{\nu}_e$, thus the decay is $\mu^- \rightarrow e^- + \overline{\nu}_e + \nu_\mu$. It must involve the weak interaction since this is the only interaction felt by neutrinos (assuming they have no mass).

6 a By conservation of Q, X must have a charge of $-1e$.

b Baryon number is only conserved if X is a baryon ($B = +1$) since the K^-, K^0 and K^+ are all mesons with $B = 0$.

c Examining strangeness, K^- has $S = -1$, K^0 and K^+ have $S = +1$ and p has $S = 0$. So $(-1) + 0 \rightarrow 1 + 1 + S_X$. For strangeness to be conserved X must have $S = -3$.
This particle is the omega-minus which was discovered in 1964 (see Box 5.3, page 166).

7 b i) This is an electromagnetic interaction. An $e^+ e^-$ pair have annihilated to produce a $\mu^+ \mu^-$ pair.

ii)

Chapter 7

1 a i) uud   **ii)** udd

b Using Coulomb's Law

$$F = \frac{1}{4\pi\epsilon_0} \frac{q_1 q_2}{r^2} =$$

$$\frac{1}{(4\pi \times 8.85 \times 10^{-12} \, \text{Fm}^{-1})} \frac{(-\tfrac{1}{3}e \, \text{C})(-\tfrac{1}{3}e \, \text{C})}{(2.6 \times 10^{-15} \, \text{m})^2}$$

$$= \frac{2.8 \times 10^{-39} \, \text{C}^2}{7.5 \times 10^{-40} \, \text{Fm}} = 3.7 \, \text{N}$$

2 a In diagram 1, A proton and a π^- have interacted electromagnetically to produce a π^+ and a K^-. The K^- is a strange meson since it contains a $\overline{s}$ quark. The strangeness on the LHS of the Feynman diagram is zero whereas the strangeness on the RHS is non-zero. Since strangeness is conserved in electromagnetic interactions, this process is not possible. In addition, baryon number is not conserved in this interaction. The baryon number on the LHS is $(+1) + 0 = +1$, on the RHS it is $0 + 0 = 0$.

Diagram 2 shows a proton decaying into a neutron via a weak interaction. This decay has never been observed and the proton is considered to be stable over large periods of time. (A proton inside a nucleus can decay into a neutron, but this is not observable.) Also, on the RHS of the interaction shown, two antiparticles are created which is not possible.

b Beta decay in terms of quarks is described on page 230.

3 a i) The strong interaction cannot be responsible since S is not constant on each side of the decay. In processes governed by the strong or electromagnetic interactions, the total strangeness must remain constant. However, in processes governed by the weak interaction, the strangeness either remains constant *or* changes by one unit. Since the strangeness difference in this decay is one unit, it could go by the weak interaction.

ii) The K^- is a strange particle with a charge of $-1e$. It is also a meson which means it must be made up of a quark–antiquark pair. The strangeness of the K^- is -1. Looking at the table, the quark combination that describes the K^- is $s\bar{u}$. The s quark has $S = -1$ and $Q = -\frac{1}{3}$; the $\bar{u}$ quark has $Q = -\frac{2}{3}$. The combination $s\bar{u}$ thus gives the correct values of Q and S, and it gives baryon number $B = +\frac{1}{3} - \frac{1}{3} = 0$ which is correct for a meson.

The π mesons are not strange particles so cannot contain strange quarks, but must consist of a quark–antiquark pair. The π^- has $Q = -1e$. This is satisfied by the quark combination $\bar{u}d$ which has $Q = -1e$, $B = 0$ and $S = 0$. Similarly, the π^0 has quark combination $d\bar{d}$ which satisfies $Q = +1e$, $B = 0$ and $S = 0$.

iii) Rest mass of $K^- = 938\,\text{MeV}/c^2 \times 0.53 = 497.14\,\text{MeV}/c^2$

Total energy of each pion is therefore $0.5 \times 497.14\,\text{MeV} = 249\,\text{MeV}$

Rest mass of individual π mesons $= 938\,\text{MeV}/c^2 \times 0.15 = 140.7\,\text{MeV}/c^2$

$$E^2 = m_0^2 c^4 + p^2 c^2$$

$$\therefore pc = \sqrt{E^2 - m_0^2 c^4} =$$

$$\sqrt{(249\,\text{MeV})^2 - \left(\frac{140.7\,\text{MeV}}{c^2}\right)^2 c^4} = \sqrt{42\,204.5\,(\text{MeV})^2}$$

so $p = 205\,\text{MeV}/c$

b i) In colliding beam experiments no beam energy goes into the kinetic energy of the reaction products since the total momentum before and after the collision is zero. The Z^0 is produced by colliding beams of protons and antiprotons together. The energy of the beams can all go into producing a Z^0 so the energy of each beam $= 0.5 \times 91.2\,\text{GeV}/c^2 = 45.6\,\text{GeV}/c^2$.

ii)

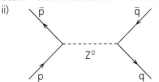

The p and $\bar{p}$ annihilate to produce a Z^0 which decays into a quark and antiquark. Note that the antiparticles are shown going backwards in time.

iii) The W particles have positive and negative charge (W^+ and W^-) and must be created in pairs in order to conserve Q. The minimum beam energy required is $2 \times 80.4\,\text{GeV} = 160.8\,\text{GeV}$.

4 a At A the incoming atom has collided with a hydrogen nucleus in the bubble chamber and three particles have been produced: two positively charged particles (since they curve in the same direction as the incoming particle) and a neutral particle which leaves no track but decays at B into two oppositely charged particles of approximately equal mass (since their radii of curvature are approximately equal).

b We first note that mesons are made up of quark–antiquark pairs. The D^+ is a meson with a charm of 1, so it must have a charmed quark in it, and also has $Q = +1e$, so the quark structure $c\bar{d}$ fits this specification. The π^+ has $Q = +1e$, $S = 0$ and is not charmed, so the quark structure $u\bar{d}$ fits this specification. The K^- meson has $Q = -1e$ and $S = -1$, so it must contain a strange quark, and the quark structure $s\bar{u}$ fits this specification. The Δ^{++} decays into a p and a π^+. The p has the quark structure uud, and the π^+ has $u\bar{d}$. Also, the Δ^{++} has zero charm and strangeness but has $Q = +2e$ and is not a meson. This means it must have the quark structure uuu.

We can write the first decay in terms of the quark structure as

$$\left(\frac{c}{\bar{d}}\right) \rightarrow \left(\frac{s}{\bar{u}}\right) + \left(\frac{u}{\bar{d}}\right) + \left(\frac{d}{\bar{u}}\right)$$

On the LHS $(q - \bar{q}) = 1 - 1 = 0$, on the RHS $(q - \bar{q}) = 3 - 3 = 0$

For the second decay

$$\left(\begin{matrix} u \\ u \\ u \end{matrix}\right) \rightarrow \left(\begin{matrix} u \\ u \\ d \end{matrix}\right) + \left(\frac{u}{\bar{d}}\right)$$

On the LHS $(q - \bar{q}) = 3 - 0 = 3$, on the RHS $(q - \bar{q}) = 4 - 1 = 3$

We know that the strong interaction is involved in Δ^{++} decay because of its very short lifetime. The D^+ has a relatively long lifetime and allows one type of quark to change into another, both of which are evidence for the weak interaction.

5 **a** If the Ω^- has $S = -3$ it must contain *three* strange quarks, each contributing to its total strangeness and making it a baryon rather than a meson (which consists of a quark–antiquark pair). In addition, it feels the strong force, providing further evidence that it is a baryon.

b The K^- has $S = -1$, so it must contain a strange quark, and has $Q = -1e$. The quark combination $s\bar{u}$ fits this specification. The proton has quark structure uud which satisfies $Q = +1e$ and $S = 0$. The Ω^- has $Q = -1e$ and $S = -3$. These conditions are satisfied if the Ω^- has a quark structure of sss. The K^+ has $Q = +1e$ and this can be achieved if it has the quark structure $u\bar{s}$, making it a strange meson with $S = +1$. Similarly, the K^0 has $Q = 0$ and this can be achieved if it has the quark structure $d\bar{s}$, making it a strange particle with $S = +1$.

c Because of the long lifetime, the weak interaction must be involved at some stage. The Ω^- with $S = -3$ has decayed into a proton and a number of π mesons, all of which have zero strangeness. So the decay does not conserve strangeness and this is further evidence for the weak interaction. (The full decay of the Ω^- is shown in Box 5.3 on page 166.) The weak interaction is mediated by W or Z exchange particles.

6 The interaction at A is

$$\pi^- + p \longrightarrow K^0 + \Delta^0$$
$$\hookrightarrow \pi^- + p$$
$$\hookrightarrow \pi^- + \pi^+$$

All mesons are made up of quark–antiquark pairs. The K^0 has $S = +1$, so it must contain a strange quark, and has $Q = 0$. From the table this gives the quark structure as $d\bar{s}$. The π^- has the quark structure $d\bar{u}$ since it has $Q = -1e$ and $S = 0$. Similarly, the π^+ has $Q = +1e$, $S = 0$, so the quark structure is $u\bar{d}$. The decay of K^0 is a weak interaction, since the K^0 lives long enough for it to be inferred from the pattern of its decay into π^- and π^+ mesons. Typical lifetime is $\sim 10^{-12}$ s.

7 **A** is true; **B** is true; **C** is true (however, in the weak interaction, strangeness can remain constant or change by one unit only); **D** is true; **E** is true; **F** is false.

Chapter 8

1 **a** i) The universe is expanding.

ii) See page 250.

iii) Being able to measure the distances to distant objects accurately.

iv) That the rate of expansion of the universe is constant, i.e. that it is not speeding up or slowing down.

b Use

$$\frac{v}{c} = \frac{\lambda - \lambda_0}{\lambda_0} = \frac{(396.820 - 396.849)\,\text{nm}}{396.849\,\text{nm}} = -0.029$$

$$\therefore v = -0.029 \times 3.0 \times 10^5\,\text{km}\,\text{s}^{-1}$$
$$= (-)8.7 \times 10^3\,\text{km}\,\text{s}^{-1}$$

Note that this answer is *negative*. This means that Alpha Centauri is moving *towards* us. This is because, in terms of astronomical distances, Alpha Centauri at just over 4 ly away is very close to us and we are observing its *local* motion in the Milky Way galaxy. The galaxy as a whole is moving away from other galaxies as the universe expands.

c That the universe is contracting.

3 **a** i) For 1500 K, $\lambda_{peak} = 1\,\mu\text{m}$; for 2000 K, $\lambda_{peak} = 0.75\,\mu\text{m}$.

ii) Using Wien's Law: $1500\,\text{K} \times 10^{-6}\,\text{m} = 0.0015\,\text{m K}$; $2000\,\text{K} \times 0.75 \times 10^{-6}\,\text{m} = 0.0015\,\text{m K}$. These two products are the same, so the data supports Wien's Law.

b $\lambda_{peak} = \dfrac{2.9 \times 10^{-3}\,\text{m K}}{2.7\,\text{K}} = 10^{-3}\,\text{m or 0.1 cm}$

This is in the microwave region.

4 **a** i) The nucleon rest energy is the same for antinucleons as it is for nucleons. So using $E = m_0 c^2$ the rest energy is

$$1.67 \times 10^{-27}\,\text{kg} \times (3 \times 10^8\,\text{m}\,\text{s}^{-1})^2 = 1.5 \times 10^{-10}\,\text{J}$$
$$= 933.8\,\text{MeV}$$

ii) To create a nucleon–antinucleon pair the energy required must be at least $2 \times 933.8\,\text{MeV} = 1867.6\,\text{MeV or } 3.0 \times 10^{-10}\,\text{J}$. Using $E = kT$,

$$T = \frac{3.0 \times 10^{-10}\,\text{J}}{1.38 \times 10^{-23}\,\text{J}\,\text{K}^{-1}} = 2 \times 10^{13}\,\text{K}$$

b There are 2 protons and 2 neutrons in $^4_2\text{He}^{2+}$, and 1 proton and 0 neutrons in $^1_1\text{H}^+$, so the total number of nucleons in He and H is 4. At 4 minutes, if the total mass of H is 3 times that of He then 75% of nucleons must be protons (i.e. the hydrogen), and 25% protons plus neutrons (the helium). This means that 12.5% of the He is protons (since there are equal numbers of protons and neutrons in the He nucleus) so the ratio of protons to neutrons must be

$$\frac{0.75(\text{protons in H}) + 0.125(\text{protons in He})}{0.125(\text{neutrons in He})}$$

$$= \frac{0.825}{0.125} = 7:1$$

5 a Refer also to Figure 8.9 on page 256. Helium formed at 10^3–10^4 s when the universe was at a temperature of about 10^9 K. Fusion needs high temperatures to overcome the electromagnetic repulsive force between protons in H and He nuclei.

b The interactions took place in the quark–lepton era. The exchange particle is a virtual photon.

6 a i) For $H_0 = 30\,\text{km}\,\text{s}^{-1}\,\text{Mpc}^{-1}$ we express H_0 as

$$\frac{30\,\text{km}\,\text{s}^{-1}}{3.1 \times 10^{19}\,\text{km}} = 9.6 \times 10^{-19}\,\text{s}^{-1}$$

so $\rho_c = \dfrac{3 \times (9.6 \times 10^{-19}\,\text{s}^{-1})^2}{8\pi \times 6.67 \times 10^{-11}\,\text{N}\,\text{m}^2\,\text{kg}^{-2}}$

$$= 1.7 \times 10^{-27}\,\text{kg}\,\text{m}^{-3}$$

ii) Similarly for $H_0 = 50\,\text{km}\,\text{s}^{-1}\,\text{Mpc}^{-1}$ or $1.6 \times 10^{-18}\,\text{s}^{-1}$,

$\rho_c = 4.6 \times 10^{-27}\,\text{kg}\,\text{m}^{-3}$

iii) For $H_0 = 75\,\text{km}\,\text{s}^{-1}\,\text{Mpc}^{-1}$ or $2.4 \times 10^{-18}\,\text{s}^{-1}$,

$\rho_c = 1.1 \times 10^{-26}\,\text{kg}\,\text{m}^{-3}$

b Since the age of the universe is equal to the reciprocal of the Hubble constant, a larger value H_0 means a younger universe.

Index

Photo credits

Thanks are due to the following for permission to reproduce copyright photographs:

Cover Lawrence Berkeley Laboratory/Science Photo Library; **p.1** *all* Science Photo Library; **p.2** Science Photo Library; **p.5** *t* Bridgeman Art Library; *b* Science & Society Picture Library; **p.8** Bridgeman Art Library; **p.9** Science & Society Picture Library; **p.11** Science Photo Library; **p.13** Science & Society Picture Library; **p.17, p.20** Corbis UK Ltd; **p.33** Science & Society Picture Library; **p.39, p.40** Science Photo Library; **p.47, p.75** Science & Society Picture Library; **p.78** Corbis UK Ltd; **p.82** Ullstein Bild; **p.92** Science Photo Library; **p.100** Science & Society Picture Library; **p.104** *t* Science Photo Library; *b* Photography and Graphics, The Royal Surrey County Hospital, Guildford; **p.109** M. Desaunay/GANIL (www.ganil.fr); **p.113** Science Photo Library; **p.119** *tl* Science & Society Picture Library; **p.120** Science Photo Library; **p.123** Science Photo Library/David Roberts; **p.151** Science & Society Picture Library; **p.152** Science Photo Library/David Parker; **p.153** *b* Corbis UK Ltd; **p.154** Science Photo Library; **p.160** Corbis UK Ltd; **p.161** Science & Society Picture Library; **p.163** Science Photo Library; **p.164** Corbis UK Ltd; **p.169** Science Photo Library; **p.185** Corbis UK Ltd; **p.194, p.218, p.229, p.234, p.244** Science Photo Library; **p.247** Corbis UK Ltd; **p.261** ICRR (Institute of Cosmic Ray Research), The University of Tokyo; **p.265** Science Photo Library; **p.271** Courtesy of Caltech/LIGO; **p.276** Science Photo Library; **p.277** © Harvard Smithsonian; **p.278** Science Photo Library.

t = top, b = bottom, l = left

Every effort has been made to contact copyright holders. The publishers apologise for any omissions which they will be pleased to rectify at the earliest opportunity.